Solid State Chemistry
Techniques

Edited by

A. K. Cheetham
University Lecturer in Chemical Crystallography
University of Oxford

and

P. Day
University Lecturer in Inorganic Chemistry
University of Oxford

CLARENDON PRESS · OXFORD

Oxford University Press, Walton Street, Oxford OX2 6DP
Oxford New York Toronto
Delhi Bombay Calcutta Madras Karachi
Petaling Jaya Singapore Hong Kong Tokyo
Nairobi Dar es Salaam Cape Town
Melbourne Auckland

and associated companies in
Berlin Ibadan

Oxford is a trade mark of Oxford University Press

Published in the United States
by Oxford University Press, New York

© Oxford University Press, 1987

First published 1987
First published in paperback 1988 (with corrections), 1990

All rights reserved. No part of this publication may be reproduced,
stored in a retrieval system, or transmitted, in any form or by any means,
electronic, mechanical, photocopying, recording, or otherwise, without
the prior permission of Oxford University Press

British Library Catagloguing in Publication Data
Solid state chemistry techniques.
1. Chemistry, inorganic
I. Cheetham, A. K.   II. Day, Peter, 1938–
546 QD151.2
ISBN 0-19-855165-7
ISBN 0-19-855286-6 (pbk)

Library of Congress Cataloging in Publication Data
Solid state chemistry.
Includes bibliographies and index:
1. Solid state chemistry. I. Cheetham, A. K.   II. Day, P.
QD478.S634 1985 541'.0421 85-3121
ISBN 0-19-855165-7
ISBN 0-19-855286-6 (pbk)

Filmset by Latimer Trend & Company Ltd, Plymouth
Printed in Great Britain by St Edmundsbury Press Ltd,
Bury St Edmunds, Suffolk

Preface

The continuing, worldwide search for new and useful materials has ensured that the Solid State is one of the major growth areas of chemistry and there is a widely perceived need for good, up-to-date text books in the area. This book is aimed at final year honours and post-graduate students who may be planning a career in the field. It deals with some of the most important techniques in solid state research, introducing the reader to both the principles of the methods and their applications. We chose a multi-author volume for this purpose in order that our account should be more authoritative, and we are delighted and encouraged by the very positive response that we received from colleagues who were invited to contribute. A book of this length cannot, of course, be comprehensive, and we are conscious that certain important techniques had to be omitted. Mössbauer spectroscopy, for example, is of considerable utility in modern solid state research, but we felt that the technique was already well covered in existing publications. Similarly, we have chosen to focus on inorganic, rather than organic, materials. We hope that readers will find this a useful and interesting book, and that Volume 2, which will deal with different classes of inorganic solids, will provide a valuable complement to it.

Oxford A. K. Cheetham
1988 P. Day

Acknowledgement

We thank Drs. J. C. Cheetham and A. K. Nowak for providing the illustration that was used for the front cover.

Contents

List of contributors   xv

1 Synthesis of solid-state materials   1
1.1 Introduction   1
    1.1.1 Reaction types   3
    1.1.2 Quality criteria and assessments   5
    1.1.3 Exclusions and other sources   7
1.2 Some choices and consequences   8
    1.2.1 Dynamic (non-equilibrium) processes   8
    1.2.2 Factors in the choice of reactants   10
    1.2.3 Phase relationships and melts in direct synthesis   17
    1.2.4 Electrolysis   20
    1.2.5 Temperature and phases of low stability   21
    1.2.6 High pressure   25
1.3 Chemical (vapour phase) transport   27
1.4 Containers   31
1.5 Recent developments   33
1.6 References   35

2 Diffraction methods   39
2.1 Introduction   39
2.2 X-rays   39
    2.2.1 Production   40
    2.2.2 Scattering of X-rays by isolated atoms   41
2.3 Crystals   42
    2.3.1 Unit cell and lattice planes   42
    2.3.2 Bragg equation   44
2.4 Powder X-ray techniques   45
    2.4.1 Debye–Scherrer method   45
    2.4.2 Other powder methods   46
    2.4.3 Applications of powder methods   47
    2.4.4 Limitations of powder methods   48
2.5 Diffraction theory   48
    2.5.1 Structure factor F_{hkl} 49

viii *Contents*

	2.5.2 Intensity of reflection I_{hkl}	50
	2.5.3 Electron density distribution	51
2.6	Single-crystal X-ray methods	51
2.7	Crystal symmetry	52
	2.7.1 Crystal systems	52
	2.7.2 Bravais lattices	53
	2.7.3 Systematic absences	54
	2.7.4 Point groups and space groups	56
2.8	Solving a structure	57
	2.8.1 The Patterson method	57
	2.8.2 Direct methods	58
2.9	Structure refinement	59
2.10	Neutron diffraction and its applications	61
	2.10.1 Wave properties	61
	2.10.2 Applications	62
	2.10.3 Deformation density studies	66
	2.10.4 Pulsed neutron sources	67
2.11	Rietveld profile analysis	68
2.12	Electron microscopy	71
	2.12.1 Electron–matter interactions	71
	2.12.2 Transmission electron microscopy (TEM)	72
	2.12.3 Scanning electron microscopy (SEM)	74
	2.12.4 Analytical electron microscopy (AEM)	75
2.13	Defective and non-crystalline materials	78
	2.13.1 Point defects in crystalline materials	79
	2.13.2 Amorphous and semi-crystalline materials	79
	2.13.3 Particle size	79
2.14	References	81
2.15	Appendix	82

3 X-ray photoelectron spectroscopy and related methods   84

3.1	Introduction	84
3.2	Techniques	87
	3.2.1 Radiation sources	88
	3.2.2 Electron spectrometers	88
	3.2.3 Sample preparation	89
	3.2.4 Typical spectra	90
3.3	Fundamental properties of core electron spectra	93
	3.3.1 Electron binding energies	93
	3.3.2 Linewidths	97

	3.3.3 Intensities	98
	3.3.4 Time-scales	99
3.4	Satellites and multiplets	99
	3.4.1 Final state configuration interaction	100
	3.4.2 Core electron multiplet splitting	101
	3.4.3 Photoionization of open-shells	102
	3.4.4 Screening in metals	104
3.5	Valence band spectra	105
	3.5.1 Some applications of valence band XPS	109
3.6	Spectroscopy of surface layers	110
	3.6.1 Clean surfaces	110
	3.6.2 Adsorbate systems	112
	3.6.3 Surface and bulk aspects of XPS	114
3.7	Related spectroscopies	114
	3.7.1 Auger electron spectroscopy (AES)	114
	3.7.2 Bremsstrahlung isochromat spectroscopy (BIS)	115
	3.7.3 X-ray absorption edge spectroscopy and extended X-ray absorption fine structure (EXAFS)	116
	3.7.4 Electron energy loss spectroscopy (EELS)	117
3.8	References	118
3.9	Bibliography	120

4 Magnetic measurements — 122

4.1	Introduction	122
4.2	Basic concepts	122
	4.2.1 Substances in magnetic fields	122
4.3	Experimental techniques	123
	4.3.1 The Gouy method	123
	4.3.2 The Faraday method	127
	4.3.3 Change-in-flux methods	128
	4.3.4 Vibrating sample magnetometer	130
	4.3.5 Evans NMR method	131
	4.3.6 Calibration standards	131
4.4	Formulas for data handling	132
	4.4.1 The Van Vleck equation	132
	4.4.2 The Curie law	133
	4.4.3 Paramagnetism	134
	4.4.4 Data analysis	135
4.5	Crystal field theory	138
	4.5.1 The octahedral crystal field	138
	4.5.2 Spin-orbit coupling	139

	4.5.3	The Zeeman effect	143
	4.5.4	Second-order Zeeman effect in Ti(III)	145
	4.5.5	Example of a low-symmetry crystal field component	
4.6	Exchange coupling	148
	4.6.1	Exchange, an orbital effect	149
	4.6.2	Vector model for exchange	151
	4.6.3	Anisotropic and antisymmetric exchange	153
	4.6.4	Dirac's permutation operator	154
	4.6.5	Exchange in linear chains	154
	4.6.6	Exchange in two-dimensional layers	158
	4.6.7	Exchange in three-dimensional systems	159
4.7	References	161
4.8	Bibliography	161

5 Optical techniques 163

5.1	Introduction	163
5.2	Experiments and their interpretation	164
	5.2.1 Linearly polarized absorption	165
	5.2.2 Circular polarization	170
	5.2.3 Other absorption techniques	179
	5.2.4 Reflectivity	180
	5.2.5 Luminescence	182
5.3	Some applications	185
5.4	References	188

6 High-resolution solid-state MAS NMR investigations of inorganic systems 190

6.1	Introduction	190
6.2	Nuclear interactions in the solid state	190
	6.2.1 The Zeeman interaction	191
	6.2.2 The dipolar interaction	191
	6.2.3 Chemical shift interaction	192
	6.2.4 Spin–spin coupling	192
	6.2.5 Quadrupolar interactions	193
6.3	Techniques for 'line narrowing' in solids: CP–MAS experiments	194
	6.3.1 Dilute spin systems	194
	6.3.2 Chemical shift anistropy: magic angle spinning (MAS) techniques	195
	6.3.3 Cross polarization (CP) techniques	197
	6.3.4 MAS spectra at high magnetic fields	200
	6.3.5 Quadrupolar nuclei with non-integral spins	204
6.4	General features of high-resolution solid-state NMR spectra	206

6.5	Investigations of crystalline inorganic systems	209
	6.5.1 Coordination compounds	211
	6.5.2 Aluminosilicates	216
6.6	Extensions of the simple CP–MAS experiment	225
6.7	References	227

7 Computational techniques and simulation of crystal structures 231

7.1	Introduction	231
7.2	Methodology	231
	7.2.1 Scope and aims of the simulation studies	231
	7.2.2 Simulation techniques	233
	7.2.3 Interatomic potentials	241
7.3	Applications	246
	7.3.1 Point defect energies	247
	7.3.2 Defect aggregation and elimination in heavily disordered solids	253
	7.3.3 Superionic conductors	265
	7.3.4 Crystal structure prediction	272
	7.3.5 Surface simulations	275
7.4	References	276

8 Transport measurements 279

8.1	Introduction	279
8.2	Conductivity	279
	8.2.1 Concentration of carriers	280
	8.2.2 Mobility of carriers	284
8.3	Measurement of conductivity	286
	8.3.1 Contacts	286
	8.3.2 Electrode arrangements	289
8.4	Some results	290
	8.4.1 Metallic compounds	290
	8.4.2 Intermediate compounds	291
	8.4.3 Semiconductors	292
	8.4.4 Organic conductors	292
8.5	Thermopower	293
	8.5.1 The Seebeck effect	293
	8.5.2 Measurement of Seebeck coefficients	296
	8.5.3 Results	297
	8.5.4 Mixed conduction	301
8.6	The Hall effect	301
	8.6.1 Theory	301

xii   *Contents*

	8.6.2 Measurement of Hall coefficients	304
	8.6.3 Results	305
8.7	Photoconductivity	306
	8.7.1 Background	306
	8.7.2 Examples	308
8.8	Conduction in amorphous and imperfectly crystalline materials	309
8.9	A.c. conductivity	312
	8.9.1 Non-ohmic contacts	314
	8.9.2 Polycrystalline materials	314
	8.9.3 Amorphous and imperfectly crystalline materials	316
8.10	Conclusions	318
8.11	References	319

9   Vibrational spectroscopy — 322

- 9.1 Introduction — 322
- 9.2 Lattice dynamics: some basic concepts — 323
  - 9.2.1 The longitudinal modes of a monatomic linear chain — 323
  - 9.2.2 Longitudinal and transverse modes of a linear chain — 324
  - 9.2.3 Long-range forces — 325
  - 9.2.4 The diatomic linear chain: acoustic and optic branches — 325
  - 9.2.5 Three-dimensional lattices — 327
  - 9.2.6 Phonons — 328
  - 9.2.7 Sampling k-space — 329
  - 9.2.8 L.o–t.o mode splitting — 331
  - 9.2.9 Dispersion of the dielectric constant — 334
  - 9.2.10 Microscopic theories — 336
  - 9.2.11 Molecular crystals — 337
- 9.3 How to obtain a spectrum — 338
  - 9.3.1 A preliminary: temperature — 338
  - 9.3.2 Particle size and shape — 339
  - 9.3.3 Transmittance and reflectance — 340
  - 9.3.4 Infrared reflectance — 341
  - 9.3.5 Spectroscopy of single crystals: the indicatrix — 343
  - 9.3.6 Spectroscopy at high pressures — 345
- 9.4 Group theory and analysis of the vibrational spectra of solids — 346
  - 9.4.1 The factor group — 348
  - 9.4.2 Wyckoff sites and the site group — 349
  - 9.4.3 Factor group analysis — 350
  - 9.4.4 L.o–t.o splitting and selection rules for Raman spectra — 350
- 9.5 Worked examples — 351
  - 9.5.1 An atomic lattice: NiAs — 351

	9.5.2	Lattice and internal modes of a molecular or complex-ionic crystal without factor group coupling	353
	9.5.3	Factor group coupling: brucite	356
	9.5.4	Line groups and polymers: orthorhombic lead oxide	357
	9.5.5	L.o–t.o splitting in Raman spectra: benitoite	360
9.6	References		360

10 Thermodynamic aspects of inorganic solid-state chemistry — 362

10.1	Introduction	362
10.2	Heat capacities, entropies, and lattice vibrations	363
	10.2.1 Basic relations and magnitudes	363
	10.2.2 Experimental techniques	364
	10.2.3 Some examples	364
	10.2.4 The interpretation of lattice heat capacities	369
10.3	Thermodynamics of solid-state reactions	371
	10.3.1 General principles	371
	10.3.2 Experimental approaches and examples	372
	10.3.3 Systematics and trends	380
10.4	Solid solutions and order–disorder phenomena	385
	10.4.1 'Simple' substitutional solid solutions	385
	10.4.2 Order–disorder and complex systems	389
10.5	Textbooks and tabulations of data	390
10.6	References	392

Index — 395

List of contributors

D. M. ADAMS
Department of Chemistry, The University, Leicester LE1 7RH, UK

C. R. A. CATLOW
Department of Chemistry, University College, 20 Gordon Street, London WC1, UK

A. K. CHEETHAM
Chemical Crystallography Laboratory, University of Oxford, 9 Parks Road, Oxford OX1 3PD, UK

J. D. CORBETT
Ames Laboratory and Department of Chemistry, Iowa State University, Ames, Iowa 50011, USA. (Ames Laboratory is operated for US Department of Energy by Iowa State University under contract No. W-7405-Eng-82.)

R. G. DENNING
Inorganic Chemistry Laboratory, University of Oxford, South Parks Road, Oxford OX1 3QR, UK

C. A. FYFE
Guelph-Waterloo Centre for Graduate Work in Chemistry, Guelph Campus, Department of Chemistry, University of Guelph, Ontario N1G 2W1, Canada

A. HAMNETT
Inorganic University Laboratory, University of Oxford, South Parks Road, Oxford OX1 3QR, UK

W. E. HATFIELD
Department of Chemistry, The University of North Carolina at Chapel Hill, Venable and Kenan Laboratories 045 A, Chapel Hill, North Carolina NC 27514, USA

A. NAVROTSKY
Department of Geological and Geophysical Sciences, Princeton University, Guyot Hall, Princeton, New Jersey 08544, USA.

R. E. WASYLISHEN
Department of Chemistry, Dalhousie University, Halifax, Nova Scotia B3H 4H6, Canada

G. K. WERTHEIM
Bell Laboratories, Murray Hill, New Jersey NJ 07974, USA

1 Synthesis of solid-state materials[†]

J. D. Corbett

1.1 Introduction

The fields of solid-state chemistry and physics are important because the chemical and physical properties of infinite non-molecular solids are so different from those associated with discrete molecules in solids or, as more frequently studied, molecules or small ions in solution. Likewise, the first requirement for a solid-state study, the synthesis and some characterization of the material of interest, usually involves techniques and concepts that are very different from those conventionally applied in molecular studies. The preparation of 'pure' and well-defined, and perhaps even novel, inorganic phases is the subject of this chapter. Some of the needs and opportunities in solid state synthesis have been outlined by Warren and Geballe.[1]

Our approach will be to describe the classes of reactions possible, the difficulties which are characteristically associated with each (and the means for avoiding some of these), and some bases for the selection of a synthetic method; although clearly we cannot go further than to categorize compounds and reactions with a few examples. Many properties of solid materials will be important in these considerations, but none will be more significant than the rate of diffusion or mass transfer within, and between, solid particles. Solid-state reactions and the successful synthesis of single-phase, homogeneous products are often very much at the mercy of these intrinsically slow (but still highly variable) processes. The use of high temperatures in synthesis is a common means of improving these rates, but a consequence of this is that much solid-state chemistry at lower temperatures is lost. One may to some extent alleviate these restrictions on reactivity through the use of liquid- or gas-phase materials with their intrinsically much higher mobilities and shorter effective path lengths, but sometimes only with some sacrifices. Vaporization equilibria which allow the transfer of solid phases through the vapour state will be especially useful here. Other means of

[†] This research was supported by the Office of Basic Energy Sciences, Materials Sciences Division, US Department of Energy.

2 Solid state chemistry: techniques

achieving reactivity as well as factors in the choice of reactants will also be considered.

Some mechanistic aspects of reactions between real solids are important for understanding later considerations. For more details the reader should see Steele (Vol. 2),[2] or the more extensive Hannay[3] or Schmalzried[4] volumes. The relevant process is the so-called chemical diffusion, that of a substitutional component under composition gradient which takes place via vacant lattice and interstitial sites. As a reaction this is strictly only pertinent to solid solutions, in exchange reactions, for example. A simple but informative description by which a third phase is formed by diffusion can be visualized as follows. For a reaction of the type $M_2Y(s) + LY(s) \rightarrow M_2LY_2(s)$, presume that the diffusion of M and L are responsible for the mass transfer, the common anion serving as a fixed reference matrix. An interface between M_2Y and LY would be transformed to the intervening product M_2LY_2 by the scheme

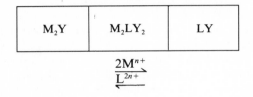

the indicated fluxes being fixed by electroneutrality. This reaction now includes the transfer of M and L across the phase boundaries together with the chemical diffusion of M and L as shown through the product phase to the opposite boundary. Unfortunately *real* synthetic reactions of this character are even more complicated, and consequently are less well studied and understood, so that making new phases in all-solid reactions is in practice more empirical. Because material transfer is facilitated by large areas of interphase contact, small diffusion pathlengths and minimum pore volume, it is customary to employ powdered reactants, usually in a compressed form. But real systems also involve the generally much more rapid diffusion on surfaces and along dislocations, and both of these also become more important in powdered samples. Of course the product is also apt to be defect-laden, which will increase reactant mobilities therein, and changes in grain boundary area and free volume usually also occur during the reaction. On the other hand, some products may form a compact and cohesive layer through which there is negligible reactant diffusion so the reaction virtually stops (tarnishing or surface blockage).

Nucleation of the new phase in most systems is not at all well understood. Nucleation in simple decomposition reactions is well known to be favoured at dislocation and defect sites. But in some oxide systems, anion defects are highly organized into crystallographic shear planes, and oxidation or reduction by H_2/H_2O mixtures occur by motion of these planes, with the reactant and product phases growing in fixed orientation (topotaxy).[5]

1.1.1 Reaction types

Our discussion of the many aspects of synthesis will be helped if we first identify classes of possible reactions and their individual advantages and problems. One common group of reactions involve gases except for one solid reactant and the desired product D(s). The most common is the metathetical or exchange reaction (1.1)

$$A(g) + B(s) \rightarrow C(g) + D(s) \qquad [\text{e.g. } CO + MnO_2 \rightarrow CO_2 + MnO] \qquad (1.1)$$

while the related combination and decomposition reactions (1.2) and (1.3), which lack C or A, are also useful,

$$A(g) + B(s) \rightarrow D(s) \qquad (1.2)$$

$$B(s) \rightarrow C(g) + D(s). \qquad (1.3)$$

The use of gaseous reactants naturally facilitates the conversions, while problems may arise with these routes from the dynamic (non-equilibrium) nature of processes (as usually carried out) as well as diffusion limitations as D forms on the surface of particles of B. Some mechanistic details of nucleation and growth of D have received substantial attention in reaction (1.2) for the oxidation of metals and for (1.3), in decomposition reactions.[3]

Synthesis with condensed phases may sometimes be conveniently performed 'neat' (i.e. on stoichiometry to yield a single phase) if one reactant is liquid,

$$A(l) + B(s,l) \rightarrow D(s) \qquad [\text{e.g. } 2NdCl_3 + Nd \rightarrow 3NdCl_2]. \qquad (1.4)$$

Not only does liquid A provide greater contact and mobility but it may also dissolve some D and prevent blockage. A particularly facile reaction occurs if the reaction can be run above the congruent melting point of D (see Section 1.2.3). Otherwise, diffusion limitations may again appear when the amount of A(l) becomes small or an intermediate solid forms, particularly if it occurs along the B→D or A→D pathway.

A more conventional solvent may also be employed to give a different version of (1.4), namely

$$A(l,s) + B(l,s) \xrightarrow{\text{solvent}} D(s). \qquad (1.5)$$

Use of a molecular or melt solvent, sometimes at elevated temperatures, speeds the reaction by bringing A and B together, presuming the solvent can be removed subsequently. Solvent-assisted reactions that are run very much below the melting point of the product will often yield a very finely divided, even amorphous material, which may be an advantage or a disadvantage depending on the intended use.

A few possibilities remain if the above reaction types are not feasible or suffer from incomplete conversions, side reactions, or contamination. Obviously, volatility of the product allows a simple separation from the contaminants. Thus the complex and incomplete reaction

$$Al_2O_3(s) + 3C(s) + 3Cl_2(g) \rightarrow 2AlCl_3(s) + 3CO(g) \qquad (1.6)$$

and its various analogues present few complications because the product $AlCl_3$ is volatile at the reaction temperature (and much below), and resublimation if necessary gives a very pure product. On the other hand, an all-solid reaction

$$A(s) + B(s) \rightarrow C(s), \qquad [\text{e.g. } CaO + TiO_2 \rightarrow CaTiO_3] \qquad (1.7)$$

will in the absence of any volatility probably be orders of magnitude slower and thence will present greater difficulty in achieving a respectable yield and purity. Reactions of this character are avoided whenever possible and (if not avoided) may utilize some combination of intimate mixing, even on the atomic scale, high temperature or high pressure or an added flux. Of course the flux remains as an impurity if not later dissolved or volatilized.

A second means of facilitating both a reaction as well as phase separation and purification amounts to the use of 'gaseous solvent', a reagent that reversibly converts a non-volatile reactant or product to a gaseous species. This process goes under the general name *chemical* or *vapour phase transport (VPT)* (see Section 1.3). Thus an all-solid reaction would be facilitated by any reagent X which carries otherwise non-volatile A to B or vice versa, that is

$$A(s) + X(g) \rightleftarrows AX(g) \qquad (1.8)$$

followed by

$$AX(g) + B(s) \rightarrow C(s) + X(g). \qquad (1.9)$$

An example is the formation of the spinel $MgCr_2O_4$ according to

$$MgO(s) + Cr_2O_3(s) \underset{}{\overset{O_2}{\rightleftarrows}} MgCr_2O_4(s) \qquad (1.10)$$

where added O_2 literally carries Cr_2O_3 to MgO through the reversible formation of gaseous CrO_3. Diffusion of this Cr_2O_3 into the MgO and nucleation of the product are still required. Obviously this process provides a means for purifying Cr_2O_3 alone by taking advantage of the temperature dependence of the last reaction, a process which would probably provide excellent single crystals of Cr_2O_3 as well.

The foregoing presentation implies that only reactions that give single-phase products, or nearly so, are found or need be considered. Though this is desired for most subsequent characterization measurements it is unfortunately often far from practice. The investigator sometimes must settle for a product that is far from ideal. Although some needs can probably be met with mixtures—catalytic and superconductivity screening, for example—others may require such measures as sorting of single crystals under a microscope even to get a powder pattern. The power of microanalytical techniques (see Section 1.1.2) and the popularity of single crystal X-ray diffraction in characterization is evident in these cases.

1.1.2 Quality criteria and assessments

Absolute criteria regarding an acceptable purity for 'single-phase' products are of course not feasible since results depend greatly on the particular compounds and routes. But the synthesizer should remember that he can usually minimize impurities by utilizing care whenever necessary, for example, through use of a good inert-atmosphere ('dry') box, high-purity gases, the best containers, a good high vacuum, and recrystallization, ignition, drying, sublimation, transport, etc. of starting materials or products whenever possible and appropriate. It will be noted that many reactions are not very forgiving in that no opportunity for fractionation or purification presents itself during or after the synthesis.

In many cases, a 99.5 per cent (by weight) recovery from quantitative analysis for all the component elements in a 'pure' compound is a reasonable minimum, although the low molecular weight impurities can still make this a poor quality index indeed. For polycrystalline materials the analysis reflects only the overall composition, and some measure of the phase purity and homogeneity, usually by X-ray powder diffraction, should be considered necessary. Such a check is often a better-quality judgment, particularly when the synthetic problems manifest themselves in the formation of separate phases rather than through substitution of impurity atoms within the product structure. However, the Debye–Scherrer method for powder patterns may give inadequate resolution[6] and can easily miss 10 per cent or more of another phase; and so the higher sensitivity (approximately 2–4 per cent) for, and better resolution of, impurity lines afforded by the Guinier technique[7,8] should be used, or a powder diffractometer in the case of an air-stable material of sufficient quantity. Non-stoichiometry or the effects of substitutional impurities are easier to recognize by this means, especially by comparison of lattice dimensions obtained for known structures through least-squares refinement of the high-resolution angular data, standard deviations generally being no more than a few parts in 10^4. Interstitial impurities, including light atoms which do not scatter X-rays well, often can be deduced from changes in lattice dimensions upon intentional variation of that element (if not already saturated). Optical microscopy, too, is often a quick and valuable check for phase purity.

Microprobe and X-ray fluorescence are useful tools for quantitative characterization with optimal detection limits of about 500 and 10 p.p.m. by weight, respectively, for elements heavier than neon, and somewhat higher below that. The former has the advantage of being able to analyse individual crystals, while the latter requires about 1 cm^2 of sample and gives better results with particle sizes greater than 200 μm. A transmission electron microscope with X-ray analysis capabilities is comparable to the microprobe in sensitivity, and its high spatial resolution (to 200 Å) allows good measure of homogeneity as well.[9] But the method also requires more robust materials which are not sensitive to the higher beam energy or the usual air transfer, and the machine is more demanding on operator experience. X-ray photoelectron spectroscopy can also provide useful semi-quantitative analysis for all elements except hydrogen, but it requires a relatively large sample.

Glassy impurities present a special problem as they contribute to most bulk analytical measures but are not revealed by X-rays. A microanalytical approach is then necessary which can examine individual crystallites, electron microscopy probably being the best because of its simultaneous diffraction check of crystallinity. A good example is the detection of '$Fe_4(P_2O_7)_3$' glass in crystalline $FePO_4$.[10]

Some additional observations on purity problems and consequences may be useful. X-ray powder patterns are often not meaningful in identifying impurities taken up at room temperature by high-melting solids because the materials so generated are not sufficiently crystalline to give recognizable diffraction maxima (coherent scattering requires a minimum of about 1000 Å crystallite or grain size). For example, the reaction of many air-sensitive halides with atmospheric moisture at room temperature is not signalled by the appearance of new lines in the powder pattern (some line broadening may be seen) *unless* the material is first strongly heated to form the oxide or oxyhalide, and to evolve hydrogen halide. The same applies to the solubility test for the rare-earth trihalides in ethanol, where the absence of cloudiness has been stated to be a purity test. And relatively low molecular weight impurities can have a large chemical effect. An extreme example[11] would be LaI_3 containing 0.5 per cent (by weight) of water. Although this would still give a product analysing a respectable 99.5 per cent (by weight) La + I and its reaction with excess La metal at elevated temperature would generate a sample analysing 99.7 weight per cent La + I, the product would be only 83 mole per cent LaI_2, the remainder being LaOI and LaH_2. The advantage of an X-ray check for the products of such low molecular weight impurities in the product is obvious. Carbonaceous material is not uncommon in commercial chemicals, and may either form unwanted materials or reduce the major compounds. For example, unsublimed commercial grade $BiCl_3$ sometimes yields a lavender melt owing to reduction to Bi^+. Traces of organic solvents may function likewise in addition to contributing products of solvolysis reactions; solvents are probably better avoided when possible if very high purity is important.

The literature also encourages healthy scepticism regarding the *real* composition of a phase, particularly with respect to the presence of the more pervasive C, N, O, and H. Some of the more striking 'mistakes' have been 'β-W' which was really[12] W_3O, and 'Ti_2S' which turned out[13] to be TiSC; while '$Zr_2H(\gamma)$' was in fact[14] $Zr_2CH_{1.5}$, 'Sr_3Sn' was actually[15] Sr_3SnO, and 'CaCl' was[16] CaHCl. The problems are compounded by the fact that some second-period elements may readily substitute for one another in crystal lattices, as illustrated by the improved formulations of the reported[17-19] 'SmO', 'Ta_2O', and 'NbF_3,' as $SmO_{0.5}N_{0.5}$, $Ta_2O_xN_{1-x}$, and $NbF_{2-y}O_{1+y}$ [$\sim 0.25 < y < 1.0$]. Many other products that have been reported rarely (or only once) presumably contain (or are stabilized by) significant impurities because of either uncritical or careless investigations, or simply the lack of better materials and methods.

Synthesis may be carried out for a variety of reasons. One motivation is to obtain new compounds and thereby broaden our knowledge of materials, properties, and the boundaries of the solid state. Another and more frequent reason is to provide a known material for a subsequent study of some physical or chemical property: diffusion, conductivity, magnetism, photoelectron or NMR or electronic spectra, catalysis, etc. But, unfortunately, a significant number of articles appear in the literature which expound or explain at length about the measured solid-state properties and even unusual features thereof, but which provide little or no information regarding the preparation or the phase purity of the materials studied. There are many variables in the synthesis of pure and single-phase compounds together with the possibility of significant intrinsic and extrinsic defects and variations in crystallinity and polytypic structure, so that some citation of pertinent details on synthesis and characterization should be given in all cases. Ideally the synthesis should be described sufficiently well that other (experienced) investigators should be able to duplicate the material. Schäfer[20] has noted that much physical and mathematical effort has doubtlessly been expended in basically meaningless attempts to correct for mistakes in synthesis.

1.1.3 Exclusions and other sources

No attempt will be made to describe a variety of methods which, although sometimes synthetic in result, are aimed at special materials, especially crystal growth from melts, epitaxial growth, thin films, and surface studies. The motivation for these specialties, which comprise a large segment of the literature on their own, is generally well removed from the chemistry. Interested readers might wish to refer to Anthony and Collongues[21], the many books on the subject, or the *Proceedings of the International Conferences on Crystal Growth*, e.g. Givargizov.[22] On the other hand, synthesis or purification may be a significant component of crystal growth by hydrothermal or vapour phase transport means and so these will be considered later. Amorphous (glassy) materials are also excluded.

Only a few portions of the present chapter have been individually treated before. This occurs principally in *Preparative Methods in Solid-State Chemistry*, edited by Hagenmuller,[23] which contains authoritative articles on high pressure, chemical transport and electrolytic methods, and on a few specific classes of compounds (fluorides, borides, carbides, ferrites). The remainder deals mainly with properties and crystal growth.

General experimental techniques are described in the introductory section to Brauer's *Handbuch der Präparativen Anorganischen Chemie*.[24] Somewhat more basic techniques for air-sensitive compounds are discussed by Shriver,[25] and Barton[26] gives a good (but old) description of the construction of and techniques in glove boxes. Brauer's *Handbuch*[24] is probably the best source for known (sometimes dated) synthetic routes to a wide variety of solid compounds, but in a terse format most useful to an experienced synthesizer. More educational aids are provided in the descriptions given in the series *Inorganic Syntheses*, but the solid state coverage is uneven.

1.2 Some choices and consequences

Many options open to the experimenter in considering synthetic routes to compounds are general in character, and are not tied to a particular reaction type. Those to be considered here include the problems and relative merits of dynamic vs. equilibrium (static) methods, some thermodynamic and kinetic matters in the choice of reactants, phase relationships, and the utilization of extreme temperatures or pressures.

1.2.1 Dynamic (non-equilibrium) processes

Reactions may be carried out far from equilibrium conditions by using a high rate of formation or conversion, or by quenching liquid or gaseous systems. Such processes may give mixtures, X-ray amorphous or inhomogeneous materials, or very finely divided compounds which may be valued for their reactivity. Intermediate phases, particularly those of some structural complexity, may be missed if crystallinity to X-rays is the criterion for their presence.

(a) Successive reactions with gases

Oxidation, reduction, or metathesis of a heated solid by a gas flowing more or less rapidly is one of the oldest processes known, and is simple to execute. Problems of consecutive equilibria may be encountered with pure gases, for example, the reduction of $CrCl_3$ by hydrogen

$$CrCl_3(s) + \tfrac{1}{2}H_2 \rightarrow CrCl_2(s) + HCl\dagger \tag{1.11}$$

†Only the solid phases are specifically identified hereafter.

which may be followed by

$$CrCl_2(s) + H_2 \rightarrow Cr(s) + 2HCl \qquad [K_{800} \sim 10^{-7}] \qquad (1.12)$$

in the early part of the reactant bed, or after a long reaction period as P_{HCl} approaches 0. In a more extreme example, CuCl is certainly not made by reduction of $CuCl_2$ by pure hydrogen because the second step forming Cu metal is also very favourable. Addition of HCl sufficient to reverse the second reaction in each case is a simple solution.

(b) Decomposition at low pressures

The thermal decomposition $CrI_3 \rightarrow CrI_2 + \frac{1}{2}I_2$ is simple and relatively free of trouble. But clearly reactions that are run far from equilibrium or at low temperatures may give products that are only poorly crystalline, particularly when a large structural arrangement is involved. For instance, disproportionation of Nb_3I_8 in a sealed tube under a large temperature gradient (600 to about 25°C)[27] gives a residue in which only the metal is evident by Debye–Scherrer methods, although, interestingly, this is signalled only when I/Nb is much less than 2.0. Nonetheless, the existence of the thermodynamically stable intermediate compound $(Nb_6I_8)I_{6/3}$ ($=NbI_{1.83}$) has since been well established.[28]

Reactions of this character were later studied much more extensively by Schäfer and Giegling,[29] who examined the disproportionation of M_3X_8, etc. to form first the $(M_6X_{12})X_2$ phases (M = Nb, Ta; X = Cl, Br, (I)) and then the metals by weight-loss measurements as the samples were slowly heated at either 10^{-7} atm. or 1 atm. In no case was the formation of the intermediate cluster evident when the reaction was carried out at 10^{-7} atm., and not even at 1 atm. for the niobium systems. The presence of clusters in the X-ray amorphous intermediates was established in some systems through solvent extraction of the product. Temperatures necessary for decomposition at 10^{-7} atm. were also observed to be 300–500°C higher than for the equilibrium processes. Of course, disproportionation reactions carried out closer to equilibrium can still give well-defined products.

(c) Discharges, plasmas, and other quenched gaseous systems

Plasmas produced by low-pressure (less than 10 torr) discharges provide a rather specific activation of electronic and vibratory modes plus moderate dissociation, while neutral gas (translation) temperatures remain relatively low. Unusual products may be isolated by a relatively rapid quenching of the plasma out of the discharge region. One application has been the preparation of pyrophoric, but evidently high-purity, ZrX_3 (X = Cl, Br, I) from a discharge-promoted reaction[30] of ZrX_4 and H_2. More recent applications[31] have emphasized the discharge capability for activating high-energy bonds to produce thin films, coatings, etc., often amorphous. Chemically useful processes include the synthesis of $(CN)_x$ and black phosphorus at moderate temperatures. Quenching of high-temperature gaseous systems, which are often structurally quite different from the solids from which they originate, has been suggested as a useful synthetic route to new compounds[1] but, so far, examples are lacking.

1.2.2 Factors in the choice of reactants

The choice of a route for the synthesis of a specific compound usually centres around either the ready availability of sufficiently pure or inexpensive reactants, or mechanistic problems, usually material transport, that inhibit a given reaction. Many means of avoiding certain difficulties have been devised. The advantages of gaseous or liquid reactants when not deleterious to the product is evident from prior discussions. Such matters as indirect routes, foreign elements, and thermodynamic factors also merit consideration.

(a) Exchange couples as illustrated by chloride syntheses

Methods for producing more or less volatile halides by reaction of a halogen-containing gas with the more common oxides are some of the oldest known, that is,

$$Fe_2O_3(s) + 6HCl \rightarrow 2FeCl_3 + 3H_2O. \quad (1.13)$$

The processes appear to be relatively facile and so a thermodynamic (equilibrium) assessment of their desirability seems reasonable. Hydrated halides might also be considered as useful sources of many anhydrous halides except that hydrolysis reactions are common for all but those with the least acidic cations, e.g.

$$LaCl_3 \cdot 7H_2O(s) \xrightarrow{\Delta} LaOCl(s) + 2HCl + 6H_2O. \quad (1.14)$$

(The hydrolysis is less when the dehydration is carried out at as low a temperature as possible.) Some of the gaseous chloride-to-oxide couples such as HCl/H_2O that might be used to repress this hydrolysis reaction will also be useful for the reverse process, the conversion of La_2O_3 to $LaCl_3$, a reaction that probably goes through the stable LaOCl.

Table 1.1 shows free-energy data for several useful chloride/oxide couples. These can be used with free-energy (or enthalpy) data for a given chloride relative to that of the corresponding oxide to ascertain which couples will be suitable for the conversion. Of course the choice should leave some appreciable spontaneity: (1) to ensure the equilibrium constant is respectable, and the amount of reactant needed minimal, (2) to overcome the extra stability of any intermediate oxychlorides, and (3) for exothermic reactions, to allow for a decrease in the equilibrium constant with increasing temperature ($d\ln K/d(1/T) = -\Delta H^0 R$. The last condition may be more likely for reactions with HCl, S_2Cl_2, and $AlCl_3$ since ΔS^0 is less favourable for these. The data in Table

Table 1.1 *Some couples for chlorination of oxides*
$$MO_{n/2}(s) + (1) \rightarrow MCl_n(s, l, g) + (2)$$

Couple†		
Reactant (1)	Product (2)	$-\Delta G^0_{298}$ (kJ/g-atom O)‡
Cl_2	$\frac{1}{2}O_2$	0
2HCl	H_2O	38
$SOCl_2$	SO_2	102
$C + Cl_2$	CO	137
CCl_4	$COCl_2$	144
$VCl_4 + \frac{1}{2}Cl_2$	$VOCl_3$	167
$\frac{1}{2}S_2Cl_2 + \frac{3}{2}Cl_2$	$SOCl_2$	182
$COCl_2$	CO_2	190
PCl_5	$POCl_3$	208
$\frac{2}{3}AlCl_3$	$\frac{1}{3}Al_2O_3$	139

†All reactants and products are gaseous except for solid C and Al_2O_3.
‡Source: Wagman *et al.*[32]

1.1 show that the common HCl is certainly not the most energetic reactant, and the classical reaction of Cl_2 with oxides mixed with carbon and a binder to give CO is better, but not the best. The first step in the latter is likely the formation of chlorinated hydrocarbons, making the subsequent reaction with oxide similar to that of CCl_4, which alone is subject to some pyrolysis. Two of the better reactants, $COCl_2$ and S_2Cl_2, are available commercially. The $SOCl_2$ and CCl_4 may also be used in bomb reactions. If possible, all of the products from these reactions (and others) should be vacuum sublimed or distilled for good purity, taking care to avoid entrainment of solid impurities.

Variations on these exchange reactions are possible. Some of the analogous couples have been applied to bromides and, with much less success, to iodides. Reactions of H_2S or CS_2 with oxides or chlorides to prepare sulfides can be similarly analysed for their favourability, e.g.

$$TiCl_4 + 2H_2S \rightarrow TiS_2(s) + 4HCl, \qquad (1.15)$$

except that conversion of a solid oxide to a non-volatile sulfide may now present more diffusion problems.

A related reaction involves simple halide exchange, for example,

$$MnCl_2(s) + 2HBr \rightarrow MnBr_2(s) + 2HCl \qquad (1.16)$$

for which $\Delta G^0_{298} = -15.4$ kJ. However, this sort of reaction is often less satisfactory in that the driving force may not be large and, worse, the reactant and product often form large or complete solid solutions, so that the activity of the reactant goes to zero asymptotically, and a good conversion is more difficult. Exchange reactions of HF with metal chlorides are useful, the lattice energy gain with the fluoride product being the principal driving force.

The 2NH$_4$X $\rightleftharpoons$ H$_2$O + 2NH$_3$ couple has long been used for conversion of oxide to chloride or bromide, especially for the rare-earth, and similar, elements. The oxide is intimately mixed with a good excess of NH$_4$X and slowly heated at 1 atm. under a slow gas flow (to remove the H$_2$O product), ultimately subliming off the excess NH$_4$X. The reaction is actually quite unfavourable thermodynamically when solid NH$_4$X is the reactant. ($\Delta H^0 = +224$ kJ mole^{-1} LaCl$_3$), and common wisdom has been that NH$_4$X merely functions as a good solid source of excess HX(g) on heating. However, recent studies[33] have shown that more stable ternary rare-earth metal chlorides are actually formed first, e.g.

$$La_2O_3(s) + 12NH_4Cl(s) \rightleftharpoons 2(NH_4)_3LaCl_6(s) + 3H_2O + 6NH_3. \quad (1.17)$$

Subsequent decomposition of these starting at about 350° in vacuum gives the desired MCl$_3$ and sublimed NH$_4$Cl, the reaction often passing through the intermediate NH$_4$M$_2$Cl$_7$ or (NH$_4$)$_2$MCl$_5$. The conversion does not appear to be nearly as effective for iodides. A much better preparation of ScCl$_3$ is obtained if (NH$_4$)$_3$ScCl$_6$ is first prepared by an aqueous route (see Section 1.2.5a), a workable procedure in other cases as well.

(b) Direct or indirect routes?

The conversion of another compound to more or less normal-valent binary halides, sulfides, etc. that were considered in the previous section usually require somewhat complex flow systems. However, more high-purity metals are now commercially available than when many of the 'standard' procedures were developed, and purification by sublimation, vapour phase transport, etc. is more frequent. Under these circumstances direct reactions of the metal with non-metal (element or compound) are often less troublesome and give better products. This is particularly true for sulfides and selenides,[34] bromides and iodides, and even some ternary phases. On the other hand some greater care regarding containers may be necessary with metal reactants (see Section 1.4).

Direct syntheses from the elements are not often carried out exactly on stoichiometry but rather with excess non-metal for a normal-valent product, depending on a difference in volatility between the product and the non-metal for a subsequent separation. A few materials, AlI$_3$ for example, strongly retain excess non-metal (I$_2$), and excess metal must be used followed by sublimation to remove the latter. General use of excess metal should be avoided with many halides and chalcogenides if the product is to be molten during the reaction; and a subsequent separation is not possible because many molten compounds dissolve significant amounts of metal even though a reduced solid does not exist.[35] Examples are CdX$_2$, SbX$_3$, Sb$_2$S$_3$, PbCh, and SnCh (X = Cl, Br, I; Ch = S, Se, (Te)).

Syntheses in sealed containers are usually intrinsically simpler. However, these may require elevated temperatures, especially with the more refractory metals, so that reaction in an open system may at first thought appear to be the only safe way to control the pressure of the non-metal. Though this is generally the case with the more volatile O_2, F_2, etc., reactions of sulfur, selenium, bromine, iodine, and many compounds which would generate a high pressure at the temperature of the metal can be easily controlled in a sealed hot–cold reactor. Figure 1.1(a, b) shows what is meant by this description. The metal in one arm is heated as strongly as necessary, often sufficient to remove a protective layer of solid product, while the volatile reactant (and perhaps product) in the other arm is maintained at a temperature necessary for a safe pressure. Even very vigorous reactions (Al + Br_2) can be safely handled as long as the volatile reactant (and product) can freely move to a cooler zone. Of course the apparatus *must* be well evacuated.

Direct routes to ternary compounds such as

$$3CsCl + 2ScCl_3 \xrightarrow{\text{fusion}} Cs_3Sc_2Cl_9 \qquad (1.18)$$

have similar advantages; however, phase relationships (see Section 1.2.3) and alternate low-temperature routes (see Section 1.2.5) should also be considered.

(c) Reductant choices and variations

The traditional H_2 is only a moderately strong reductant suitable for the production for the more easily accessible lower oxidation states in the presence of the more electronegative non-metals, $CrCl_2$, $EuCl_2$, $TiBr_3$, V_2O_3,

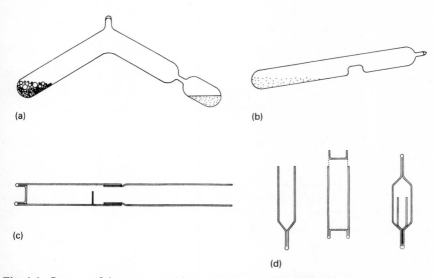

Fig. 1.1. Some useful apparatus. (a) A sealed, hot–cold reaction vessel for a volatile reactant (right side). (b) A similar vessel for a volatile product. (c) A two-piece tantalum sublimation apparatus. (d) Some tantalum container designs.

NbO_2, WO_2, for instance. The stability of the hydrogen–non-metal product of the reduction decreases relatively rapidly with increasing period and decreasing group so that hydrogen reduction of a metal selenide or iodide is ineffective. For example, gaseous HI is never less than 40 per cent dissociated to H_2 and I_2, and the reduction of SmI_3 to SmI_2 with H_2 is in effect not greatly different from the thermal dissociation of I_2 achieved with the same flow of an inert gas. It should be kept in mind that H_2 may in effect promote disproportionation of slightly stable reduced compounds of metals that also form particularly stable hydrides, such as

$$3NdCl_2(s) + H_2 \rightarrow 2NdCl_3(s) + NdH_2(s), \qquad (1.19)$$

and other well-reduced compounds may form hydrides.[36]

A wide variety of metals may be used as reducing agents for higher-valent metal compounds, but in general the optimum choice is the same metal as in the compound being reduced. This is not just a principle of maximum simplicity but is a thermodynamic necessity if the product is of borderline stability with respect to disproportionation into a higher salt and the metal in question. Thus in comparison with the process $Nd + 2NdCl_3 \rightarrow 3NdCl_2$, aluminium will be a significantly weaker and therefore less effective reducing agent in giving a good yield of $NdCl_2$, while barium can be imagined to first reduce the $NdCl_3$ to Nd metal after which the reaction proceeds as before (except that the product is now contaminated with $BaCl_2$ or some double salt with $NdCl_x$).

Aluminium has nonetheless proven to be a feasible reducing agent for other halides because the volatility of its three heavier halides allows a subsequent separation, for example

$$3HfCl_4 + Al(s) \xrightarrow{\Delta} 3HfCl_3(s) + AlCl_3(g), \qquad (1.20)$$

the reaction perhaps going through the intermediate of finely divided hafnium, a useful step when the bulk metal is especially unreactive. Similarly, Kuhn and McCarley[37] found an improved yield of Ta_6Br_{14} could be achieved in only two days in the sealed tube reaction

$$16Al(s) + 18TaBr_5 \rightarrow 3Ta_6Br_{14}(s) + 16AlBr_3 \text{ (g, l)}. \qquad (1.21)$$

The reaction was run in a hot–cold tube (450/280°) with the slight deficiency of Al (and the product) in the hot end, and $TaBr_5$ and the $AlBr_3$ (boiling points 320°C and 268°C, respectively) in the cooler zone. The reaction was completed by heating the hot end to 280°C, increasing to 570°C with the other at room temperature. The transport method of synthesis (see Section 1.3) takes much longer. The utilization of a foreign metal reductant may be helpful if the product halide serves as a flux. Thus the reaction[38] of $ZrX_4(X = Cl, Br, I)$ with Al in the presence of excess AlX_3 as a solvent for both ZrX_4 and reaction intermediates yields ZrX_3 at a relatively low temperature: 220–300°C.

Reactions run with a foreign metal reductant may be complicated by the formation of stable intermetallic phases between the two metals. Both the $NdCl_3$–Al and $HfCl_4$–Al reactions described above are apt to involve the formation of the more stable $NdAl_4$, $NdAl_2$, ..., or $HfAl_3$, $HfAl_2$, ... phases, respectively.[39] This will be particularly likely if fairly high reaction temperatures are employed, as in the reduction of HfI_3 by Al at 500°C where the formation of a thin layer of $HfAl_3$ on the metal inhibited further reaction.[40] On the other hand, intermetallics do not seem to interfere in the $TaBr_5$–Al and ZrX_4–Al reactions above, probably because the refractory nature of the metals and their aluminides allow a negligible formation at these temperatures or, possibly, because these intermetallics are still good reducing agents for $TaBr_x$ or ZrX_4. This problem is of no concern in the reduction of NbF_5 to NbF_4 by silicon, the volatile SiF_4 providing a clean separation of products.[41]

Intermetallic by-products are possible in another type of mixed metal synthesis procedure, the reaction of the pure metal of interest with an easily dried and reducible halide of another metal like HgX_2 or AgX, for example

$$2La(s) + 6HgI_2 \rightarrow 2LaI_3(s) + 3Hg_2I_2. \quad (1.22)$$

According to Carter and Murray,[42] the isothermal reaction of a threefold excess of liquid HgI_2 with lanthanum metal in a thick-walled Pyrex tube at 300–330°C is followed by sublimation of Hg_2I_2 and HgI_2 to the empty end of the container at room temperature (Fig. 1.1(b)). Evidently all of the known Hg–La phases remain good reducing agents for excess HgI_2, and are not found in the product. But a large driving force is probably necessary in cases where excess oxidant cannot be used. Thus the production of InX and (the presumed) InX_2 by reaction of In and Hg_2X_2 in the proper proportions, followed by physical removal of the mercury phase[43] likely yields a less pure product, not from interfering intermetallics but because the concentration of the Hg_2X_2 and In reactants dissolved in the liquid InX_x and Hg phases respectively do not decrease very close to zero as required.

Sometimes the rate of oxidation of the metal is very slow, for example Cu to CuO (which stops at Cu_2O), and Ir to IrO_2, both at 1000°C in oxygen. On the other hand, use of an easily oxidizable salt may allow a ready transformation to the desired product, i.e. CuI to CuO at 300–400°C, and $IrCl_3$ to IrO_2 at 600°C.[20] Likewise CuF_2 and NiF_2 are not easily obtained on fluorination of the metals owing to the formation of a protective coating of the fluorides, but the reaction of F_2 with $CuCl_2$, $NiCl_2$ or AgCl are convenient routes to the difluorides.[24]

Some other 'tricks of the trade' have been discussed by Hoppe.[44] One of the oldest 'tricks' for the synthesis of a ternary fluoride is to oxidize another ternary compound containing the metals in the right proportion, i.e.

$$2F_2 + LiInO_2(s) \rightarrow LiInF_4(s) + O_2. \quad (1.23)$$

and, when possible, a larger anion as well:

$$Ba(Ni(CN)_4)(s) \xrightarrow{F_2} BaNiF_5(s). \quad (1.24)$$

In contrast to the second, fluorination of a mixture of $BaCl_2$ and NiF_2 gives only BaF_2 and NiF_2. The high-pressure fluorination of Cs_2KNdF_6 to Cs_2KNdF_7 (see Section 1.2.6a) likewise starts with Cs_2KNdCl_6. Suitable mixtures work in other cases, as in the Klemm fluorination[45]

$$2KCl(s) + NiCl_2(s) \xrightarrow{F_2} K_2NiF_6(s) \quad (1.25)$$

and with

$$2(NH_4)_2PbCl_6(s) + Ag_2SO_4(s) \xrightarrow{F_2} AgPbF_6(s). \quad (1.26)$$

'Magic' additives such as O_2, BrF_3, H_2O, Ar may also be included to increase reactivity or to improve the growth of crystals (via $O_2^+Mn_2F_9^-$ analogues perhaps). Moist fluorine is known to form $KClF_4$ and KHF_2 as intermediates in the above synthesis of K_2NiF_6. Finally, direct oxidation of the intermetallic LiSb provides a convenient route to $LiSbO_3$, noting that moisture may be a catalyst in metal–O_2 and oxide reactions, perhaps via hydroxides.

(d) Counterions

At times only one ion of a compound is important, and the counterion is a matter of some choice. The predictable effects that this may have on the stability of the compound of interest are also worth noting.

The stability of a cation in a low oxidation state with respect to disproportionation increases with the size of the anion, as first generalized in terms of lattice energies for

$$Cd(s, l) + CdX_2(s, l) \rightleftarrows Cd_2X_2(s) \quad (1.27)$$

where AlX_4^- is much better[46] than X^- for the stability of Cd_2^{2+}. A traditional and parallel stability trend for disproportionation is $I > Br > Cl > F$ although this is not without exceptions, perhaps because of covalency. Exactly the converse relationship applies to cations in high oxidation states which are stabilized by small anions, for example, $CeF_4 > CeO_2 > CeCl_4$. And finally, the stability of complex anions with respect to decomposition to simpler (smaller) anions increases with size and decreases with charge of the cation, for example, $Cs > K > Li$ for either stability of carbonates, sulfates, peroxides, etc. with respect to the oxides, or M_2MnCl_6 against the loss of Cl_2. Johnson[47] gives useful explanations for these as well as for halide exchange reactions in terms of semiempirical lattice energy differences, although the general applicability appears to extend well beyond the more polar compounds.

1.2.3 Phase relationships and melts in direct synthesis

Reactions between solids that are inconveniently or impossibly slow become much more tractable if one of the reactants or an intermediate can be fused. The conditions necessary for a successful synthesis then depend upon either known or estimated phase relationships. Conversely, the search for and synthesis of new compounds may at times demand the concurrent determination of certain phase relationships.

Figure 1.2 presents a hypothetical diagram to illustrate some of the circumstances. The synthesis of a congruently melting compound such as A_4B is simple. Even though reaction of finely divided solids A and B below 500°C may succeed it will become significantly faster just above 500°C since the presence of the eutectic between A and A_4B gives liquid mobility to both of these. But the reaction becomes very direct when run above the melting point of A_4B (665°C), perhaps requiring only a little stirring or agitation to assist solution of B in the liquid of A plus A_4B, whereafter cooling gives the desired product.

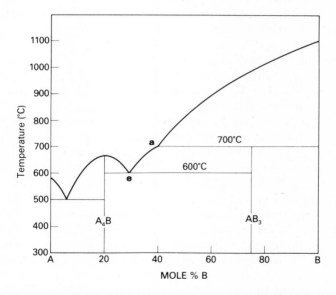

Fig. 1.2. A hypothetical solid–liquid phase diagram for the system A–B. (e eutectic point; a liquid composition in equilibrium with AB_3 and B at 700°C.)

The compound AB_3 presents more problems since *the phase AB_3 does not exist above 700°C*, its incongruent melting point. No attempt to prepare AB_3 by reaction of components above that point followed by fairly rapid cooling will be successful. In other words, some prior knowledge of this fact and perhaps of the lower (600°C) eutectic relationship are necessary for its synthesis. Relative rapid cooling of the stoichiometric melt from 1000°C first gives B, then at 700°C AB_3 starts to separate, and finally at 600°C: A_4B and more AB. Conversion of the final mixture of three phases to the equilibrium product requires that the reaction of $11B + A_4B \rightarrow 4AB_3$ take place either with A_4B as a solid below 600°C, or in the melt **ea** in the range $600 < T < 700°C$. Furthermore, the previous cooling would probably yield substantial physical segregation of the three solids within the container and, thereby, greater problems in their subsequent complete reaction. The best way to avoid these is by rapid quenching of the melt from 1000°C or above to give a finely particled, homogeneous mixture of the three phases which is then reacted by annealing just below 600°C if the solid mobilities are adequate or, if not, in the range 600–700°C where one liquid phase with composition defined by **ea** is present. Of course, any accidental cycling of the reaction temperature above 700°C will destroy AB_3. Naturally any barrier that AB_3 forms over the B-reactant will hinder complete conversion, and will require regrinding, preferably with compaction, followed by re-equilibration.

One way the complete reaction can be aided, especially if blockage of a metal B is a problem, is through the use of excess B surface in the form of a strip or foil since this can be subsequently physically separated from the AB_3 product. The same can be very useful when oxidation of a metal by $A_4B(l, g)$ is slow, and incomplete reactions are a problem, e.g. in the preparation of ZrI_2 from $ZrI_4(g)$ or $ZrCl$ from $ZrCl_2(s)$ or $ZrCl_3(g)$.[48,49] The ability to make a final separation makes this technique far better than the use of powdered metal with its higher surface area.

Some degree of suspicion and caution regarding the results, and the conclusions from them, is still merited. Synthesis of an incongruently melting (or vaporizing) phase which has an intermediate rather than the lowest composition provides greater problems, e.g. AB_3 when another phase AB_4 intrudes, Fig. 1.2. Also, synthesis with the phase relationships shown becomes more difficult the greater is the compositional difference between AB_3 and the liquid **ea**, that is, the more incongruently melting is AB_3. This in effect represents something of the difference between the solid and the solute species in **ea**. For example, many of the molten rare-earth trihalides react with metal at elevated temperatures to form only dilute solutions of the M^{2+} ions.[50] Thus the practical one-step synthesis of the fibrous Gd_2Cl_3, a phase with infinite chains of condensed metal octahedra, by reaction of melt **ea** ($\sim GdCl_{2.94}$) with excess metal for 10–20 days is limited to about a 65 per cent yield, although the product is perfectly stable.[51,52] Synthesis of a new compound under such circumstances is apt to be more successful if the metal is reacted with a gaseous oxidant so as to avoid the physical retention of unreacted liquid by the product. The quantitative reduction[53] of $ScX_3(g)$ with metal to Sc_2X_3 (X = Cl, Br) is a good example.

The implications of Fig. 1.2 pertain only to portions of binary or pseudobinary systems where all compositions can be achieved from two components. A transition to a ternary or higher-order phase diagram is necessary for consideration of many 'neat' syntheses of more complex compounds, and this is accompanied by somewhat greater difficulty in their visualization. (The inexperienced may wish to study Ricci.[54]) The series edited by Alper[55] contains many examples of the applications of phase diagrams to materials science.

Generally only a portion of the proper ternary diagram will be of interest for synthesis. Thus, for the reaction of CsCl with $TmCl_3$ and Tm to give $CsTmCl_3$, one need be concerned only with that ternary field $CsCl-TmCl_3-$Tm within the general Cs–Tm–Cl ternary system. Even in a binary liquid–solid system, the addition of a third component may lower the eutectic temperature significantly and may allow either better reactivity (more liquid phase) or access to other binary compounds which are stable only at lower temperatures. This may be of special value with incongruently melting substances. For instance the addition of a KCl 'flux' to $Y-YCl_3$ reactions allows much better yields of Y_2Cl_3 to be achieved than without flux.[56]

The addition of small amounts of various fluxes, even water, is frequently used to promote reactions between solids (sometimes called *reactive sintering*), including reactions run under pressure. These are similarly employed as solvents and have the advantage of mobilizing one reactant and lowering reaction temperatures considerably. The reaction

$$Li_2CO_3(s) + 5Fe_2O_3(s) \rightarrow 2LiFe_5O_8(s) + CO_2 \quad (1.28)$$

is aided considerably by the use of a molten flux in place of a lengthy milling, firing, and re-grinding process, specifically by the $Li_2SO_4-Na_2SO_4$ eutectic composition which at 800°C dissolves the Li_2CO_3 reactant and carries it to the iron oxide very effectively. Similar fluxes allow ready access to MFe_3O_4 [M = Mg, Ca, Co, Ni, Zn] evidently by dissolving both the Fe_2O_3 reactant and the product.[57] In another case,[58] the formation of post-transition-metal chalcogenides and pnictides (GaP, CdSe, ZnTe) directly from the elements is greatly inhibited by a thin film of product which forms on the metal, while the addition of the iodides of the same metals lowers the reaction temperature by about 500°C. The iodides act as solvents for many of the metals and probably some of the products, perhaps analogous to their function in chemical transport (see Section 1.3).

A wide variety of non-reactive fluxes like PbO, PbF_2, B_2O_3, M_2MoO_4, Bi_2O_3, Na_2S_x, and their mixtures have also been utilized for crystal growth of oxides such as yttrium–iron and yttrium–aluminium garnets, $BaTiO_3$, $PbZrO_3$, spinels, $KFeS_2$, etc. A large list is provided by Laurent.[59] The techniques have been described by White[60] as well as Anthony and Collongues.[21]

Certain liquid metals have been found to be valuable solvents for the isothermal syntheses of refractory compounds such as MnP_4, RuP_2, IrP_2, and

CuP$_2$ which are obtained from the elements using liquid tin,[61,62] and MnSi from molten copper.[63] A necessary condition is that the solvents form no (or only slightly stable) phosphides, silicides, and so forth.

Other syntheses of complex oxides, sulfides, etc. should be aided by non-reactive molten salts or metals, but these routes have not been widely explored. Except for the difficult cases there may be easier and more direct routes if good-size crystals of the product are not a principal objective.

1.2.4 Electrolysis

The synthesis of crystals of some useful materials by electrolytic reduction of melts has been known for over fifty years, but the technique and its capabilities remain rather empirical and not well explored. A typical reaction is

$$2La_2O_3 + 12B_2O_3 \rightarrow 4LaB_6(s) + 21O_2 \qquad (1.29)$$

utilizing gold electrodes and a LiBO$_2$/LiF melt to dissolve the reactants.

The success of a new reduction appears to be somewhat a matter of luck rather than design or control of more than current and voltage. The process effectively sorts out the more conducting from among the possible reduced products, and is capable of yielding good crystals at relatively low temperatures compared with direct reactions. Thus, pure crystalline phosphides of many 3d-elements can be obtained from solutions of the oxides in mixtures of the alkali metal phosphates and halides, whereas the direct reaction of the elements tends to give inhomogeneous products and poor crystals. The large amount of chemistry associated with the solvent reduction (e.g. phosphate to phosphide), and the possibility of interesting intermediates have generally not been explored. Traditional solvents have been those with high decomposition potentials: alkali metal borates, phosphates, molybdates, fluorides, and chlorides; but few systematics regarding the effect of their obvious acid–base chemistry have been considered. A given process often may be limited or complicated by the change in melt composition, and thence the change in products, as the electrolysis proceeds. Recent reviews are by Wold and Bellavance,[64] and by Feigelson.[65]

Examples of the variety of compounds so prepared include M(II)V$_2$O$_4$ spinels, TiB$_2$, CrSi$_2$, MoC, and the monosulfides of Ce, Th, Gd, U, the latter being produced from a solution of normal-valent sulfide and chloride in a KCl/NaCl eutectic. Probably the best-known products of melt electrolysis are the bronzes, for example: M$_x$WO$_3$ [M = Li, Na, Cu, and Gd] from solutions of the metal oxide in WO$_3$. Unpublished research by McCarroll[66] includes the electrolytic preparation of the metal cluster phases M$_2$Mo$_3$O$_8$ [M = Mg, Fe, Co, Ni, Zn], M$'_5$Mo$_3$O$_{16}$ [M$'$ = La, Nd], and Y$_2$MoO$_5$.

A large range of possibilities remain uninvestigated, especially for non-oxide melts and at lower temperatures. However, Masse and Simon[67] recently showed that $K_4Nb_6Cl_{18}$ is obtained at a carbon cathode by electrolysis of a solution of Nb_3Cl_8 in molten KCl/LiCl, with niobium being produced with a higher current. The most remarkable product they described is $Gd_5Cl_9C_2$ which forms from liquid $GdCl_3$ at a graphite crucible cathode, the phase containing C_2 units centred in metal octahedra which share edges. The double-metal-layered ZrCl was first, and accidentally, prepared in 1960 by R. S. Dean during electrolysis of a mixed chloride melt (cited by Struss and Corbett[49]).

The noteworthy mobility of alkali metal and other cations between sheets or in channels in many intercalated compounds permits room-temperature routes to be used for their insertion (host reduction) or removal (oxidation). Most of these processes are discussed in the next section as examples of low-temperature syntheses, but it is appropriate to note here the electrochemical means for the intercalation (and the converse) of MCh_2, MPS_3, MOX, V_2O_5, and MoO_3 phases. An oxidation (or reduction) such as

$$x\text{Li}^+ + x\varepsilon^- + MS_2(s) \rightleftarrows Li_xMS_2(s) \qquad (1.30)$$

may be readily accomplished using a lithium (or other) anode and a lithium salt in a non-aqueous solvent.[68] Several particularly favourable examples have been proposed for battery systems. The intercalation process may or may not be accompanied by a phase change, and some of the products obtained may be metastable. In addition, it has been recently demonstrated that the molybdenum chalcogenides Mo_6Ch_8 can be reversibly intercalated at room temperature by a variety of uni- and dipositive ions to form the so-called *Chevrel phases*. Coulometric titrations in CH_3CN show that materials such as $Cu_xMo_6S_8$ and $Li_xMo_6S_8$ ($0 < x < 4$), $Zn_2Mo_6S_8$ and the mixed species $Ag_{0.7}Na_xMo_6S_8$ ($0 \leqslant x \leqslant 3.2$), and many others may be so attained.[69]

1.2.5 Temperature and phases of low stability

The general problem of diffusion-limited reactions in the solid state is frequently ameliorated by the generous application of heat, and this may be quite adequate. However some general effects of high temperature should be noted: phases formed at high temperatures tend to have lower densities, higher symmetries, and more open structures, and to contain higher levels of defects and disorder. Phase boundaries also widen, and compounds become more nonstoichiometric at high temperatures. Methods of attaining very high temperatures, and some solid-state applications, have recently been described by Müller-Buschbaum.[70] Even in this extreme regime, kinetics may still be important; compounds such as $SrLa_2O_4$ and BeY_2O_4 may be obtained using focused solar radiation and electric arcs as sources of heat, while the high power (10^6 W cm^{-2}) transferred by laser irradiation, followed by the rapid quenching achieved thereafter, allows the preparation of metastable phases such as $Ca_2Ho_2O_5$ and $Sr_3La_4O_9$.

An important consequence of a high-temperature synthesis is that the products usually represent the thermodynamically most stable phases. All phases that are stable only at lower temperatures are of course lost by such a brute-force approach. A great deficiency of solid-state chemistry at present appears to be the pervasive lack of understanding of the reaction mechanisms, without which one is unable to predict, control, or exploit a reaction with some sophistication, and thereby to access phases that are of low- or even meta-stability.[20] (Imagine an organic chemistry which dealt only with the thermodynamically most stable compounds!) Other deficiencies also include poorly understood surface, nucleation, and crystallization effects, most particularly when a vapour or liquid path is not available. As pointed out by Brewer,[71] there must be many unknown compounds for which the rates of formation by ordinary means over the temperature ranges of their stabilities are very low to negligible on our time scale. And others will form at feasible rates only over a limited temperature range just below their disproportionation temperatures which serve as upper limits for their synthesis (see Section 1.2.3). Although no systematic body of knowledge pertains to 'low-temperature' routes, some diverse examples are available.

(a) Low-temperature routes

A classical means of 'lower-temperature' synthesis involves precipitation from aqueous solution followed by drying or heating, if necessary. Diversity may be obtained through the use of non-aqueous solvents and sol-gel routes. Chianelli and Dines[72] found they could precipitate the layered disulfides of transition groups IV–VI by reaction of M_2S (M = Li, Na, etc.) or $(NH_4)_2S$ with the corresponding transition-metal halides in tetrahydrofuran or acetonitrile. The products ranged from amorphous to crystalline.

Other relatively low temperature syntheses can also be achieved in solid–solid reactions when the large diffusion path lengths intrinsic to powdered reactants are reduced by the use of co-precipitated reactants, a suitable chemical compound, or a solid solution of reactants. Reaction rates in the first case are correspondingly improved since particle sizes may be <0.1 μm, but relatively high temperatures are still necessary for completion. Paulus[57] discusses several variations.

A similar, convenient low-temperature route to moisture-sensitive ternary halides of the rare-earth (and other) elements has been developed.[73-5] The route is remarkable because it starts with aqueous solutions and achieves oxyhalide-free (to Guinier diffraction) products, evidently in part because of the driving force to form the ternary products from intimately mixed reactants. Ordinarily a phase such as $Cs_3(RE)_2X_9$ would be obtained by a direct route provided the phase relationships were known or propitious, namely, by fusion of CsX and REX_3 in the proper proportions following the usual trouble of first obtaining pure REX_3. The better procedure for chlorides and bromides starts with solutions of the oxides or halides in the proper proportions in concentrated aqueous HX which are then evaporated to dryness. The residue at this point may be a mixture but generally contains the rare earth element in a ternary halide, possibly as a hydrate (e.g. $Cs_2RECl_5 \cdot H_2O$ and $RECl_3 \cdot nH_2O$). This is then slowly heated to 400–500°C under a flow of anhydrous HCl or HBr without melting. A good number of other ternary halides of the alkali metals can also be obtained this way, including those of the metals In, Bi, Sc, Cr, Fe.

A lower synthesis temperature is achieved when the diffusion path length is of the order of atomic dimensions. Thermal decomposition of an appropriate compound is one possibility, as illustrated by[76]

$$(NH_4)_2Mg(CrO_4)_2 \cdot 6H_2O(s) \rightarrow MgCr_2O_4(s) + 2NH_3 + 7H_2O + \tfrac{3}{2}O_2. \qquad (1.31)$$

The comparable oxidation of mixed metal compounds has already been described in Section 1.2.2c. But the choice of the metallic cations, their proportions, and (in the above example) the anion are all restricted by the availability of starting compounds. Solid-solution precursors provide the same advantages and also allow ready variation of the stoichiometry although, on the other hand, the initial solution can usually be obtained only for two reactants of the same charge type. One illustration[77] is the preparation of the spinels MFe_2O_4 by ignition and oxidation of solid solutions $(M, 2Fe)(C_2O_4)_3 \cdot 6H_2O$ [M = Mg, Mn, Ni, Co, Zn]. More recently, Longo and co-workers[78] found that $CaCO_3 \cdot MnCO_3$ solid solutions (calcite structure) precipitated from aqueous solutions could be decomposed in oxygen at 800–1000°C to a number of pure new phases, for instance

$$5(0.4Ca, 0.6Mn)CO_3(s) + \tfrac{3}{2}O_2 \rightarrow Ca_2Mn_3O_8(s) + 5CO_2. \qquad (1.32)$$

In contrast, the comparable reaction of the mixed components required 1300°C with frequent re-grinding for completion, and it yielded a material with much lower surface area and higher crystallinity. The initial heating of the solid solutions should generally be rapid to avoid preferential decomposition, oxidation, or segregation of one of the components ($MnCO_3$ in this case). This route allows the preparation of four mixed oxides which are not stable above 975°C in 1 atm. O_2, and which had not been obtained by conventional methods, or were observed only in multiphase mixtures.

(b) Control by structure

The reaction control afforded by the host in many intercalation (often topotactic) reactions provides good examples of useful 'low-temperature' routes. In one type a metal cation is inserted into the van der Waals gap between non-metals in layered or chain chalcogenides of group IV–VI transition metals,[79] a process that is best known for alkali metal ions and disulfides such as

$$xNa + MoS_2(s) \rightarrow (Na^+)_x MoS_2^{x-}(s) \qquad (1.33)$$
$$[\ldots S-Mo-S \ldots] \qquad [\ldots Na-S-Mo-S \ldots]$$

where the layering sequences are shown below the formulas. Some compounds of this type are unstable at high temperatures (with respect to Na_2S, MoS_2, and Mo_2S_3 in the above illustration) so that a successful synthesis requires a low-temperature route. A useful one[80,81] is the reaction of butyllithium with TiS_2 (layers) or TiS_3 (chains) suspended in hexane at 20–60°C. Other sulfides, as well as the selenides and tellurides, give analogous reaction products, often stoichiometric. Chevrel phases may also be so lithiated, e.g. to $Li_{0.8}PbMo_6S_8$, or exchanged with molten salts[82] ($ZnMo_6S_8 \rightarrow PbMo_6S_8$ at 550°C). Success obviously depends on a good lithium mobility between the non-metal layers at the reaction temperature and, when this is so, the insertion is generally reversible. Although this method is essentially limited to lithium, all of the alkali, and some alkaline-earth, metals may be incorporated into many hosts using liquid ammonia as solvent. However, non-stoichiometry and solvent incorporation may also then occur, and these stronger reducing agents may decompose tellurides, polychalcogenides, etc. Electrochemical metallation with the aid of the alkali-metal salt in propylene carbonate is another alternative (see Section 1.2.4). Products of these low-temperature routes are at best microcrystalline, and they often have sufficient stacking and intercalation defects to give only broad-line X-ray patterns.

The reverse of this process may also yield a new compound: $LiVS_2$ prepared at high temperatures ($CS_2 + LiVO_2$) may be delithiated at room temperature to give[83] the otherwise unknown layered VS_2. The host lattice control afforded by $CaMnO_3$ is illustrated[84] by a useful low-temperature topotactic route to the oxygen-deficient perovskite,

$$6CaMnO_3(s) + 2NH_3 \xrightarrow{475°C} 6CaMnO_{2.5}(s) + N_2 + 3H_2O. \qquad (1.34)$$

Two other types of metastable phases generated by reactions of a strongly bound parent structure at low temperature are (1) ZrXH [X = Cl, Br] where hydrogen is intercalated into the interstices between the double metal layers in ZrX at 25–400°C, the phases decomposing[85] above 600°C to ZrH_2 and $Zr_6X_{12}(H)$, and (2) M_6O, M_4O, and M_2O phases [= Nb, Ta] where oxygen reacts with M at 300–600°C, basically to form M-like structures.[86]

(c) Other examples

A few other illustrations of the successful reduction of reaction barriers, and thence of reaction temperature, through the proper choice of reactants are described by Hoppe.[44] Exchange reactions may access new compounds or metastable phases, as illustrated by

$$2KInO_2(s) + Na_2O(s) = K_2O(s) + 2NaInO_2(s) \qquad (1.35)$$

as well as some routes to new ternary lead oxides. Obviously Na_2O is a more effective reactant in the above than would be Na_2CO_3 or $NaNO_3$ which decompose to Na_2O at higher temperatures. 'Nascent' reactants may be even more effective; the formation of 'fresh' Na_2O and O_2 in another reaction apparently allows the oxidation of Ag metal to form the new Na_3AgO_2 at sufficiently low temperatures that the O_2 decomposition pressure is not great. In another example,[87] the metastable KYb_2Cl_7 forms when finely divided $K(NH_4)_2YbCl_6$ or $(NH_4)_3YbCl_6$ decomposes at about 400°C in vacuum to produce 'active' $YbCl_3$ in the presence of KCl.

1.2.6 High pressure

Synthetic methods utilizing high pressures may be subdivided according to the phases present into those involving a high pressure of a reactive gas, a high hydrostatic pressure on a solid, and the high solvent pressures associated with hydrothermal methods. Experimental methods for all three are relatively similar but are not germane of this chapter—the interested reader should consult Rooymans[88] and references therein. This brief description will again exclude crystal growth where these methods are much applied, but limit the discussion to synthesis capabilities. For some materials there is no alternative to the high-pressure route, but the necessary equipment is not widely available.

(a) Reactive gas

This technique invariably involves the synthesis of binary or ternary oxides, fluorides, etc. of metals in high oxidation states, reactions that often require elevated temperatures for solid reactivity and a high gas pressure, usually at or below 80 kbar, because of the limited stability of the product. Joubert and Chenavas[89] and Bougon et al.[90] give some experimental details and examples. Of course a sufficient gas pressure may also provide the characteristic benefits of high pressure on reactions between solids to be described later.

The necessary oxygen pressure is sometimes generated by decomposition of CrO_2, MnO_2, $KClO_3$, etc. within a sealed cell rather than coming from an external source. For illustration, the iron(IV) perovskite is obtained by the reaction

$$Sr_2Fe_2O_5(s) + \tfrac{1}{2} O_2 \xrightarrow{500°C} 2SrFeO_3(s) \qquad (1.36)$$

at 340 atm. oxygen whereas only the defect $SrFeO_{2.86}$ is obtained at 1 atm.[91]

Fluorine can now be used up to 4.5 kbar and 600°C in Monel autoclaves so that higher temperatures (for better crystal growth) and higher oxidation states can be obtained. Hoppe[44] describes in some detail the attainment of compounds including Cs_2CuF_6, $BaAuF_7$, and $SrNiF_6$; the higher cationic charge in the last two reduces the stability of the high oxidation state (relative to, say, Cs^+ compounds), necessitating the high fluorine pressure for synthesis.

(b) Solid reactions

Although reactions involving only solids characteristically have small entropy changes, many are accompanied by decreases in volume, and therefore the free-energy changes become more favourable with increasing pressure. This may cause structural transitions in a wide variety of phases, increase delocalization or metal–metal bonding, or stabilize new compounds. An increase in coordination number is frequently involved. Moreover, the effect of pressure alone will favour many spontaneous reactions through improved intergranular contact and the stresses generated. (Although the number of points defects is in principle diminished by increased pressure this does not appear to dominate.) For example, the reaction time necessary for

$$Re_2O_3(s) + Fe_2O_3(s) \rightarrow 2ReFeO_3(s) \tag{1.37}$$

can be reduced from several days at ambient pressure with intermittent re-grinding to one-half hour at 50 kbar.[92] High pressure may also be applied prior to reaction in order to improve interphase contact between the reactants. Many other reactions require high pressure to stabilize products which may then be isolated by cooling under pressure. Thus the distorted perovskite $PbSnO_3$ is obtained by[93]

$$SnO_2(s) + Pb_2SnO_4(s) \xrightarrow[>70\,\text{kbar}]{400°C} 2PbSnO_3(s) \tag{1.38}$$

Of course, pressure may also cause decomposition, as with the spinel[94]

$$Mg_2TiO_4(s) \xrightarrow[15\,\text{kbar}]{1000°C} MgO(s) + MgTiO_3(s). \tag{1.39}$$

Structural transformations of pure materials under high pressure have been widely studied and described (e.g. by Goodenough *et al.*[92]). Generally, those that involve the larger structural rearrangements are more apt to be quenchable.

Most of the reactants and products obtained to date by high-pressure means appear to have been handled in the atmosphere without specific precautions. Some have even used fluxes that were subsequently washed out with water. However, many of the techniques could easily be adapted to reactions of materials that are sensitive to air and moisture, if the reactants are enclosed in an autoclave or welded ampoule under an inert atmosphere box prior to reaction.

(c) Hydrothermal synthesis

This technique has received a great deal of attention as an aid for crystal growth, especially for mineralogy. It depends on the improved solvent properties and high mobilities obtained in supercritical water ($>373°C$). A large list of results is given by Demianets and Lobachev,[95] while apparatus and techniques are well described by Anthony and Collongues.[21] A wide variety of complex oxides have been synthesized, some at up to 100 kbar and 1500°C. One example is[89]

$$M_2O_3 + 2B(OH)_3 \xrightarrow[400 \text{ kbar, 1 h}]{1000-1200°C} 2MBO_3 + 3H_2O \quad [M = Al, Ga, Y, Lu]. \tag{1.40}$$

Syntheses in NH_3 and HF have also been reported, for instance MoO_2F (ReO_3 structure) from HF at 3 kbar and 700°C.[96]

1.3 Chemical (vapour phase) transport

Probably no development in the past twenty to thirty years has had a greater impact on solid state synthesis, purification, and crystal-growth techniques, than that of chemical (vapour phase) transport, largely developed by Schäfer.[20, 97-9] In the context of synthesis, this process of 'gaseous solution' by selected and often predictable chemical reagents enables otherwise non-volatile products or reactants (or both) to be moved along an activity (and usually temperature) gradient at temperatures which are low compared with those for direct volatilization. (The latter is excluded by the name *chemical tranport*.) The value of such a process is self-evident from the synthetic problems encountered in previous sections of this chapter. In addition, transport may be used to convert otherwise immobile or easily decomposed phases into excellent single crystals which are sometimes an embarrassment to synthetic chemists using other methods.

The principles of chemical transport will be illustrated first for a single phase in a process which would allow not only its separation from impurities, and crystallization, but also its synthesis. A classical example follows from the presence of the *reversible* equilibrium

$$ZnS(s) + I_2 \rightleftarrows ZnI_2 + \tfrac{1}{2} S_2 \tag{1.41}$$

in which I_2 is the transporting agent and *all products are gaseous*. A useful temperature is 900°C where $K \ll 1$. Since the reaction is *endothermic*, the equilibrium constant and therefore the partial pressures of ZnI_2 and S_2 *decrease* with *decreasing* temperature. This means that ZnS is transported to (i.e. crystals of ZnS reform in) a lower temperature region. The following schematic describes the system:

I_2	$\xrightarrow{ZnI_2}$ $\xrightarrow{S_2}$	I_2
original ZnS		product ZnS
900°C (T_2)		800°C (T_1)

Under a large (0.1–1.0 atm.) pressure of I_2 or inert gas, local equilibrium will generally pertain since the heterogeneous reaction (the above equilibrium) is more rapid than diffusion of ZnI_2 and S_2. Diffusion of the I_2 released by the back reaction $[ZnI_2 + \frac{1}{2} S_2 \rightarrow ZnS(s) + I_2]$ to the high-temperature region is not significant in this case since the background gas is principally or entirely I_2. The rate-determining transport of ZnI_2 and S_2 depends on the difference in their equilibrium pressures at T_2 and T_1 ($T_2 > T_1$) plus the obvious parameter of diffusion coefficients and the cross-sectional area and length of the container. (Generally the transport rate is low, yielding a few milligramme per hour or per day, although this can be increased by convective gas transport obtained with large inclined tubes and gas pressures > 3 atm.) As a rule, $\Delta P_i/\Sigma P > 10^{-4}$ is desirable for the diffusing species, which in effect means that K_p should not be too large or two small at the operating temperature. Furthermore, a reasonable $\Delta H^0(\pm)$ is needed in order to achieve a sufficient pressure gradient, which also puts restrictions on ΔS^0 for the better processes (see Schäfer[97]). Clearly, transporting agents that will accomplish such a process can to some extent be selected to give favourable equilibria.

Another and very impressive result is the ruby-red Cu_2O obtained by the transport reaction

$$Cu_2O(s) + 2HCl \underset{900°C}{\overset{600°C}{\rightleftharpoons}} \tfrac{2}{3}Cu_3Cl_3 + H_2O, \qquad (1.42)$$

the transport taking place from 600 to 900°C because the reaction is exothermic. (The temperatures shown with the arrows emphasize the change taking place in the different regions, although this usage is uncommon.) Only traces of HCl are necessary. Interestingly, ΔH^0 (and ΔS^0) change sign for this reaction at higher temperatures owing to dissociation of Cu_3Cl_3, so the transport

$$Cu_2O(s) + 2HCl \rightleftharpoons 2CuCl + H_2O \qquad (1.43)$$

now takes place in the opposite direction (1100 to 900°C). Consequently Cu_2O distributed over the extreme temperature range within a tube will all migrate to the central zone near 900°C where the vapour concentration is a minimum. As a further variation, if the Cu_2O is contaminated with either of the adjoining phases, Cu or CuO, the latter will move to a cooler zone by the endothermic transport processes

$$Cu(s) + HCl \rightleftharpoons \tfrac{1}{3} Cu_3Cl_3 + \tfrac{1}{2} H_2 \qquad (1.44)$$

or

$$CuO(s) + 2HCl \rightleftharpoons CuCl_2 + H_2O. \qquad (1.45)$$

The number of compounds that have been so transported is very large, and the agents which have been used, diverse. For purposes of synthesis it is important to realize that *anything that can be transported can be synthesized under transport conditions*. For illustration, the reaction

$$Zn(s) + S \rightarrow ZnS(s) \qquad (1.46)$$

near 800°C gives a low conversion because a skin of product on the liquid metal greatly inhibits its oxidation. Obviously the ZnS layer can be moved, or prevented from forming, by the addition of some I_2 to generate the transport process first described above. (One cannot distinguish whether I_2 first attacks the metal to form ZnI_2, followed by: $ZnI_2 + \frac{1}{2} S_2 \rightarrow ZnS(s) + I_2$ [$K \gg 1$]; but the equilibrium reaction responsible for the transport is clearly the reverse.) The effective syntheses of many other metal sulfides and selenides proceed similarly. Note that here, as in many other cases, the clean separation of the product that is achieved makes the process selective, and stoichiometric amounts of the reactants need not be used. The synthesis and transport of Cu_2O (see above) could be accomplished in one tube, starting with say Cu, CuO, and a trace of concentrated aqueous HCl, the transport reactions shown for Cu, CuO, and Cu_2O ensuring that Cu and CuO in effect react to form Cu_2O.

Materials of considerable technological importance, for example GaAs, are formed, transported, and grown epitaxially in static or flow systems[100, 101] utilizing equilibria of the character

$$GaAs(s) + HCl \rightleftarrows GaCl + \tfrac{1}{2} H_2 + \tfrac{1}{4} As_4. \qquad (1.47)$$

In this instance, convenient starting materials are $AsCl_3$, Ga, and H_2 which react to establish this equilibrium.

Many halides and oxyhalides may be transported, for example

$$TaOCl_2(s) + TaCl_5 \rightleftarrows TaOCl_3 + TaCl_4 \qquad (1.48)$$

from 500°C to 400°C starting with, for instance, Ta_2O_5, $TaCl_5$, and Ta, not necessarily in the stoichiometric amounts. Similarly, clusters are synthesized and transported by a (probable) process such as

$$Nb_6I_{11}(s) + \tfrac{13}{2}I_2 \rightleftarrows 6NbI_4 \qquad (1.49)$$

starting with Nb_3I_8 and excess Nb, and by

$$Sc_5Cl_8(s) + 2ScCl_3 \rightleftarrows 7ScCl_2 \qquad (1.50)$$

where Sc and $ScCl_3$ are convenient reactants.[102] It will be noted that transport of a *reduced* phase requires that the gaseous product in a higher oxidation state should still be reduced (or easily dissociated), a condition which appears difficult to meet in some binary systems.

A transport aid to synthesis will also follow if one of the reactants is mobilized instead. Diffusion path lengths in an intimate mixture need then be only a few microns, and $\Delta P_i/\Sigma P$ may be as low as 10^{-6}. One example is

$$2\text{CaO(s)} + \text{SnO}_2(\text{s}) \rightarrow \text{Ca}_2\text{SnO}_4(\text{s}) \qquad (1.51)$$

which is promoted under *isothermal* conditions by H_2 or CO as these partially reduce and therefore transport SnO_2 via the volatile SnO. It is important to note that the *total* surface area of the CaO is now available for reaction, not just the points of contact. The synthesis of NbO is a well-known illustration.[97] The direct reaction

$$3\text{Nb(s)} + \text{Nb}_2\text{O}_5(\text{s}) \rightarrow 5\text{NbO(s)} \qquad (1.52)$$

is very difficult under clean conditions (as in a vacuum), the finest powders possible in pressed pellets requiring 1500–1700°C for complete reaction. However, a small amount of water in a sealed tube (perhaps from the silica container) will produce H_2 on reaction with Nb, after which the H_2/H_2O couple transports the surplus oxygen from Nb_2O_5 to Nb in a few hours at 900°C. The reaction stops when either Nb_2O_5 (and the NbO_2 intermediate) or Nb is consumed, so that separate piles of Nb and Nb_2O_5 are converted without metal transfer to NbO in one place, and NbO mixed with the excess reactant at the other. These are now isothermal reactions driven by a gradient in chemical potential of oxygen rather than by a temperature-derived difference. The Nb–Nb_2O_5 reaction may be catalysed by a variety of other agents that transport oxygen, for example

$$\text{Nb}_2\text{O}_5(\text{s}) + 3\text{NbCl}_5 \rightleftarrows 5\text{NbOCl}_3 \qquad (1.53)$$

which would come about following the addition of Cl_2, $NbCl_x$, $NbOCl_y$, or (in part) HCl to the reactants. Iodine or iodides will similarly function via $NbOI_3$, but in this case the NbO is transported as well (950 up to 1100°C) leaving any excess NbO_2 behind. Naturally the volatile transporting agents in all of these reactions are condensed elsewhere, and are removed on completion of the reaction.

Many elements can also be transported, particularly as halides or oxides and this may provide synthetic advantages. In addition, a great many halides are mobilized at temperatures where they are otherwise not very volatile through the formation of gaseous complexes with AlX_3, GaX_3, FeX_3 etc., e.g.

$$\text{VCl}_2(\text{s}) + \tfrac{3}{2}\text{Al}_2\text{Cl}_6 \rightleftarrows \text{VAl}_3\text{Cl}_{11}, \qquad (1.54)$$

thereby accelerating halide reactions. Similarly, AlX_3 stabilizes gaseous halides of Au and Pd so that the latter may be transported,[103] and it can be expected to do the same for more conventional gaseous species responsible for transport by HCl, I_2, etc., such as $HgCr_2Se_4$.[104]

Rather than enumerate more examples in detail, a collection of transportable phases is listed in Table 1.2 in order to give the reader a better idea of the variety of possibilities. In some cases, the gaseous products responsible for the transport are not known. Wolf *et al.*[105] give detailed consideration to

Table 1.2 Additional examples of chemically transportable compounds†

Compound	Agent	Compound	Agent
TiO_2	$I_2 + S_2$	$CrTaO_4$	Cl_2
AlOCl	$NbCl_5$	MWO_4	Cl_2
CrOCl	Cl_2	(M = Mg, Mn, Fe, Ni, Zn)	
SiO_2‡	HF	MFe_2O_4	HCl
IrO_2, RuO_2	O_2	(M = Mg, Mn, Co, Ni)	
Be, WO_3	H_2O	MNb_2O_6	Cl_2 or HCl
BP	HCl	(M = Ca, Mg, Co, Ni, Zn)	
Nb_5Sb_4	I_2	ZrOS, ZrSiS	I_2
TiS_2, NbS_2, TaS_3, WS_2	S	MIn_2S_4	I_2
Cr_2S_3, FeS_2, PtS		(M = Mn, Co, Zn, Cd)	
$RbNb_4Cl_{11}$	$NbCl_5$	Cu_3MS_4	I_2
NbS_2Cl_2	$NbCl_4$	(M = Nb, Ta)	
BiSBr	Br_2	Cd_4SiSe_6	I_2
$CuCr_2S_3Cl$	Cl_2	$ZnSiP_2$	I_2
$MGeO_3$	HCl	$LaTe_2$, La_2Te_3	I_2
(M = Mn, Fe, Co)		Be_2SiO_4, Zn_2SiO_4	Li_2BeF_4
$MTiO_3$	Cl_2	$HgCr_2Se_4$	$AlCl_3 + HgCl_2$
(M = Mg, Ni)		$V_nO_{2n-1}[n = 4\text{–}8, \infty]$	$TeCl_4$
		$FeGeO_3$, $MnGeO_3$	NH_4Cl

† From Schäfer[20,97,98] and Emmenegger[106]. For transport of elements, see Schäfer.[20,97,98]
‡ See Section 1.4 for additional examples.

the transport of non-stoichiometric phases, particularly FeS_{1+x} and vanadium oxides.

1.4 Containers

One of the larger problems has been left until last, the matter of containment. The literature naturally relates what has been successful, though perhaps only by ignoring small amounts of container attack, or 'getting-by' because of some fortunate fractionation. Presumably many useful reactions cannot be realized because a satisfactory container is lacking.

No general description of containers suitable for different reactions is available. Fused silica ('quartz') is widely used although it is important to recognize its limitations. Obviously the more active metals reduce SiO_2, and form very stable oxides and silicides. Because of limited contact and poor diffusion between metal and silica, many of these reactions may be so slow as to be insignificant, for instance with Zr, Nb, Ta, and Mo. On the other hand, certain materials will chemically transport either the metal or the SiO_2, and lead to substantial contamination. For example, the process

$$SiO_2(s) + H_2 \rightarrow SiO + H_2O \qquad (1.55)$$

at high temperatures (1000°C) leads to effective transport from the wall to other materials, including metals. The H_2 may come from degassing of metals or, more frequently, by reduction of traces of water that are evolved from silica on heating. Hydrolysis of $FeCl_2$ at 600°C has been noted even with SiO_2 which has been previously flamed under vacuum.[107] Halogen is also deleterious in the presence of certain metals. With $TaCl_5$, the SiO_2 transport is in the same direction

$$SiO_2(s) + 2TaCl_5 \rightleftarrows SiCl_4 + 2TaOCl_3. \tag{1.56}$$

Contamination from fingerprints even appears responsible for some transport. Many metals are transported to the silica wall as the iodides, with the same result. Fused silica also appears marginal with many metal sulfides and selenides at high temperature (see comments in Lieth[34]) although attack by normal-valent sulfides can be reduced by a coating or boat of pyrolytic or glassy carbon.[108]

Basic oxides are clearly not satisfactory in silica because of the stability of silicates. Less well-recognized metathetical reactions also occur at elevated temperature between SiO_2 and metal halides, especially iodides of metals that form stable oxyhalides or basic oxides, for example

$$2PrI_3(l, g) + SiO_2(s) \rightarrow 2PrOI(s) + SiI_4. \tag{1.57}$$

This reaction is easily displaced to the right in a vacuum or by an active metal which reacts with the SiI_4. In addition, such salts should not be allowed to condense on the silica wall during sublimation or distillation, but rather on a tantalum condenser (e.g. see Fig. 1.1(c)). The heavier rare-earth metal iodides, ThI_4, etc., appear to be the worst in this respect.[109-11] Further reaction of a basic oxide produced by these reactions with the wall will also promote the reaction. Thus for

$$2MgI_2(l) + SiO_2(amorph.) \rightarrow 2MgO(s) + SiI_4 \tag{1.58}$$

ΔG^0_{1100} is 143.1 kJ but this drops to only $+10.7$ kJ when the subsequent formation of $MgSiO_3$ is included. The reaction

$$4ScCl_3(s, g) + SiO_2(s) \rightleftarrows 2Sc_2Si_2O_7(s) + 3SiCl_4 \tag{1.59}$$

is sufficient above 750°C that the equilibrium may be studied in a closed diaphragm cell.[112]

General experiences regarding ceramic alternatives to glass are not documented except that the introductory chapter in Brauer[24] gives a good summary of the properties, uses, and availability (in Europe) of about twenty ceramic materials, plus tabulations of the suitability of a few ceramics and metals toward certain elements and salts at high temperatures.

Metal containers fall into two classes: reactive and inert. The use of Cu, Ni, Co, Ti, Nb, Ta, Mo, and so forth as both containers and reductants in the study of the lowest oxidation state compounds of these elements is well known. Sizeable temperature gradients should be avoided over long periods if volatile metal species are present for which decomposition can lead to metal transport and penetration of the wall. Containers of the platinum and coinage metals are traditional for their inertness to some melts, hydrothermal reactions, etc. Note should be made of the fact that platinum is transported by Cl_2, O_2, or $CO + Cl_2$, and Mo and W by water, too. The noble metals are generally unsatisfactory as containers for many of the more reactive metals and their reduced compounds because of the formation of very stable intermetallic phases, e.g. Cs–Au, Re–Au, Zr–Pt.

Among the traditional refractory materials—Nb, Ta, Mo, and W—the first two are the most versatile in terms of ductility, strength and ease of fabrication and welding.[110] For instance, tantalum containers are inert to metals and their halides which lie to the left in the periodic table, and are suitable to ⩾1000°C and, with care, 30 atm. internal pressure. Some useful apparatus was shown in Fig. 1.1(d). The less expensive niobium is not quite as strong, but has been used extensively with alkaline-earth and rare-earth metal–metal halide systems. Of course the group VI metals Mo and W are still more inert, and the former can be weld-sealed. Some gas-tight crucibles with threaded lids have been described by Cullmann and Schuster.[113] Chalcogenides are not generally satisfactory in Nb or Ta. When higher temperatures are needed in the synthesis of metal-rich chalcogenides, it is sometimes useful to first react the starting materials in silica as far as possible without contamination, so as to reduce their vapour pressures, and then to switch to a tungsten container.[114]

1.5 Recent developments

A number of important articles have appeared since this chapter was first written. Regarding the effects of impurities (Section 1.1.2), the tendency of the lighter nonmetals to bond within certain metal-rich halides, often to stabilize new phases, has become increasingly evident. Thus, the reported clusters Zr_6I_{12} and $CsZr_6I_{14}$ actually contain carbon (or other) atoms centred in all metal octahedra, the carbon originally deriving from adventitious impurities.[115] Similarly, certain 'binary' metal chain compounds are judged to probably be carbides, $Gd_6(C_2)Br_7$ for instance, and the rare-earth metal monohalides appear to be hydrides, e.g. $Gd(H_{0.9})Cl$.[116] In these cases, the typically inconsistent and low yield synthesis of the supposedly simpler phase is usually changed into a high yield reaction when the correct impurity is added in quantitative amounts. Such possibilities should be generally considered for metal-rich compounds of at least the early transition elements; a considerable variety of analogous compounds have been discovered, $Gd_3(C)Cl_3$,[117] $Sc_7(C_2)Cl_{10}$,[118] and $Zr_6(B)Cl_{13}$,[119] to name a few. Substitutional oxygen has also been detected in the Chevrel phases $SnMo_6S_8$ and $PbMo_6S_8$.[120]

A valuable review of phase relationships (Section 1.2.3) as well as other problems associated with the high temperature synthesis of Chevrel phases has been published by Flükiger and Baillif,[121] while the synthesis and crystal growth of transition metal borides, carbides, silicides, and phosphides via their solutions in liquid metals has been summarized by Lundström.[122]

Some of the most promising results in recent years have come in the area of lower temperature routes (Section 1.2.5) to new phases, structures, or more reactive materials, an approach that has been termed '*chemie douce*' or 'soft chemistry'. Many examples fall in the category of intercalation,[69] namely oxidation, reduction, or ion exchange under the structural control of the host matrix. For example, such reactions have been used to produce an increasing variety of metastable derivatives of the Chevrel phases.[123] The remarkable hydration, hydrolysis, condensation and thermolysis products derived from $K_2Ti_4O_9$ have been described by Marchand et al.,[124] these including new acid derivatives, the phase $K_2Ti_8O_{17}$ build from condensed sheets and, by dehydrration of '$H_2Ti_4O_9$', a new metastable and open form of TiO_2 that is related to the bronze $Na_xTi_4O_8$. In another vein, room temperature reaction of the three-dimensional structure of $(Nb_6I_8)I_3$ with CH_3NH_2 gives a remarkable transformation, first to a layered intermediate, evidently through solvolysis of strained iodine bridges between clusters, and then to crystalline $Nb_6I_8(NH_2CH_3)_6$.[125] New routes to better or more reactive ceramics or their precursors are also receiving considerable attention. For example, a low temperature hydrolysis or pyrolysis of metal alkoxides or their mixtures have been found to lead to oxide powders that are useful precursors for important ceramics,[126,127] while pyrolyses of a variety of polymers have been studied as routes to SiC, Si_3N_4, BN, etc.[128,129] Such approaches should certainly allow access to new phases that are not stable at the higher temperatures used in traditional preparations. Also, devitrification of a glass of appropriate composition is another means of achieving an adequate reaction on a microscopic scale, as has been shown in the synthesis of the rare tube silicate litidionite.[130] A lower temperature route to oxy- and thiohalides of the rare-earth elements has also been found. Intermediates such as $(NH_4)_3YCl_6$ (see Section 1.2.2.a) react with H_2O or Y_2O_3, or H_2 or Y_2S_3, at 300–500°C to give YOCl, or YSCl, respectively, in a new structure.[131,132] Also, recrystallization of the oxychlorides from various chloride melts affords new structures for these for several metals, the results being clearly dependent on the nature of the melt.

Finally, a valuable review on transport reactions (Section 1.3) has appeared, covering additional data from the literature together with careful evaluation and thermodynamic modelling of transport experiments, especially for the binary and ternary niobium oxides.[133]

1.6 References

1. Warren, J. L. and Geballe, T. H. *Mat. Sci. Eng.* **50**, 149 (1981).
2. Steele, B. C. H. In *Inorganic Solids*, Vol. 2 (to be published).
3. Hannay, N. B. (ed.). *Treatise on solid-state chemistry*, Vol. 4: *Reactivity of solids*. Plenum, New York (1976).
4. Schmalzried, H. *Solid-state reactions* (2nd edn). Verlag Chemie, Weinheim (1981).
5. Nagel, L. and O'Keeffe, M. In *MTP International Review of Science*, Vol. 10, p. 1. Butterworths, London (1972).
6. Wadsley, A. D. In *Nonstoichiometric compounds* (ed. L. Mandeleorn) p. 101. Academic Press, New York (1964).
7. Westman, S. and Magnéli, A. *Acta Chem. Scand.* **11**, 1587 (1957).
8. Simon, A. *J. appl. Crystallogr.* **3**, 11 (1970).
9. Cheetham, A. K. and Skarnulis, A. J. *Anal. Chem.* **53**, 1060 (1981).
10. Long, G. J., Cheetham, A. K., and Battle, P. D. *Inorg. Chem.* **22**, 3012 (1983).
11. Corbett, J. D. In *Preparative inorganic reactions* (ed. W. L. Jolly) Vol. 3, p. 1. Interscience Publishers, New York (1966).
12. Hägg, G. and Schönberg, N. *Acta Cryst.* **7**, 351 (1954).
13. Kudielka, H. and Rohde, H. *Z. Kristallogr.* **114**, 447 (1960).
14. Rexer, J. and Peterson, D. T. *Nucl. Metallurg. Series* 10, 334 (1964).
15. Widera, A. and Schäfer, H. *J. less-common Metals* **77**, 29 (1981).
16. Ehrlich, P., Alt, B., and Gentsch, L. *Z. anorg. allg. Chem.* **283**, 58 (1956).
17. Felmlee, T. L. and Eyring, L. *Inorg. Chem.* **7**, 660 (1968).
18. Wasilewski, R. J. *Trans. Am. Inst. Mech. Engrs* **221**, 647 (1961).
19. Schäfer, H., Schnering, H.-G., Niehues, K.-J., and Nieder-Vahrenholz, H. G. *J. less-common Metals* **9**, 95 (1965).
20. Schäfer, H. *Angew. Chem., Int. Educ. Engl.* **10**, 43 (1971).
21. Anthony, A. M. and Collongues, R. In *Preparative methods in solid-state chemistry* (ed. P. Hagenmuller) p. 147. Academic Press, New York and London (1972).
22. Givargizov, E. I. (ed.). *Proc. 6th int. Conf. on crystal growth*, North Holland Publishing Co., Amsterdam (1980).
23. Hagenmuller, P. (ed.). *Preparative methods in solid-state chemistry*. Academic Press, New York and London (1972).
24. Brauer, G. *Handbuch der Präparativen Anorganischen Chemie*, (3rd edn) 3 volumes. Ferdinand Enke Verlag, Stuttgart (1975–81).
25. Shriver, D. F. and Drezdzon, M. A. *The manipulation of air-sensitive compounds* (2nd edn). Wiley, New York (1986).
26. Barton, C. J. In *Technique of inorganic chemistry* (eds H. B. Jonassen and A. Weissberger) Vol. 3, p. 259. Interscience, New York (1963).
27. Seabaugh, P. W. and Corbett, J. D. *Inorg. Chem.* **4**, 176 (1965).
28. Simon, A., von Schnering, H.-G., and Schäfer, H. *Z. anorg. allg. Chem.* **355**, 295 (1967).
29. Schäfer, H. and Giegling, D. *Z. anorg. allg. Chem.* **420**, 1 (1976).
30. Newnham, I. E. and Watts, J. A. *J. Am. Chem. Soc.* **82**, 2113 (1960).
31. Veprek, S. *Chimica* **34**, 489 (1980).
32. Wagman, D. D. et al. *Selected values of chemical thermodynamic properties*, NBS Technical Note 270–3, US Government Printing Office, Washington DC (1968).

33. Meyer, G. and Ax, P. (1982). *Mat. Res. Bull.*, **17**, 1447 (1982).
34. Lieth, R. M. A. (ed.). *Preparation and crystal growth of materials with layered structures*. D. Reidel, Dordrecht, Holland (1977).
35. Corbett, J. D. In *Fused salts* (ed. B. R. Sundheim) Chap. 6. McGraw-Hill, New York (1964).
36. Struss, A. W. and Corbett, J. D. *Inorg. Chem.* **17**, 965 (1978).
37. Kuhn, P. J. and McCarley, R. E. *Inorg. Chem.* **4**, 1482 (1965).
38. Larsen, E. M., Moyer, J. W., Gil-Arnao, F. and Camp, M. J. *Inorg. Chem.* **13**, 574 (1974).
39. Hansen, M. and Anderko, K. P. *Constitution of binary alloys* (2nd edn). McGraw-Hill, New York (1958).
40. Struss, A. W. and Corbett, J. D. *Inorg. Chem.* **8**, 227 (1969).
41. Gortsema, F. P. *Inorg. Syn.* **14**, 105 (1973).
42. Carter, F. L. and Murray, J. F. *Mat. Res. Bull.* **7**, 519 (1972).
43. Clark, R. J., Griswold, B., and Kleinberg, J. *J. Am. chem. Soc.* **80**, 4764 (1958).
44. Hoppe, R. (1981). *Angew. Chem., Int. Educ. Engl.* **20**, 63 (1981).
45. Klemm, W. and Huss, E. *Z. anorg. allg. Chem.* **258**, 221 (1949).
46. Corbett, J. D., Burkhard, W. J., and Druding, L. F. *J. Am. chem. Soc.* **83**, 76 (1961).
47. Johnson, D. A. *Some thermodynamic aspects of inorganic chemistry* (2nd edn). Cambridge University Press (1982).
48. Corbett, J. D. and Guthrie, D. H. *Inorg. Chem.* **21**, 1747 (1982).
49. Struss, A. W. and Corbett, J. D. *Inorg. Chem.* **9**, 1373 (1970).
50. Corbett, J. D. *Rev. chim. Minerale* **10**, 239 (1973).
51. Mee, J. E. and Corbett, J. D. *Inorg. Chem.* **4**, 88 (1965).
52. Lokken, D. A. and Corbett, J. D. *Inorg. Chem.* **12**, 536 (1973).
53. McCollum, B. C., Camp, M. J., and Corbett, J. D. *Inorg. Chem.* **12**, 778 (1973).
54. Ricci, J. E. *The phase rule and heterogeneous equilibrium*. Dover, New York (1966).
55. Alper, A. M. (ed.). *Phase diagrams: materials science and technology*, 5 volumes. Academic Press, New York (1970–8).
56. Corbett, J. D., Ford, J., and Hwu, S.-J. Unpublished research (1982).
57. Paulus, M. In *Preparative methods in solid-state chemistry* (ed. P. Hagenmuller), p. 487. Academic Press, New York (1972).
58. Kwestroo, W. In *Preparative methods in solid-state chemistry* (ed. P. Hagenmuller) p. 564. Academic Press, New York (1972).
59. Laurent, Y. *Rev. Chim. Minerale* **6**, 1145 (1969).
60. White, E. A. D. In *Technique of Inorganic Chemistry* (eds H. B. Jonassen and A. Weissberger) Vol. 4, p. 31. John Wiley, New York (1965).
61. Jeitschko, W., Ruhl, R., Krieger, U., and Heiden, C. *Mat. Res. Bull.* **15**, 1755 (1980).
62. Kaner, R., Castro, C. A., Gruska, R. P., and Wold, A. *Mat. Res. Bull.* **12**, 1143 (1977).
63. Johnson, V. *Inorg. Syn.* **14**, 182 (1973).
64. Wold, A. and Bellavance, D. In *Preparative methods in solid-state chemistry* (ed. P. Hagenmuller) p. 279. Academic Press, New York (1972).
65. Feigelson, R. S. *Adv. Chem. Ser.* **186**, 243 (1980).
66. McCarroll, W. H. Private communication, unpublished research (1982).
67. Masse, R. and Simon, A. *Mat. Res. Bull.* **16**, 1007 (1981).

68. Whittingham, M. S. and Jacobson, A. J. (eds). *Intercalation Chemistry*. Academic Press, New York (1982).
69. Schöllhorn, R. In *Inclusion compounds* (eds. J. L. Atwood, J. E. D. Davies, and D. D. MacNicol) Vol. 1, p. 249. Academic Press, London (1984).
70. Müller-Buschbaum, Hk. *Angew. Chem., Int. Educ. Engl.* **20**, 22 (1981).
71. Brewer, L. *J. chem. Ed.* **35**, 153 (1958).
72. Chianelli, R. R. and Dines, M. B. *Inorg. Chem.* **17**, 2758 (1978).
73. Meyer, G. and Schönemund, A. *Mat. Res. Bull.* **15**, 89 (1980).
74. Meyer, G. *Prog. solid-state Chem.*, **14**, 141 (1982).
75. Meyer, G. *Inorg. Syn.* **22**, 1 (1983).
76. Whipple, W. and Wold, A. *J. inorg. nucl. Chem.* **24**, 23 (1962).
77. Kleinert, P. and Funke, A. *Z. Chem.* **1**, 155 (1961).
78. Longo, J. M., Horowitz, H. S., and Clavenna, L. R. *Adv. Chem. Ser.* **186**, 139 (1980).
79. Rouxel, J. In *Intercalated Layered Materials* (ed. F. Levy) p. 201. Reidel Publishing, Dordrecht, Holland (1979).
80. Murphy, D. W., DiSalvo, F. J., Hull, G. W., and Waszczak, J. V. *Inorg. Chem.* **15**, 17 (1976).
81. Chianelli, R. R. and Dines, M. B. *Inorg. Chem.* **14**, 2417 (1975).
82. Behlok, R. J., Kulberg, M. L., and Robinson, W. R. *Proc. 4th International Conf. on Chemistry and Uses of Molybdenum*. Climax Molybdenum Co., Golden, Colorado (1982).
83. Murphy, D. W., Carides, J. N., DiSalvo, F. J., Cros, C., and Waszczak, J. V. *Mat. Res. Bull.* **12**, 825 (1977).
84. Poeppelmeier, K. R., Leonowicz, M. E., and Longo, J. M. *J. solid-state Chem.*, **44**, 89 (1982).
85. Imoto, H., Corbett, J. D., and Cisar, A. *Inorg. Chem.* **20**, 145 (1981).
86. Niebuhr, J. *J. less-common Metals* **10**, 312; **11**, 191 (1966).
87. Meyer, G., Ax, P., Cromm, A., and Linzmeier, H. *J. Less-common Metals*, **98**, 323 (1984).
88. Rooymans, C. J. M. In *Preparative methods in solid-state chemistry* (ed. P. Hagenmuller) p. 71. Academic Press, New York (1972).
89. Joubert, J. C. and Chenavas, J. In *Treatise on solid-state chemistry* (ed. N. B. Hannay) Vol. 5, p. 463. Plenum Press, New York (1975).
90. Bougon, R., Ehretsmann, J., Portier, J., and Tressaud, A. In *Preparative methods in solid-state chemistry* (ed. P. Hagenmuller) p. 401. Academic Press, New York (1972).
91. MacChesney, J. B., Sherwood, R. C., and Potter, J. F. *J. chem. Phys.* **43**, 1907 (1965).
92. Goodenough, J. B., Kafalas, J. A., and Longo, J. M. In *Preparative methods in solid-state chemistry* (ed. P. Hagenmuller) p.1. Academic Press, New York (1972).
93. Sugawara, F., Syono, Y., and Akimoto, S. *Mat. Res. Bull.* **3**, 529 (1968).
94. Akimoto, S. and Syoro, Y. *J. chem. Phys.* **47**, 1813 (1967).
95. Demianets, L. N. and Lobachev, A. N. In *Current topics in materials science* (ed. E. Kaldis) Vol. 7, p. 563. North-Holland Publishing Co., Amsterdam (1981).
96. Sleight, A. W. *Inorg. Chem.* **8**, 1764 (1969).
97. Schäfer, H. *Chemical transport reactions*. Academic Press, New York (1964).
98. Schäfer, H. In *Preparative methods in solid-state chemistry* (ed. P. Hagenmuller) p. 252. Academic Press, New York, London (1972).

99. Schäfer, H. *Agnew. Chem., Int. Ed. Engl.* **15**, 713 (1976).
100. Jones, M. E. and Shaw, D. W. In *Treatise on solid-state chemistry* (ed. N. B. Hannay) Vol. 5, p. 283. Plenum Press, New York (1975).
101. Arizumi, T. In *Current topics in materials science* (ed. E. Kaldis) Vol. 1, p. 343. North-Holland Publishing Co., Amsterdam (1978).
102. Poeppelmeier, K. R. and Corbett, J. D. *J. Am. Chem. Soc.* **100**, 5039 (1978).
103. Schäfer, H. and Trenkel, M. *Z. anorg. allg. Chem.* **414**, 137 (1975).
104. Schäfer, H., Binnewies, M., Domke, W., and Karbinski, J. *Z. anorg. allg. Chem.* **403**, 116 (1974).
105. Wolf, E., Opperman, H., Krabbes, G., and Reichett, W. In *Current topics in materials science* (ed. E. Kaldis) Vol. 1, p. 697. North-Holland Publishing Co., Amsterdam (1978).
106. Emmenegger, F. P. *J. Crystal Growth* **17**, 31 (1972).
107. Rustad, D. S. and Gregory, N. W. *Inorg. Chem.* **21**, 2929 (1982).
108. Flahaut, J. Private communication (1982).
109. Corbett, J. D. *Inorg. nucl. Chem. Lett.* **8**, 337 (1972).
110. Corbett, J. D. *Inorg. Syn.* **22**, 15, 31, 39 (1983).
111. Mironov, K. E., Abdullin, R. V., Popova, E. D., and Vasil'eva, I. G. *Izv. Sib. Otd. Akad. Nauk.* USSR, *Ser. Khim. Nauk.* **3**, 96 (1981).
112. Polyachenok, L. D., Nazarov, K., and Polyachenok, O. G. *Russ. J. phys. Chem.* **52**, 1021 (1978).
113. Cullmann, H.-O. and Schuster, H.-U. *J. thermal Anal.* **24**, 187 (1982).
114. Franzen, H. F. Private communication (1982).
115. Smith, J. D. and Corbett, J. D. *J. Am. chem. Soc.*, **107**, 5704 (1985).
116. Simon, A. *J. solid-state Chem.* **57**, 2 (1985).
117. Warkentin, E. and Simon, A. *Rev. chim. Minerale* **20**, 488 (1983).
118. Hwu, S.-J., Corbett, J. D., and Poeppelmeier, K. R. *J. solid-state Chem.* **57**, 43 (1985).
119. Ziebarth, R. P. and Corbett, J. D. *J. Am. chem. Soc.* **107**. 4571 (1985).
120. Hinks, D. G., Jorgensen, J. D., and Li, H.-C. *Phys. Rev. Lett.* **51**, 1911 (1983).
121. Flükiger, R. and Baillif, R. In *Superconductivity in ternary compounds* (eds. Ø. Fischer and M. B. Maple) Vol. 1, p. 113. Springer-Verlag, Heidelberg (1982).
122. Lundström, T. *J. less-common Metals* **100**, 215 (1984).
123. Potel, M., Gougen, P., Chevrel, R., and Sergeant, M. *Rev. chim. Minerale* **21**, 509 (1984).
124. Marchand, R., Broham, L., M'Bedi, R. and Tournoux, M. *Rev. chim. Minerale* **21**, 476 (1984).
125. Simon, A. and Stollmaier, F. *Inorg. Chem.* **24**, 168 (1985).
126. Mazdiyasni, K. S. *Ceram. Intl.* **8**, 42 (1982).
127. Uhlmann, D. R., Zelinski, B. J. J., and Wnek, G. E. *Proc. Mat. Res. Soc. Symp.* **32**, 59 (1984).
128. Wynne, K. J. and Rice, R. W. *Ann. Rev. mat. Sci.* **14**, 297 (1984).
129. Seyferth, D. and Wiseman, G. H. *J. Am. ceram. Soc.* **67**, C132 (1984).
130. Hefter, J. and Kenney, M. E. *Inorg. Chem.* **21**, 2810 (1982).
131. Garcia, E., Corbett, J. D., Ford, J. E., and Vary, W. J. *Inorg. Chem.* **24**, 494 (1985).
132. Meyer, G., Staffel, T., Dötsch, S., and Schleid, T. *Inorg. Chem.*, **24**, 3504 (1985).
133. Gruehn, R. and Schweizer, H.-J. *Angew. Chem. Int. Ed. Eng.* **22**, 82 (1983).

2 Diffraction methods

A. K. Cheetham

2.1 Introduction

The evolution of our understanding of chemistry in the solid state has relied heavily upon structural information obtained from a wide range of physical methods. Spectroscopic and resonance techniques have played an important rôle, of course, but the most definitive structural data has largely been acquired using diffraction methods. For example, the determination of precise interatomic distances and bond angles for inorganic solids is performed almost exclusively by single-crystal, X-ray diffraction, and since a thorough understanding of electronic properties and bonding in solids requires a sound structural basis, it is clear that diffraction studies will continue to be pre-eminent among structural methods in solid-state chemistry.

The aim of this chapter is to introduce the reader to the principles of structure determination using X-rays, and to explore the areas in which the use of alternative radiations, in particular neutrons or electrons, may be preferred. There will also be some emphasis on the extent to which structural information can be obtained from polycrystalline samples, either by Rietveld profile analysis of powder diffraction data or by electron microscopy. In the final section, there is a brief discussion of techniques that can be used to study defective and non-crystalline solids.

2.2 X-rays

Early investigations into the structure of matter at the micron level utilized optical microscopy, but in the nineteenth century Abbé showed that this approach is intrinsically wavelength-limited and that significant improvements in resolution could only be achieved by using a radiation with a much smaller wavelength. In particular, a wavelength of about 1 Å (10^{-10} m) is required to probe structure at the atomic level, and the discovery of X-rays in 1895 happily provided a radiation that met this requirement. The following discussion of X-ray diffraction methods is necessarily brief, and for further

40   *Solid state chemistry: techniques*

details the reader is referred to the books by Stout and Jensen,[1] and Woolfson.[2]

2.2.1 Production

X-rays for diffraction experiments are normally produced by bombarding a metal target, often Cu or Mo, with a beam of electrons emitted from a heated filament (Fig. 2.1(a)). The typical spectrum of X-ray wavelengths is shown in Fig. 2.1(b). The incident electron beam will ionize electrons from the K-shell (1 s) of the target atoms, and X-rays are emitted as the resultant vacancies are filled by electrons from the L (2 p) or M (3 p) levels. This gives rise to the intense $K\alpha$ and $K\beta$ lines:

$$L \rightarrow K: \quad K\alpha_1, K\alpha_2$$
$$M \rightarrow K: \quad K\beta_1, K\beta_2.$$

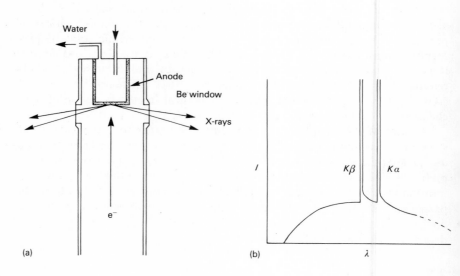

Fig. 2.1. (a) X-ray tube; (b) X-ray emission spectrum, showing characteristic $K\alpha$ and $K\beta$ lines.

As the atomic number Z of the target element increases, the energy of the characteristic emissions increases and the wavelength decreases:

Line	Wavelength λ(Å)	Energy (keV)
Cu $K\alpha$	1.54178	8.04
Mo $K\alpha$	0.71069	17.44

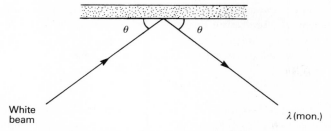

Fig. 2.2. Reflection of X-rays from a crystal monochromator.

We can select out a monochromatic beam of, say, the $K\alpha$ wavelength by reflecting an X-ray beam from a crystal monochromator (Fig. 2.2) according to the Bragg equation, $\lambda = 2d\sin\theta$ (see Section 2.3.2). Alternatively, we may use a filter. Absorption of X-rays is governed by the equation $I = I_0 e^{-\mu\tau}$, where μ is the linear absorption coefficient and τ is the path length through the solid. The variation of μ with λ takes the form $\mu = k\lambda^3$, but there are discontinuities or absorption edges at energies that are just sufficient to knock an electron out of an atomic orbital. The K absorption edge is just on the short-wavelength (high-energy) side of the $K\beta$ line (Fig. 2.3). A filter of elements $(Z-1)$ or $(Z-2)$ will normally absorb the $K\beta$ emission for an element of atomic number Z; we use a Ni filter for Cu radiation and a Zr filter for Mo radiation.

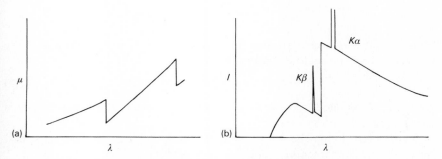

Fig. 2.3. (a) Variation of linear absorption coefficient, μ, with wavelength; (b) the effect of a suitable filter on the X-ray emission spectrum.

2.2.2 Scattering of X-rays by isolated atoms

X-rays are scattered by their interaction with atomic electrons, and interfer-

42 Solid state chemistry: techniques

ence takes place between X-rays scattered from different parts of an atom. As a consequence, the scattering power or scattering factor f_X decreases with increasing scattering angle, 2θ (Fig. 2.4).[3] The shape of the scattering factor curve can be calculated from atomic wave functions:

$$f_X = \tfrac{1}{2}(1+\cos 2\theta)\frac{e^2}{mc^2}\int \Psi\Psi^* e^{ik \cdot r} dT. \tag{2.1}$$

At $\sin\theta/\lambda = 0$, f_X is proportional to atomic number and is therefore very small for light atoms such as hydrogen and lithium. To a good approximation, f_X is

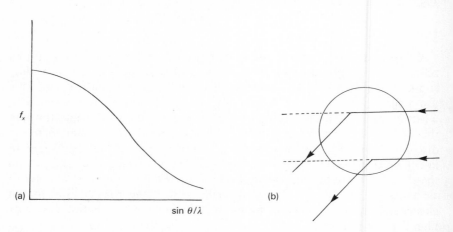

Fig. 2.4. (a) A typical X-ray form factor, f_x; (b) interference takes place between X-rays scattered from different parts of an atom.

independent of wavelength, but for accurate work it is necessary to make corrections for anomalous scattering:[4]

$$f_X^{\text{corr.}} = f_X + \Delta f' + \Delta f''. \tag{2.2}$$

$\Delta f'$ and $\Delta f''$ depend upon λ, and are important when the X-ray energy is close to an absorption edge.

Before considering the scattering from an array of atoms, we must first discuss some fundamental properties of crystalline materials.

2.3 Crystals

2.3.1 Unit cell and lattice planes

The *unit cell* is the simplest repeating unit of a crystalline structure and is defined by three translations a, b, and c, and three angles α, β, and γ. The cubic unit cells of the NaCl and CsCl structures, in which $a=b=c$ and $\alpha=\beta=\gamma=90°$, will be familiar. At the origin of each cell is a *lattice point*.

Lattice planes can be understood in terms of a two-dimensional example. A set of lattice planes must be parallel and equally spaced, and every lattice point must lie on some member of the set. The planes cut the cell axes in fractional parts: $\frac{1}{1}, \frac{1}{2}, \frac{1}{3}, \ldots \frac{1}{n}$, or at infinity (Fig. 2.5). The reciprocals of the intercepts are the Miller indices, and in three dimensions each plane is identified by three indices, h, k, l. The spacing between adjacent members of the hkl planes is designated d_{hkl}. Figure 2.6 shows an example.

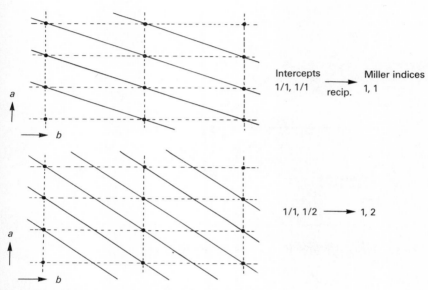

Fig. 2.5. Miller indices for a two-dimensional lattice.

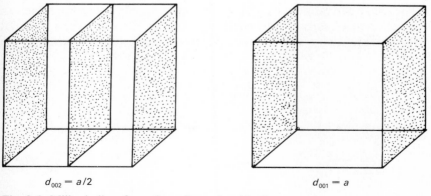

$d_{002} = a/2 \qquad\qquad d_{001} = a$

Fig. 2.6. Miller indices for a three-dimensional lattice.

2.3.2 Bragg equation

Scattering from a crystal can be described in terms of reflection from a set of lattice planes. In Fig. 2.7,

$$\text{deviation of beam} = 2\theta,$$

$$\text{path difference} = 2x = 2d_{hkl}\sin\theta.$$

For constructive interference, $n\lambda = 2d_{hkl}\sin\theta$.

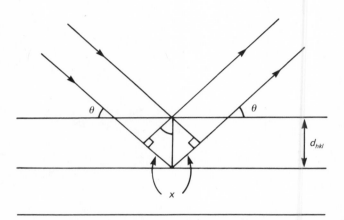

Fig. 2.7. Bragg reflection from crystal planes with spacing d_{hkl}.

High-order ($n = 2, 3$, etc.) scattering from planes d_{hkl} is indistinguishable from first-order scattering ($n = 1$) from planes $d_{hkl}/2$, $d_{hkl}/3$, etc. Hence, $\lambda = 2d_{hkl}\sin\theta$, known as the *Bragg equation*, represents the condition for diffraction to take place.

The above derivation is over-simplified, but the result is rigorous and can be applied to crystals of any symmetry. Taking, for example, a material with cubic symmetry, we have the following relationship between d_{hkl}, a, h, k, and l:

$$\frac{1}{d_{hkl}} = \left(\frac{h^2 + k^2 + l^2}{a^2}\right)^{\frac{1}{2}} \tag{2.3}$$

whence, substituting in the Bragg equation:

$$\sin^2\theta = \frac{\lambda^2}{4a^2}(h^2 + k^2 + l^2). \tag{2.4}$$

2.4 Powder X-ray techniques[5]

2.4.1 Debye–Scherrer method

A powder sample for X-ray work should ideally contain an infinite number of randomly oriented crystallites. Each set of lattice planes *hkl* will scatter at the appropriate 2θ angle, according to the Bragg equation, and, since all possible orientations of crystallite should be present, a cone of scattering will be formed at each value (Fig. 2.8). For example, a simple cubic material with a lattice parameter $a = 5\,\text{Å}$ will scatter Cu $K\alpha$ X-rays, ($\lambda = 1.54\,\text{Å}$) at the following angles (according to eqn 2.4):

hkl	d(Å)	2θ (degrees)
100	5.00	17.72
110	3.54	25.15
111	2.89	30.94 ...

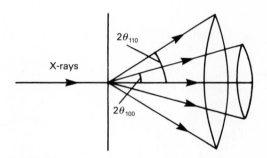

Fig. 2.8. Scattering of X-rays into cones by a powder sample.

We can intercept the cones at the different 2θ values on an X-ray film, as in the Debye–Scherrer method (Fig. 2.9). The sample is mounted on a glass fibre, or in a capillary tube, which is rotated during the exposure.

The procedure for analysing a film of a cubic material is as follows:

(i) Obtain 2θ values from the film; for a camera of diameter 114.6 mm, 1 mm = 1 degree for 2θ.
(ii) Calculate $\sin^2\theta$ values.
(iii) Index lines, i.e. assign *hkl* values. $\sin^2\theta = C(h^2 + k^2 + l^2)$, where $C = \lambda^2/4a^2$ from eqn (2.4).
(iv) Find the mean value of *a*.

46 *Solid state chemistry: techniques*

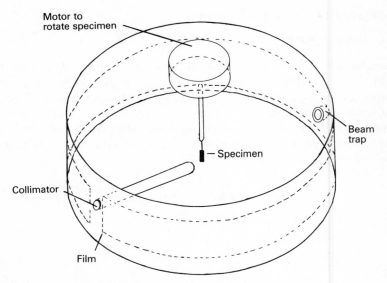

Fig. 2.9. The Debye–Scherrer X-ray powder method.

For example, the following data was obtained for NaCl with Cu $K\alpha$ radiation:

$\sin^2\theta$	$(h^2+k^2+l^2)$	C	a (Å)
0.0560	3	0.01867	5.642
0.0746	4	0.01865	5.644
0.1492	8	0.01865	5.645
0.2052	11	0.01865	5.644
0.2239	12	0.01866	5.644
0.2984	16	0.01865	5.645

If a sample contains heavy elements, low-angle lines will be shifted to higher 2θ values by *absorption*, and a correction may be necessary to determine an accurate lattice parameter. A common procedure is to plot the values of a against the Nelson–Riley function:[6]

$$f(\theta) = \tfrac{1}{2}\left(\frac{\cos^2\theta}{\theta} + \frac{\cos^2\theta}{\sin\theta}\right) \tag{2.5}$$

and extrapolate to $\theta = 90°$.

2.4.2 Other powder methods

Other important powder techniques include the Guinier–Haag focusing camera, which gives very high-resolution patterns by arranging for the scattered beams to come to a focus on the film, and the diffractometer method in which a detector is used to measure the scattered intensity of X-rays as a function of angle (Fig. 2.10). This method yields both the 2θ value and *intensity* of each reflection.

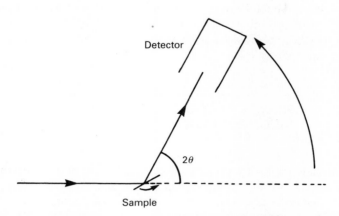

Fig. 2.10. Schematic diagram of a simple X-ray powder diffractometer.

2.4.3 Applications of powder methods

(a) Routine identification of materials

The X-ray powder pattern of a compound provides a convenient and characteristic fingerprint which can be used in qualitative analysis. Precise values of the d-spacings (which are properties of the material itself, independent of the X-ray wavelength) and of estimated line intensities can be compared with those listed in the Powder Diffraction File,[7] which contains entries for approximately 30 000 inorganic compounds.

(b) Quantitative analysis of mixtures

Conventional chemical analysis will yield the bulk composition of a mixture, but it does not reveal the identity and proportions of the component phases. With powder X-ray diffractometer methods, the phases can be identified using the procedure described above, and their proportions can be determined by comparing the intensities of characteristic lines from each phase; standards are usually necessary. For example, we could readily analyse a dust containing several forms of SiO_2: quartz, cristobalite, and tridymite.

(c) Precise determination of lattice parameters

This contributes towards the overall characterization of a compound. The lattice parameters are readily obtained for high symmetry (e.g. cubic,

48   Solid state chemistry: techniques

hexagonal) compounds using procedures of the type described in Section 2.4.1, but the indexing of low-symmetry patterns may be difficult. For example, for an orthorhombic compound with $a \neq b \neq c$, $\alpha = \beta = \gamma = 90°$:

$$\sin^2\theta = \frac{\lambda^2}{4}\left(\frac{h^2}{a^2} + \frac{k^2}{b^2} + \frac{l^2}{c^2}\right). \tag{2.6}$$

If we can obtain the lattice parameters, either by hand or by using one of the computer programs that has been developed for this purpose,[8] the structure may become obvious from a comparison with a known compound which is isomorphous, e.g.

	a(Å)	b(Å)	c(Å)
$KMnO_4$	7.41	9.09	5.72
$KClO_4$	7.24	8.84	5.65

(d) Phase diagrams

Powder measurements are very useful in the elucidation of phase diagrams, both for identifying the compositions of line phases by measuring X-ray patterns for samples at regular intervals of composition, and for determining the solubility limits of non-stoichiometric phases. In the CaF_2–YF_3 system, a measurement of the lattice parameter of the cubic fluorite phase will give the solubility of YF_3 in CaF_2 (Fig. 2.11).[9] To avoid working with quenched samples, a variable temperature attachment can be used to study the variation of the X-ray pattern with temperature.

2.4.4 Limitations of powder methods

The above examples give some measure of the scope and utility of X-ray powder measurements in solid-state chemistry, but powder data do not normally yield enough information to give a complete structure determination. For example, a measurement of the lattice parameter of cubic pyrites, FeS_2, does not define the lengths of the Fe–S or S–S bonds. In order to determine the atomic positions, we must use the *intensity* information, ideally from measurements on single crystals.

2.5 Diffraction theory

For the determination of an unknown structure, we must understand the relationship between the atomic arrangement and the intensity I_{hkl} of an X-ray reflection from the planes hkl. We shall deal only with centrosymmetric structures in which the unit cell contains a centre of symmetry; for the general case, the $\cos(x)$ terms are replaced by $\exp(x)$.

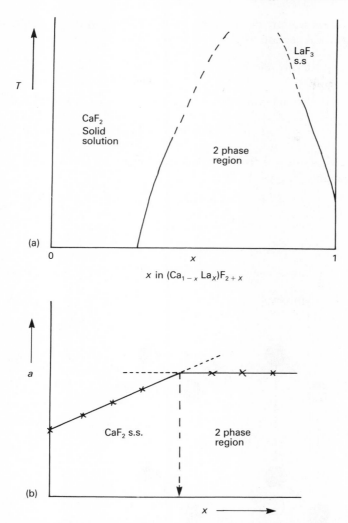

Fig. 2.11. (a) Phase diagram for the system CaF_2–YF_3; (b) lattice parameter, a, of the CaF_2-rich solid solution as a function of composition, x; the arrow shows the solubility limit of YF_3 in CaF_2.

2.5.1 Structure factor F_{hkl}

The scattering power of an isolated atom f_x was discussed in Section 2.2.2. For an arrangement of several atoms, we have to consider interference

50 *Solid state chemistry: techniques*

between waves scattered by different atoms. For example, the scattering by a unit cell for a reflection hkl:

$$\text{scattering amplitude} = \sum_n (f_n \times \text{phase factor}) \quad (2.7)$$

for n atoms in the unit cell. This is the structure factor F_{hkl} and takes the form:

$$F_{hkl} = \sum_n f_n \cos 2\pi (hx_n + ky_n + lz_n) \quad (2.8)$$

where x_n, y_n, z_n are the fractional coordinates of the nth atom: e.g. for the CsCl structure (Fig. 2.12):

$$F_{100} = f_{Cl}\cos 2\pi(1\cdot 0 + 0\cdot 0 + 0\cdot 0) + f_{Cs}\cos 2\pi(1\cdot \tfrac{1}{2} + 0\cdot \tfrac{1}{2} + 0\cdot \tfrac{1}{2})$$

$$= f_{Cl}\cos(0) + f_{Cs}\cos(\pi)$$

$$= f_{Cl} - f_{Cs}.$$

Similarly, $F_{110} = f_{Cl} + f_{Cs}$; chlorine and caesium are scattering in phase for the 110 reflection.

For centrosymmetric structures, F_{hkl} may be positive or negative; the structure factor amplitude is $|F_{hkl}|$.

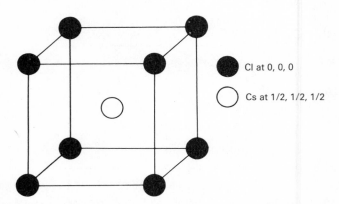

Fig. 2.12. The caesium chloride structure.

2.5.2 Intensity of reflection I_{hkl}

The intensity of a reflection is something that we can measure, and it is related to the structure factor according to:

$$I_{hkl} = sLpF_{hkl}^2 \quad (2.9)$$

where s is a scale factor, L is the Lorentz (geometrical) correction,

($L = (\sin 2\theta)^{-1}$ for single crystal data), and p is a polarization correction, $p = (1 + \cos^2 2\theta)/2$.

Essentially

$$I_{hkl} \propto F_{hkl}^2. \tag{2.10}$$

In CsCl, therefore:

$$I_{100} \propto (f_{Cl} - f_{Cs})^2$$
$$I_{110} \propto (f_{Cl} + f_{Cs})^2$$
$$\therefore I_{110} > I_{100}.$$

2.5.3 Electron density distribution

If we know the structure factors F_{hkl} for a complete set of X-ray reflections, we can calculate the electron density ρ at any position xyz in the unit cell (and, therefore, the atomic positions):

$$\rho(xyz) = \frac{1}{V} \sum_h \sum_k \sum_l F_{hkl} \cos 2\pi (hx + ky + lz). \tag{2.11}$$

This is an example of a Fourier summation; the electron density is the *Fourier transform* of the X-ray diffraction pattern. In order to calculate $\rho(xyz)$ we require the values of F_{hkl}, but unfortunately we can only measure F_{hkl}^2; we can obtain $|F_{hkl}|$, but not the sign of F_{hkl}. This dilemma is known as the *phase problem*. The + or − signs of F_{hkl} correspond to phases of 0 and 180 degrees, respectively. Non-centrosymmetric structures can have phases anywhere in the range 0–360 degrees.

For a data set containing m structure factors, each of which may be positive or negative, we have 2^m possible combinations of phases; e.g. for only 300 reflections we have 2×10^{90} possibilities! Clearly, the solving of the phase problem by trial-and-error methods is not very practicable.

It is instructive to compare the X-ray experiment with an optical diffraction experiment. The diffraction pattern obtained by passing a beam of monochromatic visible light through a grating can be transformed into an image using an optical lens; a magnetic lens will effect the same transformation in an electron microscope. Lenses suitable for X-rays are not available, and it is therefore necessary to measure the diffraction pattern and carry out the transformation mathematically. Unfortunately, there is no direct way of measuring the phases of the diffracted beams; only their amplitudes $|F_{hkl}|$ can be measured. The solution to the phase problem is normally performed using sophisticated computational methods which will be described in Section 2.8.

2.6 Single-crystal X-ray methods

A wide range of methods is available for the examination of single crystals with X-rays, but only a brief outline of a typical procedure can be given here. After a preliminary photographic examination of the crystal in order to establish its perfection and to determine lattice parameters and symmetry information, intensity data can be collected on a computer-controlled diffractometer of the type shown in Fig. 2.13. Each set of planes hkl can be brought into the reflecting position using three (φ, χ, and ω) of the four degrees of freedom, 2θ, φ, χ, and ω. The counter can then scan across the reflection using the fourth degree of freedom, 2θ. A data set containing several thousand reflections can be measured in a period of a few days. Equivalent reflections are then merged, and corrections are made for Lorentz and polarization effects, as described in eqn (2.9). The resulting list of $|F_{hkl}|$ values provides the basis for the determination of the structure.

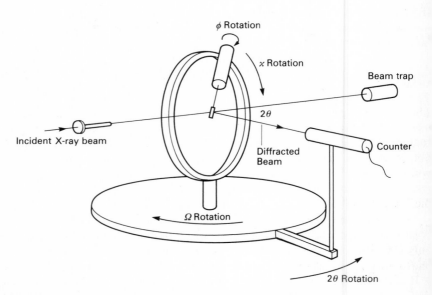

Fig. 2.13. Schematic diagram of a four-circle diffractometer.

In order to simplify the job of solving the phase problem, we must first examine some aspects of crystal symmetry. For example, the unit cell of sodium zeolite-A contains approximately 480 atoms, but if we take advantage of the symmetry within the unit cell, we only need to locate 9 atoms in order to place all the others.

2.7 Crystal symmetry

2.7.1 Crystal systems

There are seven basic shapes of unit cell, known as the *seven crystal systems;* these are listed in Table 2.1.

Table 2.1 *The seven crystal systems*

Crystal system	Lattice parameters		Lattice symmetry
Triclinic	$a \neq b \neq c$;	$\alpha \neq \beta \neq \gamma$	$\bar{1}$
Monoclinic	$a \neq b \neq c$;	$\alpha = \gamma = 90°$; $\beta \neq 90°$	$2/m$
Orthorhombic	$a \neq b \neq c$;	$\alpha = \beta = \gamma = 90°$	mmm
Tetragonal	$a = b \neq c$;	$\alpha = \beta = \gamma = 90°$	$4/mmm$
Rhombohedral	$a = b = c$;	$\alpha = \beta = \gamma \neq 90°$	$3m$
Hexagonal	$a = b \neq c$;	$\alpha = \beta = 90°$; $\gamma = 120°$	$6/mmm$
Cubic	$a = b = c$;	$\alpha = \beta = \gamma = 90°$	$m3m$

Some common examples are cubic NaCl, tetragonal CaC_2, and hexagonal graphite.

2.7.2 Bravais lattices

A primitive (P) cell contains one repeating motif, which may be an atom or group of atoms, per cell; a centred cell contains 2 or 4. Centred cells have translational symmetry *within* the cell. A body-centred cell (*I*) contains identical motifs at its origin (0,0,0) and at its centre $(\frac{1}{2},\frac{1}{2},\frac{1}{2})$, whereas a face-centred cell contains motifs at its origin and on one (A or C) or all (F) of its faces. The seven crystal systems yield fourteen Bravais lattices, shown in Table 2.2. They may be primitive or centred.

Table 2.2 *The fourteen Bravais lattices*

Crystal system	Bravais lattices
Triclinic	P
Monoclinic	P, C
Orthorhombic	P, C, I, F
Tetragonal	P, I
Rhombohedral	R
Hexagonal	P
Cubic	P, I, F

For a body-centred cell, we require that for every atom at *xyz*, there is an identical atom at $x+\frac{1}{2}$, $y+\frac{1}{2}$, $z+\frac{1}{2}$. Body-centred cubic metals, e.g. Nb, represent a special case with *xyz* = 0,0,0, and the identical atom placed at $\frac{1}{2},\frac{1}{2},\frac{1}{2}$. CsCl is primitive, not body-centred, because the atom at $\frac{1}{2},\frac{1}{2},\frac{1}{2}$ is not identical to the one at 0,0,0 (Fig. 2.14).

Centred cells are readily apparent in the diffraction pattern because the translational symmetry leads to *systematic absences*.

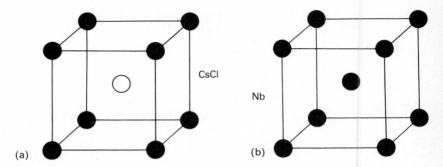

Fig. 2.14. (a) The primitive cubic caesium chloride structure; (b) body-centred cubic niobium metal.

2.7.3 Systematic absences

In a centred cell, certain classes of reflection will be systematically absent because atoms in one motif will scatter exactly out of phase with atoms in another:

I cell: reflections with $h+k+l$ odd are absent.
A(C) cell: reflections with $k+l$ ($h+k$) odd are absent.
F cell: reflections with $h+k$, $k+l$, or $h+l$ odd are absent, i.e. only reflections with h, k, and l all odd or all even are observed.

e.g.

Body-centred cubic (Nb metal)	Face-centred cubic (NaCl)
100 absent	100 absent
110	110 absent
111 absent	111
200	200
210 absent	210 absent
211	211 absent
220	220

These absences help us to define the crystal symmetry. There are two other translational symmetry elements, found only in crystals, which lead to systematic absences.

Screw axis

Here, one part of a unit cell is related to another by a rotation (e.g. two-fold plus a translation. For example, a two-fold screw axis, 2_1, represents a two-fold rotation plus a translation of $\frac{1}{2}$ cell (Fig. 2.15); 3_1: a three-fold rotation plus a translation of $\frac{1}{3}$ cell. Absences again occur: e.g. a two-fold screw along the b-axis leads to $0k0$, k odd absences.

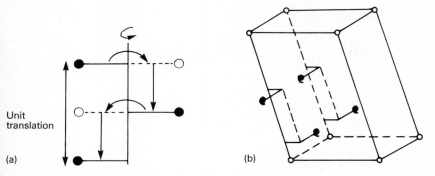

Fig. 2.15. Schematic representations of (a) a two-fold screw axis and (b) a glide plane.

Glide plane

Here, two parts of the cell are related by a reflection across a plane plus a translation, e.g. a c-glide might be a reflection across a plane perpendicular to b plus a translation ($c/2$) along c; this gives $h0l$, l odd absences.

Systematic absences *only* arise from translational symmetry within the cell, i.e. centred cells or cells containing screw axes or glide planes. Some important symmetry elements and their systematic absences are shown in Table 2.3.

Table 2.3 *Translational symmetry elements and their systematic absences*

Symmetry element	Affected reflection	Condition for systematic absence
Centred-cells:		
Body-centred (I)	hkl	$(h+k+l)$ odd
Face-centred (A)	hkl	$(k+l)$ odd
Face-centred (B)	hkl	$(h+l)$ odd
Face-centred (C)	hkl	$(h+k)$ odd
Face-centred (F)	hkl	$(h+k)$ odd
		$(h+l)$ odd
		$(k+l)$ odd
Screw axes:		
Two-fold screw (2_1) along a	$h00$	h odd
2_1 along b	$0k0$	k odd
2_1 along c	$00l$	l odd
Glide planes $\perp b$:		
Translation $(a/2)$ (a-glide)	$h0l$	h odd
Translation $(c/2)$	$h0l$	l odd
Translation $(a/2+c/2)$(n-glide)	$h0l$	$(h+l)$ odd
Translation $(a/4+c/4)$(d-glide)	$h0l$	$(h+l)=4n+1, 2, 3$

2.7.4 Point groups and space groups

In the same way that the symmetry of an isolated molecule can be described by a point group, crystal symmetry is described by a space group. Only thirty-two point groups are compatible with the restrictions imposed by the translational symmetry of a crystal; for example, five-fold rotation axes cannot be accommodated in a crystal. We can assign the thirty-two permissible point groups among the seven crystal systems as shown in Table 2.4.

Table 2.4 *Permissible point groups assigned to crystal systems*

Crystal system	Point groups
Triclinic	$1, \bar{1}$
Monoclinic	$2/m, m, 2$
Orthorhombic	$mmm, mm2, 222$
Tetragonal	$4/mmm, \bar{4}m, 4mm, 422, 4m, \bar{4}, 4$
Rhombohedral	$\bar{3}m, 3m, 32, \bar{3}, 3$
Hexagonal	$6/mmm, \bar{6}m2, 6mm, 622, 6/m, \bar{6}, 6$
Cubic	$m3m, 43m, 432, m3, 23$

The two-hundred-and-thirty space groups, which describe all the permutations of symmetry that are available to a crystal, can be generated from the thirty-two point groups by introducing the possibility of the translation elements: centring, glide planes, and screw axes. For example, we can generate six monoclinic space groups from the point group $2/m$:

(1) The cell may be primitive or C-centred; thus we have $P2/m$ and $C2/m$.
(2) The two-fold rotation axis may be degraded to a two-fold screw axis; thus $P2_1/m$ and $C2_1/m$ ($\equiv C2/m$).
(3) The mirror plane may be degraded to a c-glide plane; thus $P2/c$ and $C2/c$.
(4) A screw axis and c-glide may occur together; thus $P2_1/c$ and $C2_1/c$ ($\equiv C2/c$).

What possibilities occur for point groups m and 2?

It is sometimes possible to assign the space group uniquely on the basis of the systematic absences. In other instances, it is only possible to narrow the field.

The two-hundred-and-thirty space groups are described in the International Tables for Crystallography, Volume I[10]. Each entry summarizes the symmetry relationships between identical atoms within a unit cell. For example, in space group $P2_1/c$: $x,y,z; \bar{x},\bar{y},\bar{z}; \bar{x}, (\frac{1}{2}+y), (\frac{1}{2}-z); x, (\frac{1}{2}-y), (\frac{1}{2}+z)$.

Thus, for every atom at a general position xyz, there are three identical atoms at $\bar{x},\bar{y},\bar{z}$, and so on. If an atom lies on a symmetry element, then the number of equivalent atoms will be reduced. For example, the *special position* $\frac{1}{2},0,\frac{1}{2}$ has only one equivalent ($\frac{1}{2},\frac{1}{2},0$).

A knowledge of the space group is important in solving the phase problem.

2.8 Solving a structure

2.8.1 The Patterson method

The Patterson technique is the most common method for solving the phase problem in inorganic chemistry. Let us assume that we have collected a set of intensity data as described in Section 2.6, and that we have generated a list of $|F_{hkl}|$ values. The crystal system will be known from preliminary photographic work and the data collection; let us consider a monoclinic structure for which the systematic absences confirm the space group $P2_1/c$. Our problem remains that we do not know the absolute values of the F_{hkl} for the calculation of the electron density, according to eqn (2.11). In the Patterson method, we generate a three-dimensional map from a Fourier summation using $|F_{hkl}|^2$ as the coefficients, rather than F_{hkl}:

$$P(uvw) = \frac{1}{V}\sum_h\sum_k\sum_l |F_{hkl}|^2 \cos2\pi(hu+kv+lw). \quad (2.12)$$

The resulting Patterson map will look like a three-dimensional electron density map, but the peaks (high values of P) will correspond to *vectors* between pairs of atoms in the unit cell.

For example, atom 1 (atomic number Z_1) at $x_1 y_1 z_1$, and atom 2 (atomic number Z_2) at $x_2 y_2 z_2$, will yield a vector at: $u = x_1 - x_2$, $v = y_1 - y_2$, $w = z_1 - z_2$. Furthermore, the height of the peak will be proportional to $Z_1 Z_2$, so that vectors between heavy atoms will be very prominent. The Patterson method therefore provides a powerful means of locating heavy atoms in, say, metal oxides, coordination compounds, or organometallics. The procedure is described in the appendix (Section 2.15). We can then use our knowledge of the heavy-atom positions to locate the lighter atoms. The heavy atoms have the largest X-ray scattering factors f_X, and their contribution will dominate the structure factor summation (eqn 2.8). If we calculate partial structure factors F_{hkl}^{heavy} using just the coordinates of the heavy atoms in the unit cell, the magnitudes of the F_{hkl} values will only be approximate, but the phases are likely to be correct; i.e. the sign of F_{hkl} is unlikely to be wrong. For a single heavy atom M at a position x', y', z', we would have simply:

$$F_{hkl}^{\text{heavy}} = f_M \cos 2\pi(hx' + ky' + lz').$$

We can use these phases with the *observed* $|F_{hkl}|$ values in order to calculate the electron density using eqn (2.11), and the resultant map should reveal the light atom positions. We shall return in Section 2.9 to the optimization of the atomic coordinates.

2.8.2 Direct methods[11]

The Patterson method described above is very powerful for solving structures in which the unit cell contains a small proportion of heavy atoms and a large proportion of light ones, but the Patterson map can become very congested if all the atoms have approximately the same atomic number, as for example in most organic compounds. Under these circumstances, the 'direct-methods' approach is preferred. This is a statistical method for predicting phases and ideally requires:

(1) that all atoms have the same X-ray scattering power;
(2) that the distribution of the atoms within the unit cell is quasi-random.

The first step in the procedure is to place all the structure factors on an absolute scale, normalized for the effects of the angular dependence of f_X and the variations in f_X from one compound to another. *Normalized structure factors E_{hkl} are calculated according to*:

$$|E_{hkl}|^2 = \frac{|F_{hkl}|^2}{\varepsilon \Sigma_n f_n^2}. \tag{2.13}$$

As in eqn (2.8), the summation is over all the n atoms in the unit cell; ε is normally unity, but may differ for certain classes of reflection. The resultant $|E|$ values typically fall in the range 0–5, and for Fourier calculations E_{hkl} has the same phase as F_{hkl}.

The prediction of phases is largely based upon the Sayre probability relationship:[12]

$$S(hkl) \sim S(h'k'l') \cdot S(h-h', k-k', l-l') \tag{2.14}$$

where, for a centrosymmetric structure, $S(hkl)$ indicates the sign of E_{hkl} and $\sim$ means 'probably equal to', e.g.

$$\begin{array}{ll} h, k, l & 442 \\ h', k', l' & 110 \\ h-h', k-k', l-l' & 332. \end{array}$$

Thus, if E_{442} and E_{110} are both of negative phase, E_{332} is likely to be positive. The probability P that this relationship holds depends upon the $|E_{hkl}|$:

$$P = \tfrac{1}{2} + \tfrac{1}{2}\tanh\left\{\frac{1}{N}|E_{hkl} \cdot E_{h'k'l'} \cdot E_{h-h',k-k',l-l'}|\right\}. \tag{2.15}$$

N is the number of atoms in the cell, $\tanh x = (e^x - e^{-x})/(e^x + e^{-x})$.

Table 2.5 shows the probability P as a function of N and the average $|E|$

Table 2.5 *Probabilities calculated according to eqn 2.15*

			P	
$\|E_{av}\|$	$\|E_{av}\|^3$	$N=20$	$N=60$	$N=100$
3.0	27.0	1.0	1.0	0.99
2.6	17.4	1.0	0.99	0.97
2.3	12.2	1.0	0.96	0.92
1.8	5.8	0.93	0.81	0.76
1.5	3.4	0.82	0.71	0.66

value in the triple product. It can be seen that for a given value of $|E_{av}|^3$, the predictions become less precise as the number of atoms in the cell increases.

Our objective is to start with a set of (say) six phases, and to use the probability relationship to predict the phases of other reflections. The six starting reflections should have $|E|$ values greater than 3.0, so that the initial predictions will be reliable. A computer program[13] can be used to select the best starting reflections from the larger $|E|$s; three of the six will be arbitrarily assigned + phases in order to define the position of cell origin along the three principal crystallographic axes, but the other three may be + or −. There are eight possible combinations of the three unknown phases (+ + +, + + −, + − −, etc.), one of which must be correct; and the phase prediction will be performed for each of the eight possibilities.

A successful Fourier calculation will require approximately ten E values per atom to be located, so for fifteen atoms we would normally take the 150 largest Es, and generate phases for them using each of the eight starting sets. Some of the Es will be as small as 1.5, and many of the predictions will be incorrect. A range of sophisticated methods is available, for example tangent refinement, to ensure that the optimum set of phases is obtained.

The starting model that gives the most self-consistent set of predicted phases is assumed to be correct, in the first instance, and this set of phases is used to calculate an E-map

$$F(xyz) = \frac{1}{V} \sum_h \sum_k \sum_l E_{hkl} \cos 2\pi(hx + ky + lz). \qquad (2.16)$$

This is essentially equivalent to the electron density map (eqn 2.11) and should reveal the atomic positions. The direct-methods technique can also be applied to non-centrosymmetric structures, albeit with more difficulty.

2.9 Structure refinement

The procedure outlined in Section 2.8 will furnish approximate coordinates for all the atoms in the cell, with the exception of hydrogen atoms, and the positions can then be optimized by a *least-squares* refinement. The coordinates are adjusted to minimize the function:

$$D = \sum_{hkl} w_{hkl}(|F^{obs.}_{hkl}| - |F^{calc.}_{hkl}|)^2$$

where w_{hkl} is the weight given to the observation $|F_{hkl}|$. At this stage it is also necessary to include parameters that take account of the thermal motion of the atoms:

$$F^{calc.}_{hkl} = \sum_n f_n \cos 2\pi (hx + ky + lz) e^{-B_n \sin^2\theta/\lambda^2}. \tag{2.17}$$

The isotropic temperature factor B_n is equal to $8\pi^2 \overline{u_n^2}$, where $\overline{u_n^2}$ is the mean square amplitude of vibration of atom n. In many refinements, the thermal motion is described anisotropically by an elipsoid, with six parameters.

The goodness-of-fit for a refinement of a structure is measured in terms of a 'reliability factor' or R-factor:

$$R = \frac{\Sigma_{hkl}|F^{obs.}_{hkl} - F^{calc.}_{hkl}|}{\Sigma_{hkl}|F^{obs.}_{hkl}|} \times 100\% \tag{2.18}$$

or, a weighted R-factor:

$$R_w = \frac{\Sigma_{hkl} w_{hkl}(|F^{obs.}_{hkl} - F^{calc.}_{hkl}|)}{\Sigma_{hkl} w_{hkl} F^{obs.}_{hkl}} \times 100\%. \tag{2.19}$$

The R-value would typically be in the range 20–30% after locating heavy atoms in a Patterson map, or locating atoms approximately by direct methods; but, after refinement of the model, the R-value usually falls to 5% or less. In many cases it is necessary to correct for systematic errors such as absorption. In order to detect any omissions in the final model, a *difference Fourier* map is normally computed:

$$\Delta(xyz) = \frac{1}{V}\sum\sum\sum(F^{obs.}_{hkl} - F^{calc.}_{hkl})\cos 2\pi(hx + ky + lz). \tag{2.20}$$

This should reveal any errors and should also indicate the locations of hydrogen atoms in, for example, organic molecules. It is often not possible to refine the positions of hydrogen atoms.

The structure determination is now complete and the final set of atomic coordinates, with their estimated standard deviations, may be used to calculate bond lengths and bond angles of chemical interest. The precision of a good X-ray structure determination is extremely high and the errors of bond lengths are typically only a few thousandths of an angstrom. It has now become routine to characterize new materials that can be obtained in single crystal form by this method.

2.10 Neutron diffraction and its applications

2.10.1 Wave properties

The vast majority of crystal structure determinations are performed using X-rays, but there are circumstances in which the use of an alternative radiation, particularly neutrons or electrons, may be advantageous. The wave nature of the neutron leads in principle to the possibility of doing diffraction experiments, and this was first demonstrated in 1936 using a radium–beryllium source.[15] At the present time the most common source of neutrons for diffraction purposes is the nuclear reactor, but recently the production of neutrons by *spallation* from metal targets bombarded with high-energy (about 800 MeV) protons has received much attention (see Section 2.10.4).[16]

For neutrons produced in a reactor and moderated using heavy water, we have the following condition:

$$\text{kinetic energy, } \tfrac{1}{2}mv^2 = \tfrac{3}{2}kT. \tag{2.21}$$

Substituting the de Broglie relationship, $\lambda = h/mv$, we can show that the mean wavelength of neutrons moderated to 273 K should be 1.55 Å, and their energy equal to 40 meV. In practice, the moderation is imperfect, and the spectral distribution of neutrons is rather broad (Fig. 2.16). However, the mean wavelength is close to 1 Å, confirming that neutrons should be suitable for probing matter at the atomic level. The beam is normally monochromated using a crystal (e.g. germanium).

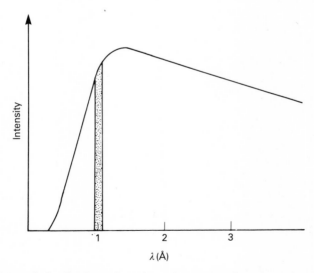

Fig. 2.16. Spectral distribution of moderated neutrons from a nuclear reactor. A narrow band of wavelengths can be selected for diffraction experiments.

Unlike X-rays, which are scattered by the atomic electrons, the neutron is scattered by the nucleus. Consequently, the interference effects that cause X-ray scattering to diminish with $\sin\theta/\lambda$ (Section 2.2.2) are absent with neutrons, and the scattering is isotropic. Because neutron scattering is a nuclear property, it may vary dramatically from one element to the next, and is indeed different for different isotopes, e.g. hydrogen and deuterium. The neutron also has a magnetic moment, and magnetic scattering is observed, in *addition* to nuclear scattering, from paramagnetic ions. Neutron absorption is usually negligible.

2.10.2 Applications

Many important applications arise from the dependence of neutron scattering amplitudes, usually known as *scattering lengths,* upon atomic number (Fig. 2.17). The scattering lengths show a rather slow increase with atomic number, with the consequence that light atoms, which scatter X-rays weakly, make a substantial contribution to the neutron diffraction pattern. This leads to a higher precision in the location of light atoms using neutrons, an advantage that is especially useful in work on hydrogen-containing materials, and there has been a great deal of work on metal hydrides, hydrogen-bonded systems, and biological macromolecules. Cerium hydride is a typical example; X-rays are sensitive only to the cerium atoms, but neutron diffraction measurements on CeD_2 (deuterium has better neutron scattering properties than hydrogen) show that it adopts the face-centred cubic fluorite structure with deuterium atoms occupying tetrahedral sites.[1]

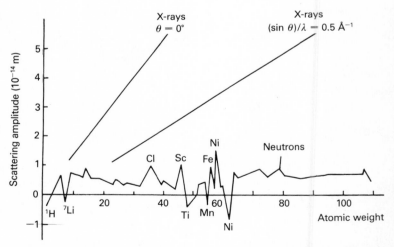

Fig. 2.17. Neutron scattering lengths of the elements (after Bacon[17]); negative values arise when there is an anomalous phase change on scattering.

In CeD$_{2+x}$ $(0 < x < 1)$, neutrons reveal not only the placement of the extra deuterium in octahedral sites, but also long-range ordering of the additional deuterium atoms to generate a tetragonal superstructure (Fig. 2.18). This is apparent from the appearance of new reflections that are not observed in CeD$_2$ (Fig. 2.19).

A second type of application stems from variations in scattering lengths between adjacent elements in the periodic table, a feature that is particularly

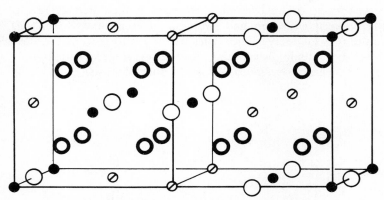

Fig. 2.18. The crystal structure of CeD$_{2.50}$; only alternate octahedral interstices in the fluorite structure are occupied (small circles, cerium; thick-lined large circles, tetrahedral deuterium; thin-lined large circles, octahedral deuterium).

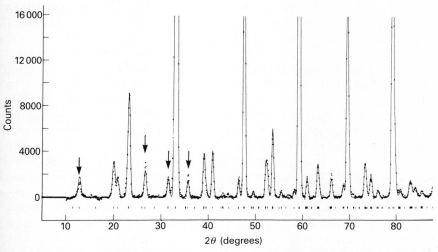

Fig. 2.19. The neutron powder pattern of CeD$_{2.50}$ with arrows indicating some of the superstructure reflections arising from long range ordering of interstitial deuterium atoms.

marked in the first transition series. When this is the case, it is possible to distinguish between pairs of elements that have almost identical X-ray scattering amplitudes. Iron and cobalt, with scattering lengths of 0.96 and 0.22×10^{-14} m, respectively, provide a good illustration. A neutron diffraction study on the spinel $NiCo_2O_4$ has shown that cobalt is present on both the tetrahedral and octahedral sites, whereas X-rays are totally insensitive to the degree of inversion.[19] The same property has also been used to study order–disorder transitions in many alloys, for example FeCo.[20]

An even more striking example is afforded by the difference in scattering lengths between hydrogen and deuterium (-0.36 and $+0.68 \times 10^{-14}$ m, respectively). A single crystal neutron study of YH (oxalate)$_2$·2H$_2$O revealed a non-random isotopic distribution in the hydronium ion $H_5O_2^+$, with hydrogen showing a preference for site 2 rather than site 1:[21]

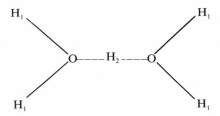

This effect, which must be vibrational in origin, is strongly temperature-dependent.

The other principal area in which neutron diffraction has played an important rôle is in the study of magnetically ordered materials: ferromagnets, antiferromagnets, and ferrimagnets. In the same way that neutron scattering from a regular array of atoms leads to a normal diffraction pattern, magnetic scattering from an ordered array of spins gives rise to magnetic Bragg peaks. Figure 2.20 shows the powder neutron diffraction profiles of $Fe_{0.96}O$ above and below its magnetic ordering temperature (about 200 K).[22] Note the magnetic peaks at low temperatures, and the splittings that arise from a structural phase transition that accompanies magnetic ordering in this case. From the d-spacings and intensities of the magnetic reflections, it is possible to determine the arrangement of the iron spins and their magnitude. This information gives a valuable insight into the bonding. Single-crystal studies are particularly advantageous because the Fourier transform of the magnetic scattering yields a map of the spin density (by analogy with eqn 2.11). The use of a *polarized* neutron beam is also advantageous since it increases the sensitivity to magnetic scattering.[23]

Two further properties of the neutron give rise to advantages in certain areas. The neutron beam energy, 40 meV for $\lambda \sim 1.5$ Å, is orders of magnitude lower than that of a typical X-ray beam (8 KeV for Cu $K\alpha$), and

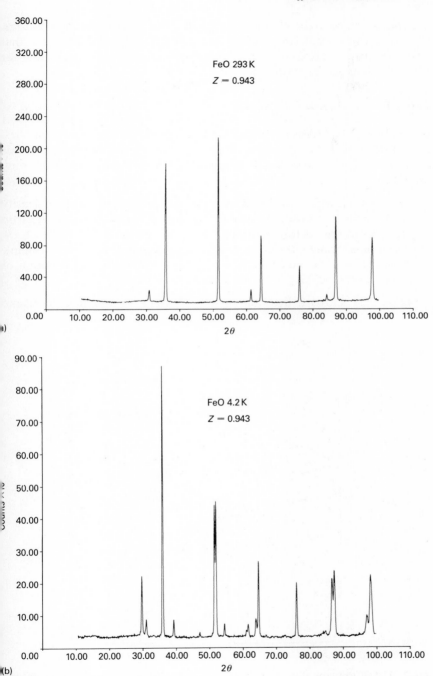

Fig. 2.20. Powder neutron diffraction patterns of $Fe_{0.943}O$ at (a) room temperature and (b) 4.2 K.

66 *Solid state chemistry: techniques*

radiation damage to the sample is therefore less likely when using neutrons. Neutron absorption cross-sections are also very low, a property that facilitates data collection from samples held in special environments, for example furnaces and cryostats.

2.10.3 Deformation density studies

By combining the results of precise, single-crystal X-ray and neutron studies, we can determine the way in which electrons are redistributed when bonds are formed; this is known as the *deformation density*. Consider the following procedure:

(1) Locate nuclear positions with neutrons and produce an electron density map for the non-bonded crystal by placing calculated free-atom electron densities at the nuclear positions (N-map).
(2) Determine the actual electron density distribution using X-rays (X-map).
(3) Calculate a difference map (X − N map) to show the redistribution of electrons that takes place when bonds are formed.

Figure 2.21 shows a deformation density map for a simple inorganic

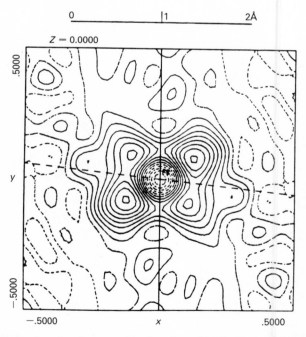

Fig. 2.21. Deformation density map showing the redistribution of electrons around iron in FeS_2.[24]

compound.[24] The precision of this type of study has now reached the point where useful comparisons with theory can be made.

2.10.4 Pulsed neutron sources[16]

Up to the present time, most neutron diffraction experiments have been performed with constant wavelength beams from reactor sources, but a new generation of accelerator-based sources promises to play an important rôle in the future. In the case of a spallation source, pulses of high-energy (about 800 MeV) protons from a synchrotron are directed on to a heavy-metal target. A high flux of neutrons (about thirty per proton) is produced by a process known as spallation, and, after moderation to increase the proportion of thermal neutrons, the neutron beam, which retains the original pulsed structure of the protons, is available for scattering experiments. In the Linac, the neutrons are produced from a pulsed electron beam.

Pulsed beams can conveniently be used to collect diffraction patterns by *time-of-flight* methods. In a simple case, the counter is placed at a fixed 2θ angle, and the different d-spacings are sampled by measuring the time that neutrons with different wavelengths (and, therefore, different energies) take to reach the counter (Fig. 2.22); the neutron velocity is sufficiently slow to permit this (about 4×10^5 cm s^{-1}). In terms of the Bragg equation, we have the following situation:

$$\lambda = 2d_{hkl}\sin\theta.$$
$$\text{(variable)} \quad \text{(fixed)}$$

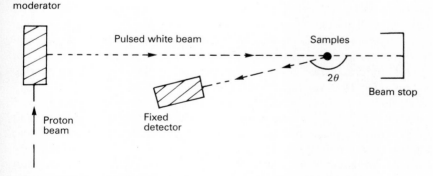

Fig. 2.22. Schematic diagram of a time-of-flight powder diffractometer.

In a conventional experiment, λ is fixed and θ is varied. The appearance of a typical time-of-flight pattern is shown in Fig. 2.23.[25]

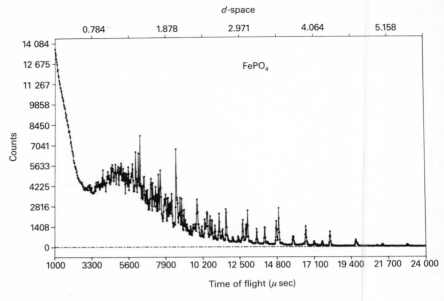

Fig. 2.23. Time-of-flight neutron powder pattern of FePO$_4$.[25]

2.11 Rietveld profile analysis[26]

The relatively low flux of a neutron beam compared with that of an X-ray source results in a need for rather large crystals (at least 1 mm^3) for single-crystal neutron experiments. The difficulty in preparing such crystals has stimulated the development of powder neutron diffraction techniques to a point where they can now be used to refine complex, low-symmetry structures. A conventional refinement using structure factors from single, integrated powder intensities is quite feasible for high-symmetry structures, but is clearly impracticable for low-symmetry materials because of overlap between adjacent Bragg reflections (Fig. 2.24).

Rietveld has devised a method for analysing the more complex patterns by means of a curve-fitting procedure in which the least-squares refinement minimizes the difference between the observed and calculated profiles, rather than individual reflections. In order to do this, it is assumed that the reflections are gaussian in shape, and the calculated intensity at each point (say, 0.1° 2θ steps) on the profile is obtained by summing the contributions from the gaussians that overlap at that point. In addition to the conventional parameters in the least-squares (i.e. atomic coordinates and temperature factors), additional parameters are required: the lattice parameters (which determine the positions of the reflections), a zero-point error for the counter setting, and three parameters that describe the variation of the gaussian half-width (full width at half maximum intensity) with scattering angle. The

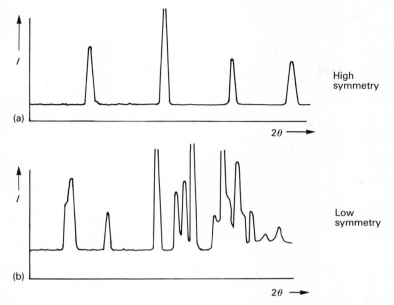

Fig. 2.24. Powder diffraction patterns of samples with (a) high symmetry and (b) low symmetry.

technique has been applied to a wide range of solid-state problems[27] and an example is shown in Fig. 2.25.

Profile analysis procedures have also been developed for the analysis of time-of-flight powder data.[29] The procedure is essentially the same as that used in constant-wavelength experiments except that the peak shape function is more complex. One advantage of time-of-flight powder methods is that the whole diffraction pattern is collected simultaneously (since the counter remains stationary), making it an attractive way of following changes with time, temperature, or pressure. In addition, the incident and scattered beams can pass through small apertures in, say, a high-pressure apparatus; the design of special environments is clearly easier for such measurements.

The extension of profile refinement to X-ray data has been achieved,[30] although the method does not appear to be as powerful as its neutron counterpart. One disadvantage of using X-rays is that the fall-off in intensity with increasing $\sin\theta/\lambda$ reduces the amount of useful data, a serious drawback for a method that is intrinsically short of data compared with single-crystal experiments. Powder data can be collected by means of a conventional diffractometer or by scanning a Guinier film using a microdensitometer. The latter procedure, which appears to be preferred, requires the use of a modified Lorentzian function to describe the peak shape. Data has also been collected using 'white' synchrotron radiation in an experiment similar to the time-of-flight neutron experiment.[31] The counter is again placed at a fixed 2θ value, but the different wavelengths are resolved, not by time-of-flight, but by

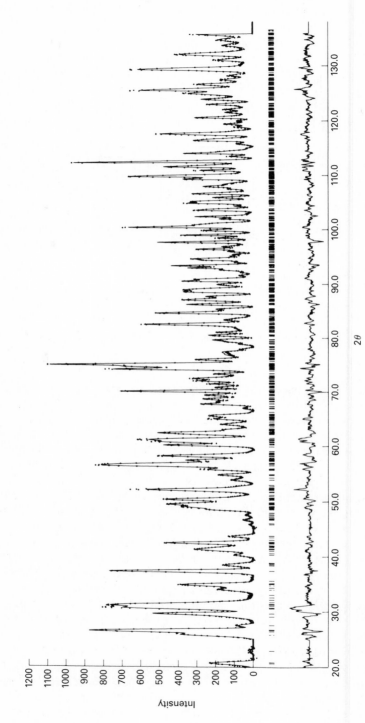

Fig. 2.25. Observed (dots) and calculated (full line) powder neutron diffraction patterns of anhydrous ferric sulphate; a difference curve and the reflection positions are also shown.

using an energy-dispersive X-ray detector (e.g. a Li-drifted silicon detector). As in the neutron experiment, this has the advantage that the whole diffraction pattern is accumulated simultaneously.

Profile analysis is limited by the same drawback that affects powder methods in general: the loss of information arising from the compression of the three-dimensional diffraction pattern into a single dimension. The inability to assign intensities to individual reflections, except in the simplest cases, also renders the use of Fourier methods rather difficult. In particular, Patterson techniques may not yield the information necessary to locate heavy atoms and solve the phase problem. Similarly, direct methods are difficult to apply to powder data. This serves to underline the fact that profile analysis, though an excellent technique for refining structures, requires a good starting model and is not a good method for *ab initio* structure determination. However, this problem may be overcome as instrumental resolution improves, giving larger numbers of individual reflections (see Postscript, section 2.14).

2.12 Electron microscopy

In many respects, the electron microscope offers the ideal solution to the difficulties presented by the resolution limitations of a light microscope. The wavelength is less than 1 Å so atomic resolution should be possible, and, as with the light microscope, the diffraction pattern can be transformed directly into an image by using a suitable lens. In this section, we shall examine the extent to which the electron microscope fulfils this promise. We shall also find that a modern electron microscope is a very versatile instrument, offering a wide range of modes in addition to diffraction and imaging.

2.12.1 Electron–matter interactions

Electrons are generated in the electron microscope by thermionic emission from a cathode filament (often tungsten) and 'monochromated' by acceleration through a potential E. For an accelerating voltage of 100 keV, the wavelength is 0.037 Å. The electrons are scattered by the atomic potentials of the atoms in the specimen, and, as with X-rays, there is a sharp fall-off of scattering power with $\sin\theta/\lambda$. Broadly speaking, the scattering factor f_e increases with atomic number. In contrast to X-rays and neutrons, the scattering of electrons by matter is very strong ($f_e \approx 10^4 f_x$), and diffraction of electrons is feasible with gaseous samples. The strength of the interaction also leads to a breakdown of the kinematic theory that we have used to treat X-ray and neutron diffraction, and the interpretation of electron diffraction intensities is very complex.

The condenser lenses produce a fine, parallel beam of electrons at the specimen position, and a range of interactions takes place, some of which are summarized in Fig. 2.26. The diffracted and unscattered electrons form the basis for the conventional imaging mode described in the next section.

72 *Solid state chemistry: techniques*

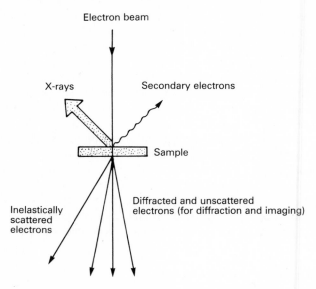

Fig. 2.26. The interaction of an electron beam with a specimen.

2.12.2 Transmission electron microscopy (TEM)[32]

Electrons that are diffracted will form a diffraction pattern which can be transformed directly into an image by magnetic lenses. Either the diffraction pattern or the image can be projected on to a viewing screen (Fig. 2.27). Single-crystal diffraction patterns, similar in appearance to zero-level, X-ray precession photographs, can be obtained by appropriately orientating a thin specimen (approximately 1 µm thick, or less). Using the high magnification of the instrument, we can therefore select individual crystallites from a powder sample, and produce single crystal diffraction patterns; a typical example is shown in Fig. 2.28.

With a great deal of care and a good electron microscope, it is possible to obtain images with atomic resolution by permitting an optimum number of diffracted beams to contribute to the image. The resolving power is dependent upon both the wavelength and the quality of the objective lens (which produces the first image)

$$d_{\min.} \propto C_s^{\frac{1}{4}} \lambda^{\frac{3}{4}}. \tag{2.22}$$

C_s is the spherical abberation coefficient of the objective lens. The image will be sensitive to the thickness of the crystal and the focusing conditions. Extremely thin crystals (<500 Å) are ideal, and it is customary to take

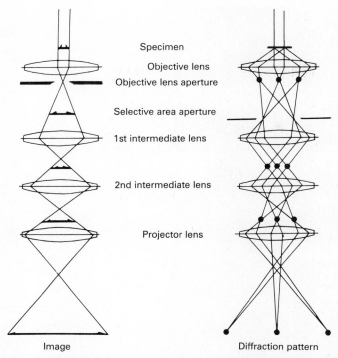

Fig. 2.27. Ray diagrams for the generation of (a) images and (b) diffraction patterns in a transmission electron microscope.

photographs at a series of focusing steps (a through-focal series); these can then be compared with theoretical images calculated by the multislice method[33] (Fig. 2.29). The image will be dominated by heavy atoms since these scatter electrons strongly and we should remember that the image will be a *projection* of the structure along the direction of the beam; this can often lead to problems in interpretation. There is no direct way of obtaining the structure from the image so we have to depend upon comparisons between the observed images and those calculated for a range of structural models. As with powder X-ray and neutron methods, we need a good starting model. In favourable cases it will be possible to deduce a reasonable model for the structure, but the optimization of atomic coordinates cannot be achieved with precision approaching that of X-ray or neutron methods. Another limitation of TEM is that many materials will not survive the high vacuum in the microscope, and the intense electron beam necessary for lattice imaging. One great strength of the microscope, however, is its ability to image defects—a theme to which we shall return in Section 2.13.

74 *Solid state chemistry: techniques*

Fig. 2.28. Selected-area electron diffraction pattern of a molybdenum–vanadium oxide.

2.12.3 Scanning electron microscopy (SEM)

The low-energy (< 50 eV) secondary electrons emitted from the surface of the specimen provide the basis for a different type of imaging. The beam can be concentrated to a small probe (say 20 Å diameter) that may be deflected across the specimen in a raster fashion using scanning coils. The secondary electrons can be detected above the specimen, and an image showing the intensity of secondary electrons emitted from different parts of the specimen can be displayed on a CRT. This scanning image is particularly useful for examining the morphology of a crystal surface (Fig. 2.30).

A related technique involves the collection of electrons transmitted *through* the specimen during scanning. These can be used to produce a scanning transmission (STEM) image, with the advantage, compared with TEM, that radiation damage is reduced because the beam is not stationary.

Fig. 2.29. High resolution transmission electron micrograph of $TiNb_{24}O_{62}$.

2.12.4 Analytical electron microscopy (AEM)

An analytical electron microscope is usually a transmission instrument that has been equipped with scanning coils to provide SEM and STEM modes, plus other ancillary equipment for studying X-ray emission, luminescence, or

Fig. 2.30. Scanning electron micrograph of a limpet's teeth, which contain biomineralized goethite, FeOOH (courtesy of Dr. S. Mann).

the spectral distribution of the transmitted electron beam (electron energy loss spectroscopy—EELS). The measurement of X-ray emission spectra is particularly valuable for providing high-resolution, chemical analysis of solids.[34]

The bombardment of a specimen with electrons gives rise to the emission of characteristic X-rays from the elements in the specimen (as in the X-ray tube—see Section 2.2.1). These can be detected and sorted into different energies using either a crystal monochromator or a solid-state, energy-dispersive detector (Fig. 2.31). X-ray detection has been used for many years on scanning instruments (the microprobe), but only recently has it found widespread use in transmission microscopes.

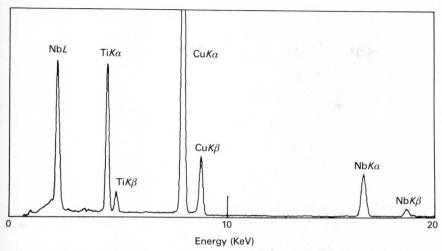

Fig. 2.31. The X-ray emission spectrum of a titanium–niobium oxide, mounted on a copper grid.

The intensities of the X-ray emission lines can be used to give quantitative chemical analysis. For thick specimens (the usual microprobe situation):

$$C_x = k\, ZAFI_x \tag{2.23}$$

where C_x is the concentration of element x, I_x is the intensity of a characteristic X-ray line of x, Z is the atomic number correction, A is the absorption correction, and F is the fluorescence correction.

For thin specimens (the TEM situation):[35]

$$\frac{C_x}{C_y} = k_{x/y} \frac{I_x}{I_y} \quad \text{(ratio method)}. \tag{2.24}$$

Both methods require standards. The intensity of the X-ray emission from a particular element can be superposed on a scanning image to show the distribution of that element in the specimen (an 'elemental map'). Analytical work is restricted to elements with $Z > 10$ on most instruments because the emissions from light elements are usually absorbed by the detector window, but a new range of thin window detectors promises to facilitate quantitative, light element analysis.[36]

Energy loss spectroscopy[37] looks at inelastically scattered electrons that have lost energy in the sample due to a range of processes, including plasmon excitations and the ionization of core electrons; the latter is the preliminary step to X-ray emission (see Section 2.2.1). A typical spectrum is shown in Fig. 2.32. Fine structure on the low-energy side of the edge stems from electric

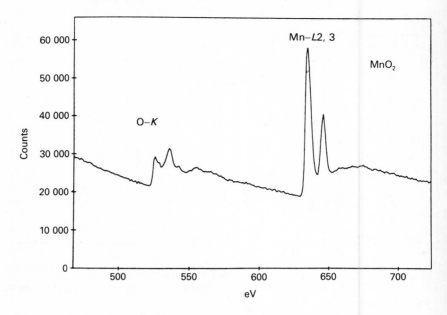

Fig. 2.32. The oxygen-K and manganese-L edges in the electron energy loss spectrum of MnO_2.

dipole transitions between discrete levels (e.g. $2p \rightarrow 3d$ in the $L_{2,3}$ lines of Mn in MnO_2), and the detail on the high energy side is similar to the EXAFS fine structure in X-ray absorption spectra. The EELS method is most sensitive in the 0–1000 eV range, a range that includes the K edges of the elements H–F, and is therefore useful for analysing light elements that are not readily detected in X-ray emission. Elemental maps can also be obtained in the form of energy-filtered transmission images. The EELS technique is now in a phase of rapid development and should become a powerful tool in solid-state chemistry during the next decade.

2.13 Defective and non-crystalline materials

Our treatment of structure determination using X-rays or neutrons has assumed that the solid under study is perfectly crystalline and ordered. All materials contain some element of disorder, albeit usually below the detection level of diffraction methods, but some contain substantial concentrations of defects; how can we obtain structural information for these grossly defective materials? Many spectroscopic methods are capable of giving information about the local environment around particular elements, e.g. Mössbauer, EXAFS, NMR, and ESR, but diffraction methods, too, can play a useful rôle.

2.13.1 Point defects in crystalline materials

For a crystal containing substantial concentrations (a few atom percentage) of point defects, for example a non-stoichiometric oxide such as UO_{2+x}, the effect of the defects on the diffraction pattern is to produce diffuse scattering around and between the Bragg reflections. If, as in a conventional structure determination, we simply measure the Bragg reflections and solve the structure, the unit cell contents that we obtain from our Fourier maps will represent an *average* cell for the whole crystal.[38] From this we can ascertain whether the defects are present as, say, interstitials or vacancies, but information about the local atomic arrangements in the vicinity of the defects (e.g. any defect clustering) is contained in the diffuse scattering and is therefore lost in the Bragg experiment. Unfortunately, even if we can measure the diffuse intensity, we have no way of phasing the data to obtain electron density maps for the defects. The electron microscope circumvents this problem because the lenses do not discriminate between diffuse and Bragg scattering; the images therefore contain the defects as well as the periodic structure (see Fig. 2.29). If we wish to use the diffuse scattering in X-ray or neutron patterns, then we must use the more indirect approach described below, for amorphous materials.

2.13.2 Amorphous and semi-crystalline materials

For an amorphous material, the absence of long-range order leads to a breakdown of the condition required for Bragg scattering, and all the scattering will be diffuse. The pattern will resemble that of a powder sample, with cones of scattering at different 2θ values, except that the cones will be broadened into haloes.

We can obtain some useful information by carrying out a Fourier transform of the scattering curve $I(s)$ since, by analogy with the Patterson method (Section 2.8.1), this should furnish a one-dimensional radial distribution function (RDF):

$$p(r) = \int sI(s)\sin(sr)ds \qquad (2.25)$$

where $s = 4\pi\sin\theta/\lambda$. The treatment is essentially the same as that used for gas-phase electron diffraction data. In Fig. 2.33, we show the scattering curve and radial distribution function for amorphous WS_3.[39] A model for the local atomic arrangement can be derived from the RDF.

If a material is partially crystalline, then the degree of crystallinity can be determined by comparing the relative intensities of the Bragg peaks and diffuse scattering.

2.13.3 Particle size

The Bragg condition can also break down when the particle size in a powder

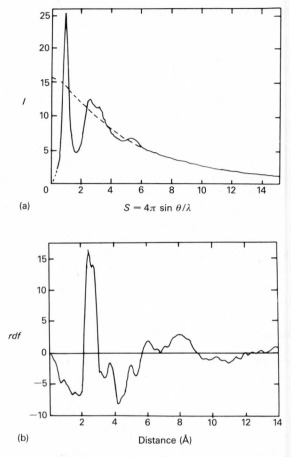

Fig. 2.33. The scattering curve (a) and radial distribution function (b) for amorphous MoS_3. The *rdf* is dominated by Mo-Mo vectors at ~ 2.5 Å.

becomes very small (say, < 500 Å). The effect of this is to broaden the reflections, and by measuring the broadening we can estimate the particle size. We use the Scherrer formula:[40]

$$t = \frac{0.9\lambda}{B\cos\theta_B} \qquad (2.26)$$

where t is the particle size, B is the full width at half maximum (in 2θ), and θ_B is the Bragg angle. Corrections must, of course, be made for instrumental broadening.

2.14 Postscript

Since this chapter was written, the *ab initio* determination of crystal structures by powder diffraction methods has been demonstrated by several groups. Examples include $FeAsO_4$[41], determined from time-of-flight neutron data, $AlPO_4-12$[42] from laboratory X-ray data, and $\alpha\text{-}CrPO_4$[43], I_2O_4[44] and $MnPO_4 \cdot H_2O$[45] from synchrotron X-ray data. This important advance has been brought about by improvements in both computational methods and instrumentation. In general, it seems that X-rays are to be preferred for structure determination, since a reasonable phasing of the data can usually be obtained if a small number of dominant scatterers can be located, but that neutrons, with their approximately constant sensitivity to all elements, are superior for refinement.

2.15 References

1. Stout, G. H. and Jensen, L. H. *X-ray structure determination. A practical guide.* MacMillan, New York (1968).
2. Woolfson, M. M. *An introduction to X-ray crystallography.* Cambridge University Press (1970)
3. Cromer, D. T. and Mann, J. B. *Acta Cryst.* **A24**, 321 (1968).
4. *International tables for X-ray crystallography* (eds C. H. MacGillavry, G. D. Rieck and K. Lonsdale) Vol. III. Kynoch Press, Birmingham (1962).
5. Klug, H. P. and Alexander, L. E. *X-ray diffraction procedures for polycrystalline and amorphous materials.* John Wiley and Sons, New York (1974).
6. Nelson, J. B. and Riley, D. P. *Proc. Phys. Soc. (London)*, **57**, 160 (1945).
7. *Powder Diffraction File—Inorganic Compounds.* JCPDS International Centre for Diffraction Data, Philadelphia (1984).
8. Visser, J. W. *J. appl. Cryst.* **2**, 89 (1969).
9. Ippolitov, E. G., Garashina, L. S., and Maklasklov, A. G. *Inorg. Mat.* **3**, 59 (1967).
10. *International tables for X-ray crystallography* (eds N. F. M. Henry and K. Lonsdale) Vol. I. Kynoch Press, Birmingham (1952).
11. Woolfson, M. M. *Direct methods in crystallography.* Oxford University Press (1961).
12. Sayre, D. *Acta Cryst.* **5**, 60 (1952).
13. Main, P., Fiske, S. J., Hull, S. E., Lessinger, L., Germain, G., Declercq, J-P., and Woolfson, M. M. *MULTAN80, A system of computer programs for the automatic solution of crystal structures from X-ray diffraction data.* Univs of York, England, and Louvain, Belgium (1980).
14. Rollett, J. S. *Computing methods in crystallography.* Pergamon Press, Oxford (1965).
15. Mitchell, D. P., and Powers, P. N. *Phys. Rev.* **50**, 486 (1936).
16. Windsor, C. G. *Pulsed neutron scattering.* Taylor and Francis, London (1981).
17. Bacon, G. E. *Neutron diffraction.* Clarendon Press, Oxford (1975).
18. Titcombe, C. G., Cheetham, A. K., and Fender, B. E. F. *J. Phys., C—Solid-state Phys.* **7**, 2409 (1974).

19. Battle, P. D., Cheetham, A. K., and Goodenough, J. B. *Mat. Res. Bull.* **14**, 1013 (1979).
20. Shull, C. G., and Siegel, S. *Phys. Rev.* **75**, 1008 (1949).
21. Brunton, G. D., and Johnson, C. K. *J. chem. Phys.* **62**, 3797 (1975).
22. Battle, P. D., and Cheetham, A. K. *J. Phys., C—Solid-state Phys.* **12**, 337 (1979).
23. Moon, R. M., Riste, T., and Koehler, W. C. *Phys. Rev.* **181**, 920 (1969).
24. Stevens, E. D., and Coppens, P. *Acta Cryst.* **A35**, 536 (1979).
25. Cheetham, A. K., and Eddy, M. M. Unpublished results.
26. Rietveld, H. M. *J. appl. Cryst.* **2**, 65 (1969).
27. Cheetham, A. K., and Taylor, J. C. *J. solid state Chem.* **21**, 253 (1977).
28. Long, G. J., Longworth, G., Battle, P. D., Cheetham, A. K., Thundathil, R. V., and Beveridge, D. *Inorg. Chem.* **18**, 624 (1979).
29. Von Dreele, R. B., Jorgensen, J. D., and Windsor, C. G. *J. appl. Cryst.* **15**, 581 (1982).
30. Malmros, G., and Thomas, J. O. *J. appl. Cryst.* **10**, 7 (1977).
31. Glazer, A. M., Hidaka, M., and Bordas, J. *J. appl. Cryst.* **11**, 165 (1978).
32. Thomas, G., and Goringe, M. J. *Transmission electron microscopy of materials.* Wiley–Interscience, New York (1979).
33. Skarnulis, A. J. *J. appl. Cryst.* **12**, 636 (1979).
34. Cheetham, A. K., and Skarnulis, A. J. *Anal. Chem.* **53**, 1060 (1981).
35. Cliff, G., and Lorimer, G. W. *J. Microsc. (Oxford)* **103**, 203 (1975).
36. Cheetham, A. K., Skarnulis, A. J., Thomas, D. M., and Ibe, K. *J. Chem. Soc., Chem. Comm.* 1603 (1984).
37. Ahn, C. C., and Krivanek, O. L. *EELS atlas.* ASU HREM facility and GATAN, USA (1983).
38. Cheetham, A. K. *Chemical applications of thermal neutron scattering* (ed. B. T. M. Willis). Oxford University Press (1972).
39. Liang, K. S., de Neufville, J. P., Jacobson, A. J., and Chianelli, R. R. *J. non-cryst. Solids* **35**, 1249 (1980).
40. Scherrer, P. *Gött. Nachr.* **2**, 98 (1918).
41. Cheetham, A. K., David, W. I. F., Eddy, M. M., Jakeman, R. J. B., Johnston, M. W., and Torardi, C. C. *Nature (London)* **320**, 46–48 (1986).
42. Rudolph, P. R., Saldarriaga-Molina, C., and Clearfield, A. *J. Phys. Chem.* **90**, 6122–6125 (1986).
43. Attfield, J. P., Sleight, A. W., and Cheetham, A. K. *Nature (London)* **322**, 620–622 (1986).
44. Lehmann, M. S., Christensen, A. N., Fjellvåg, H., Feidenhans'l, R., and Neilsen, M. *J. Appl. Crystallogr.* **20**, 6122–6125 (1987).
45. Lightfoot, P., Cheetham, A. K., and Sleight, A. W. *Inorg. Chem.* **26**, 3544–3547 (1987).

2.16 Appendix: The determination of heavy-atom coordinates from a Patterson map

Example: MoO_2, space group $P2_1/c$.

Lattice parameters: $a = 5.584$ Å, $b = 4.842$ Å, $c = 5.608$ Å, $\beta = 120.9°$.

Density measurements confirm that there are four formula units per unit cell ($n = 4$):

By inspection, we can assign the Mo–Mo vectors:

Vector A is Mo(1)–Mo(4)

$\therefore 2y - \frac{1}{2} = \frac{1}{2}$ $\quad \therefore y = 0.50$

Vector B is Mo(1)–Mo(3)

$\therefore 2x = 0.464,\quad x = 0.232$

and $2z - \frac{1}{2} = 0.568$ $\quad \therefore z = 0.534$

We can then check that Mo(1)–Mo(2), expected at $2x$, $2y$, $2z$, i.e. 0.464, 0.00, 0.068, is observed at the correct position. This is peak C. Note that a vector at $2z = 1.068$ is equivalent to one at 0.068 because of the unit cell translational symmetry. Thus, Mo is located at (0.232, 0.50, 0.534) and the three equivalent positions.

3 X-ray photoelectron spectroscopy and related methods

G. K. Wertheim

3.1 Introduction

Photoelectron spectroscopy is a deceptively simple technique. The basic principle of the photoelectric effect was enunciated by Einstein[1] in a paper which is one of the cornerstones of the quantum theory. Experimental observations had shown that there is a threshold in frequency below which light, regardless of intensity, fails to eject electrons from a particular metallic surface. Einstein recognized that this is a natural consequence of the postulate that light is quantized, i.e. that light consists of photons of energy

$$E = h\nu \tag{3.1}$$

where h is Planck's constant and ν the frequency. The threshold frequency, ν_c, then corresponds to the minimum energy, denoted $e\varphi$, required to remove an electron from the metal to empty space,

$$e\varphi = h\nu_c. \tag{3.2}$$

φ is called the work-function, and e is the magnitude of the electronic charge. Typical values of φ fall between 2 and 6 eV. Agreement between the work-functions obtained for a particular metal from photoemission and from thermionic emission provided additional support for the quantum hypothesis.

At higher frequency, the photoelectrons emerge into space with a *maximum* kinetic energy which reflects the excess of the photon energy over the work function,

$$E_{\text{kin}}^{\text{max}} = h\nu - e\varphi. \tag{3.3}$$

The maximum kinetic energy is associated with electrons emitted from the immediate vicinity of the Fermi level, i.e. from the highest lying occupied electronic states of the metal. Electrons which lie E_B below the edge of the conduction band emerge with correspondingly less energy,

$$E_{kin} = hv - E_B - e\varphi. \tag{3.4}$$

E_B is called the binding energy, see Fig. 3.1.

If the experiment is carried out with photons whose energy is much greater than the work-function, the kinetic energy spectrum will be a replica of the distribution of occupied bound states. Photoelectron spectroscopy then becomes a technique for the direct imaging of the electronic states of atoms, molecules, and solids.

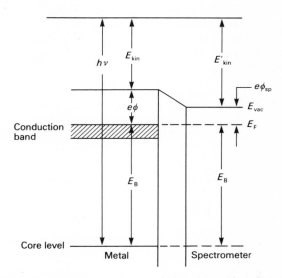

Fig. 3.1. Energy level diagram for photoemission. Note that the contact potential between the sample and the spectrometer, $e(\varphi - \varphi_{sp})$, changes the kinetic energy of the electron as it enters the spectrometer. As a result only the spectrometer work function is needed to calculate E_B from the measured kinetic energy.

Before the development of synchrotron sources, the shorter wavelength radiation required to probe the deeper-lying core levels was available only in the form of characteristic X-rays. The choice of radiation is largely dictated by the requirements of high photoelectron flux and good resolution. Photoelectron intensity depends strongly on the photoelectric cross section, which decreases with frequency as $(v)^{-3.5}$. The attainable resolution is limited by the line width of the X-ray. Both considerations point toward the use of low-energy X-rays. Since it is also essential to use a simple X-ray spectrum the choice is limited to the $K\alpha$ radiation of the low-Z elements. The need for a stable X-ray tube anode surface with good thermal conductivity narrows the choice to metals. From these considerations Mg and Al *Kα* radiation, with photon energies of 1253.6 and 1486.6 eV, emerge as the most reasonable candidates for use in X-ray photoelectron spectroscopy (XPS).

Siegbahn et al.[2,3] were the first to demonstrate the power of X-ray photoemission. He and his collaborators at the University of Uppsala in Sweden carried out the pioneering development, and showed that chemical information could be obtained by this technique, which they called *electron spectroscopy for chemical analysis*, or ESCA.

ESCA experiments have produced interesting and provocative results. A wide scan provides immediate identification of the chemical composition, because the core electron binding energies of the elements are distinctive; see Fig. 3.2. High-resolution scans yield a wealth of unexpected detail, e.g. chemical shifts, multiplet structure, and satellites. The ability to resolve small changes in binding energy associated with changes in chemical bonding has been responsible for much of the interest in this technique.

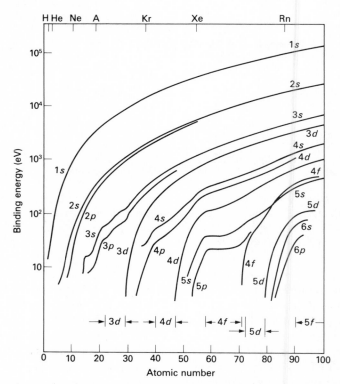

Fig. 3.2. Core electron binding energies of the elements. The regions of transition metals, rare earths, and actinides are indicated. Below 10 eV chemical effects become too large to allow a universal plot. (Plotted from Bearden and Burr,[12] and Siegbahn et al.[2])

It is important to recognize at the outset that ESCA provides information about a thin surface layer. The absorption length of the X-rays is large, 10^3 to 10^4Å, so that photoelectrons are created with significant intensity to that depth within the solid. Before their kinetic energy can be measured, they must travel to the surface and escape into the vacuum of the spectrometer. Most of them are inelastically scattered before they emerge from the solid, reducing their kinetic energy and removing them from the sharp photo-peak that is studied in ESCA. The mean-free-path of electrons with energies of the order of 1.5 keV is typically only 15 Å; see Fig. 3.3. The sharp peak in the photoemission spectrum is therefore due largely to those few photoelectrons which are created within the first few atomic layers. Electrons from deeper in the solid are inelastically scattered and produce a broad structure extending toward greater binding energy. The area of this 'loss tail' greatly exceeds that of the sharp peak itself, but is relatively unobtrusive because it is spread over a wide range of energy.

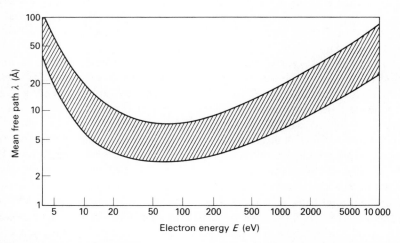

Fig. 3.3. Mean-free-paths of electrons in solids as a function of kinetic energy. The majority of reported values falls within the shaded region. The general trend is often called 'the universal curve' since the data show no systematic dependence on the nature of the solid.

3.2 Techniques

Photoelectron spectroscopy requires (1) a source of monochromatic radiation, (2) an electron spectrometer, and (3) an adequately prepared sample. For detailed discussion of the first two subjects see Barrie.[4]

3.2.1 Radiation sources

As initially conceived,[2] an ESCA spectrometer consists of an Al or Mg $K\alpha$ X-ray tube irradiating a specimen whose photoelectrons are energy-analysed in a magnetic or an electric deflection spectrometer. The resolution of such an instrument is limited by the nature of the X-rays used to excite the photoemission. The Al $K\alpha_{1,2}$ line consists of two components separated by the 0.4 eV spin-orbit splitting of the $2p^5$ final state created by the dipole transition to the $1s$ hole state. Both components have the 0.47 eV lifetime width of the $1s$ hole state. The resolution is consequently inherently no better than 0.9 eV. With Mg $K\alpha$ somewhat better resolution can be obtained. The X-ray tube emission spectrum further contains the $K\alpha_{3,4}$ and $K\beta$ radiations, as well as a broad bremsstrahlung background. Photoemission from a single electronic level with the total output of the X-ray tube produces an image of the X-ray tube spectrum, a rather unfavourable situation. A significant improvement is obtained by using Bragg diffraction from a spherically bent crystal to disperse the X-rays and focus only one component on the sample. This can improve the resolution to 0.25 eV, but reduces the X-ray flux at the sample by orders of magnitude and increases the time required to obtain satisfactory data.

An important source of radiation for ultraviolet photoemission (UPS) is found in rare-gas discharge lamps. Windowless systems using He(I), He(II), or Ne resonance radiation at 21.2, 40.8, and 16.8 eV, are extremely valuable for valence band spectroscopy, offering the best resolution currently available. See Orchard[5] and Williams[6] for more details.

The future of ESCA lies in the use of synchrotron radiation from electron storage rings which provide a continuous spectrum with intensities which exceed those of the characteristic X-ray lines. The utilization of synchrotron radiation has made rapid progress, limited only by the need to develop suitable monochromators, especially in the region above the carbon K edge at 285 eV. The availability of tunable radiation has already made possible new types of photoemission experiments, and has resulted in the discovery of threshold effects that add a new dimension to our understanding of the many-electron aspects of the photoemission process (see below).

3.2.2 Electron spectrometers

Once a photoelectron has emerged from the surface of a sample, its kinetic energy must be measured. This can be accomplished by retardation or by deflection. If the electron flux is high, retardation is adequate. In its simplest implementation, retardation yields the integral of the spectrum, which is then differentiated. A system of hemispherical retarding grids, with modulated retardation and lock-in detection can be used to effect the differentiation. If the signal is weak, this is disadvantageous because the small differences between counting rates at two closely spaced energies carry with them the large statistical fluctuations of the original spectrum.

X-ray photoelectron spectroscopy usually requires a deflection spectrometer and the detection of individual electrons. The relatively low kinetic energy, 1 keV, makes electrostatic analysers preferable to magnetic ones. Such spectrometers could be operated by scanning the voltages on the deflection elements, but this would result in a resolution which varies with electron kinetic energy. It yields the lowest resolution at the highest kinetic energy, i.e. for electrons from the vicinity of the Fermi level, where the best resolution is desired. To obtain a constant optimum resolution it is advantageous to operate at a fixed, low analyser pass-energy and employ variable retardation before the photoelectrons enter the analyser.

The most common analyser is the cylindrical mirror type. Its major advantages are its large solid angle of acceptance and its ease of manufacture. Entrance and exit slits are annular, so that the entire spectrometer has cylindrical symmetry. Such analysers can be modified for angle-resolved photoemission by blocking off all but a small part of the 2π azimuthal transmission. Electron detection in these instruments is generally by a channeltron or similar electron multiplier. The other type of analyser which is widely used is based on hemispherical deflection elements. It is double-focusing, gives an essentially linear dispersion along the radius of the sphere, and is thus well suited for multi-element detection. For more detail see Roy and Carette.[7]

3.2.3 Sample preparation

The realization that the mean-free-path of photoelectrons is typically of the order of a few atomic layers implies that special surface preparation is required to study bulk properties. The minimal procedure is to crush a crystalline material in an inert atmosphere in order to expose fresh surfaces just prior to insertion into the spectrometer.

In recent years it has become general practice to work with surfaces prepared in the vacuum of the spectrometer, and to monitor surface conditions in detail. No single technique is applicable to all materials. Without doubt the optimum procedure is cleaving. It works well for single crystals of ionic compounds with high symmetry, and in materials that have a natural layer structure. For the more covalent materials special cleaving techniques have been developed. When large single-crystals are not available, one is often content simply to fracture the specimen in vacuum, exposing a rough but clean surface. Metals are generally cleaned by argon ion sputtering, followed by high-temperature annealing to remove the disorder and the implanted argon. Sputtering is often not a satisfactory technique for compounds or alloys because lighter atoms are preferentially removed; thus metal oxides are reduced, and alloy compositions are changed. An alterna-

90 *Solid state chemistry: techniques*

tive approach is to deposit a layer of the material on a substrate by vacuum evaporation or molecular beam epitaxy in a preparation chamber attached to the spectrometer.

3.2.4 Typical spectra

It is instructive to consider the details of the spectrum associated with photoemission from a single core electron state. For simplicity we have chosen the spectrum of the 1s state of carbon in graphite. The data shown in Fig. 3.4 were taken with monochromatized Al $K\alpha$ radiation on a surface prepared by cleaving in vacuum. The sharp line at 284.5 eV is due to photoelectrons originating in the first few monolayers, which have emerged from the sample without causing excitations of the solid. The long tail

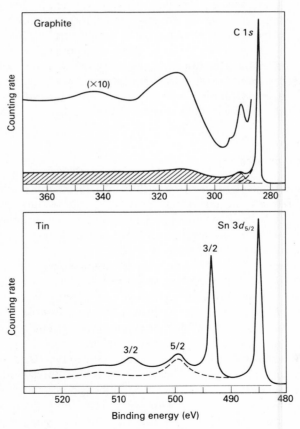

Fig. 3.4. Photoemission from graphite and tin with monochromatized Al $K\alpha$ radiation. Note the area of the energy loss tail above the background level.

towards greater binding energy is due to C 1s photoelectrons which have lost some of their kinetic energy to the solid. This energy-loss tail has structure which reflects the excitation spectrum of graphite, i.e. the interband transitions and the collective excitations of the valence band called *plasmons*. These can be seen more clearly in the vertically expanded trace

Plasmons which are excited as the energetic photoelectron propagates through the solid *after* the photoelectric process are called *extrinsic* plasmons. There is, however, also the possibility of exciting a plasmon as an integral part of the photoelectric process. These are called *intrinsic* plasmons and are part of the screening response of the solid to the sudden creation of a core hole in one of its atoms. Calculations for photoemission from simple metals have shown that the intrinsic plasmons make up almost one-half of the first plasmon loss peak. The spectrum of these two kinds of plasmons is usually indistinguishable.

Photoemission from an electronic state with non-zero orbital angular momentum produces a spin-orbit doublet; see the 3d spectrum of tin in Fig. 3.4. The two lines correspond to final states with $j_+ = l+s$ and $j_- = l-s$. The intensity ratio of these lines is given by the $(2j_- + 1)/(2j_+ + 1)$, e.g. $\frac{1}{2}$ for p subshells, $\frac{2}{3}$ for d subshells, etc. Each component has its own energy loss tail, resulting in a relatively complex spectrum.

Wide-scan X-ray photoemission spectra are useful mainly for chemical analysis, see Fig. 3.5. The photoelectron lines are readily matched up with the known inner-shell binding energies of one of the elements, in this case Nb,

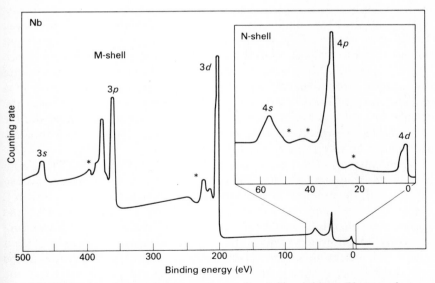

Fig. 3.5. Wide-scan photoemission spectrum of metallic niobium. Plasmon features are indicated by an asterisk.

and provide an unambiguous chemical identification of the sample. The spin-orbit splitting of sub-shells with non-zero orbital angular momentum is resolved, and can be used to confirm the chemical identification. Within each core shell the lifetime width decreases, and the cross section increases, with increasing orbital angular momentum. As a result the core electron with largest j of a given shell generally gives the signal with the largest peak height and is most useful in chemical analysis. The full extent of the energy-loss tails are seen most clearly in such wide scans, and under favourable circumstances the quantized excitations of the solid, the plasmons or interband transitions, are also resolved. In Fig. 3.5 the plasmon satellites, which have energies of 11 and 20 eV, are indicated by asterisks.

The measurables in a core electron spectrum are the positions, widths, and shapes of the lines. In many cases the position and width can be obtained by inspection, and the shape safely ignored. However, when faced with unresolved overlapping lines, numerical data analysis becomes necessary. The best approach is to fit a theoretical function to the data by least-squares optimization. This function should be a convolution of the spectrometer resolution function with a mathematical representation of the photoemission line. The spectrometer resolution function may be obtained from the derivative of the Fermi cut-off in a metal with a simple conduction band, e.g. Ag, since the width of the resolution function is usually much greater than kT. The photoemission line itself has a Lorentzian shape corresponding to the lifetime of the core hole which is created.

Binding energies are necessarily expressed relative to a reference level, but different levels are used depending on the nature of the experiment. In gas-phase photoemission, binding energies are measured from the vacuum level. The binding energy of the outermost electron is called the first ionization potential. In the study of solids it is more convenient to use the Fermi level. The Fermi levels of the spectrometer and a conducting sample coincide, and they can be determined by direct observation of the Fermi cut-off in a metallic sample. The work-function of the sample does not enter into the calculation of E_B because the contact potential between the sample and the spectrometer changes the kinetic energy of the photoelectron appropriately as it enters the spectrometer, see Fig. 3.1.

When working with insulating samples new problems arise. The Fermi level could be anywhere within the band gap. Furthermore, photoemission can result in the build-up of positive charge near the surface resulting in a dipole layer. Binding energies measured relative the Fermi level of the spectrometer then lack validity. Attempts to compensate for such charging effects by flooding the sample surface with low-energy electrons provide a partial but unreliable solution. Alternatively an external reference can be provided by depositing a standard substance like Au on the surface of the sample, and using one of its core levels to define the energy scale. The best procedure is to measure all energies from some well-defined feature of the electronic structure, e.g. the edge of the valence band, which itself can be located by XPS. For a full discussion of these problems see Evans.[8]

For valence band studies, data analysis often consists simply of a comparison of the spectrum with molecular orbital or band structure calculations, the latter preferably convoluted with the instrumental resolution function. There is little benefit in the opposite approach, that of deconvoluting the resolution function from the data; it tends to distort line shapes and introduce fictitious features into the output.

3.3 Fundamental properties of core electron spectra

3.3.1 Electron binding energies

(a) Binding energies in free atoms

The measured binding energy of a core electron in a free neutral atom is not in agreement with the calculated eigen-energy of that state. The resolution of this conflict lies in the recognition that the two energies are inherently different quantities. The calculation refers to the atom in its initial state with all orbitals occupied; the experiment to the atom in the final state, *after* photoemission, with one of its orbitals empty. The two would be equal if the outer orbitals remained frozen as the inner electron is removed. In reality the outer shells of the atom readjust when an inner electron is removed because the coulomb attraction of the nucleus is then less effectively screened. To the outer electrons after photoemission the core looks like that of the atom with the next higher atomic number. Explicit use is often made of this similarity by substituting properties of the valence-ionized $Z+1$ atom for those of the core-ionized Z atom. This is called the 'equivalent cores' approximation.

It is useful to think of the photoemission process as comprising the following events. A bound electron absorbs the photon, converting the photon energy into kinetic energy. As the electron leaves the atom some of its energy is used to overcome the coulomb attraction of the nucleus, reducing its kinetic energy by its initial state binding energy. At the same time the outer orbitals readjust, lowering the energy of the final state which is being created and giving this additional energy to the outgoing electron. The change due to outer shell readjustment is termed the intra-atomic relaxation energy.

Relaxation does not, however, necessarily proceed to the ground state of the hole-state atom. Electronic transitions between outer orbitals will produce excited final states which result in satellite structure at the high binding energy side of the main photoemission line. The process which produces these excited states is generally referred to as *shake-up*. It was shown by Lundquist[9] that in the sudden approximation the centroid of all the final states lies at the frozen-orbital energy, i.e. the energy which would be obtained if all relaxation were turned off. This is the basis of Koopmans' theorem[10] which equates the binding energy to the eigen energy. It has proved possible to measure the Koopmans' energy for free atoms by measuring the centroid of the entire spectrum; but for solids the extrinsic losses generally obscure the excited state structure. See Shirley[11] for a more detailed discussion.

Since most of the early information regarding core electron states was obtained from *X-ray emission* spectroscopy,[12] it is worth noting that the above considerations apply equally well in that case. X-ray emission energies reflect the difference between the energies of two relaxed final states, and are consequently directly comparable to differences in binding energies obtained by XPS.

(b) Binding energies in solids

Electron binding energies in molecules or solids depend on the environment of the atom. For the discussion of the resulting binding energy shifts it is advantageous to take the value of the core electron binding energy in the neutral free atom as the standard from which shifts are measured. These shifts are of coulombic origin. For a free atom the initial state core electron binding energy shift that occurs when a valence electron is missing is well approximated by $<e/r>$, where r refers to the charge distribution of the empty valence orbital. All core electrons that lie well within the valence orbital will exhibit the same shift, because the potential due to a shell of charge is constant inside the shell. Such free atom core electron shifts typically have value of the order of $1\text{ Ry} = 13.6\text{ eV}$.

In an ionic solid, which consists of a regular lattice of positive and negative ions, there is a second initial state term due to the potential of all the other charged ions in the solid. It is called the *Madelung potential*. It always opposes the single-ion shift, and typically cancels out 80 per cent of it. For simple structures the Madelung *potential* can be deduced from equations for the Madelung *energies* which have been tabulated. However, this method is applicable only when an ionic charge can be assigned to each lattice site.

The discussion of covalent materials takes isolated neutral atoms as its starting point and allows them to coalesce to form a solid. As the outer orbitals interact to form bands, charge will redistribute among the valence orbitals. Thus, one might expect an initial state increase in core electron binding energy when atomic Si with s^2p^2 configuration forms diamond structure Si with sp^3 configuration because a valence electron has moved from an s into a more extended p orbital. Similarly with $[Xe]4f^65d^06s^2$ configuration Sm atoms become $4f^55d^16s^2$ in the metal suggesting a large initial state increase in core electron binding energy.

Measured binding energies are, however, further modified by new final state relaxation effects which appear in the transition from the isolated atom to the solid. We assume that intra-atomic relaxation, which is included in the free-atom binding energy, remains unchanged when a solid is formed. Extra-atomic relaxation, the response of the environment to the ionization of an atom, is specific to the material under consideration. It always lowers the energy of the final state. In an insulator the neighbouring ions are polarized to screen the core hole. In a metal the conduction electrons respond to perform the same function. The resulting shifts are not small and must be included in any complete analysis of binding energy shifts. Unfortunately it generally requires a formidable effort to calculate these shifts from first principles. For more details on ionic systems see Citrin and Thomas;[13] for metals see Gelatt, Ehrenreich, and Watson,[14] and Williams and Lang.[15]

The above analyses clarify why *X-ray emission* lines, which result from transitions between two core hole states, in the same atom exhibit only small shifts with change in chemical state: the single-ion, charge redistribution, relaxation, and Madelung terms act almost equally on all core electrons.

(c) The Born–Haber cycle

An alternate approach to core electron binding energies in solids bypasses detailed electronic considerations and refers only to macroscopic quantities. It is based on a Born–Haber cycle which was initially explored in detail by Johansson and Martensson,[16] and is illustrated in Fig. 3.6. By traversing the

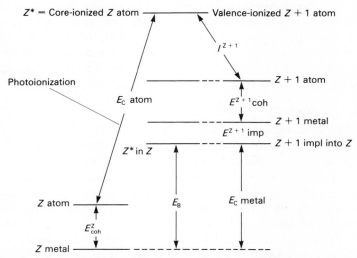

Fig. 3.6. Born–Haber cycle showing the connection between the binding energies of a core electron in the free atom and the solid, and various thermodynamic quantities.

cycle, starting at the metallic state in the lower left, one obtains a relationship between atomic ionization potentials, cohesive energies, and the free atom and metal core electron binding energies. The steps are as following:

(1) Remove one atom from the metal at the cost of the cohesive energy E_{coh}^Z,
(2) core-ionize the atom at a cost of the free-atom core electrons binding energy, $E_{\text{atom}}^{\text{core}}$,
(3) use an equivalent cores argument to replace the core-ionized Z atom with a valence-ionized $Z+1$ atom without any change in energy,
(4) add an electron to the $Z+1$ ion to form a neutral $Z+1$ atom resulting in a gain in energy equal to the ionization potential of the $Z+1$ atom, I^{Z+1},
(5) condense such $Z+1$ atoms into a metal with a gain of the $Z+1$ atom cohesive energy, E_{coh}^{Z+1},
(6) transfer a $Z+1$ atom from the $Z+1$ metal to the Z metal host, releasing the implantation energy, E_{imp}^{Z+1}.

A $Z+1$ impurity atom dissolved in a Z metal is electronically identical to a fully screened core-ionized Z atom in the Z metal. As a result one obtains the relationship

$$E_{\text{core}}^{\text{metal}} = E_{\text{core}}^{\text{atom}} + E_{\text{coh}}^Z - E_{\text{coh}}^{Z+1} - I^{Z+1} - E_{\text{imp}}^{Z+1}. \quad (3.5)$$

The only poorly known quantity is the implantation energy, which has been found to be small. The energy $E_{\text{core}}^{\text{metal}}$ can be identified with the usual core electron binding energy referenced to the Fermi level, since it represents the energy to create a fully screened state by moving a core electron to E_F.

Cycles of this nature can be constructed to gain insight into other phenomena effecting core electron binding energies, including surface atom core level shifts (see Section 3.6.1). The connection between electronic and thermodynamic quantities which emerges from such cycles shows that photoelectron spectroscopy can be used to determine such relatively inaccessible thermodynamic quantities as the implantation energy.

(d) Chemical shifts

Siegbahn et al.[2,3] have shown that core electron binding energies in molecular systems exhibit chemical shifts which are simply related to various quantitative measures of covalency. In general, the greater the electronegativity of the ligands, the greater the binding energies of the core electrons of the ligated atom. From a purely coulombic point of view, the chemical shift arises in the initial state from the displacement of electronic charge from the atom towards its ligands, reducing the electrostatic potential at the atom in question. In addition there is a final state shift due to the polarization of the ligand by the core hole on the central atom. Since the net effect always depends on the combined effects of electronegativity and polarizability of the ligand, a certain universality can be anticipated.

Chemical shifts in solids are much less predictable because the final state screening and the Madelung potential can vary profoundly from substance to substance. There may also be inconsistencies in reference level when comparing metals and their compounds. Nevertheless, the expectation that core electron binding energies of cations should increase with valence is usually borne out. Good examples are found in the oxides of Ti, V, Nb, and W, but there are some cases which violate this rule, e.g. the oxides of Sn and Pb. The retrograde shift from PbO to PbO_2 has been shown to arise from differences in the Madelung potential in the two structures. Unexpectedly small shifts have also been found for Cu_2O relative to Cu, and for Ag_2O relative to Ag.

3.3.2 Linewidths

The irreducible width Γ of an XPS line is due to the lifetime of the core hole state created in the photoemission process.

$$\Gamma = 2\hbar/\tau. \quad (3.6)$$

The resulting line shape is Lorentzian. A core hole has a number of decay channels, whose contributions to the width are additive. Radiative decay, i.e. X-ray emission, follows dipole selection rules and is dominant only for the inner shells of the heavy elements. Other shells de-excite by two-electron processes in which one electron is emitted from the atom while the other makes a transition to the initial-state core hole. They are denoted by the shell of the initial-state hole and the shells of the final state holes, e.g. $L_{III}M_{IV,V}M_{IV,V}$ indicates that a $2p_{3/2}$ hole is filled by one $3d$ electron while another is emitted. Transitions in which the initial and final state holes are in different shells are termed *Auger transitions*, after P. Auger[17, 18] who first described this process. If one of the final state holes lies in the shell of the initial state hole they are termed *Coster-Kronig transitions*.[19] The latter have far larger transition probabilities because the overlap between initial and final state wave-functions is much greater than in the Auger case. The most weakly bound electronic state of a given shell always has the smallest lifetime width, because it is not subject to a Coster-Kronig de-excitation. Partial and total transitions probabilities have been calculated for most of the inner core states of the elements.[20-25] In the few cases which have been studied experimentally, good agreement has been reported, especially when the dominant decay channels involve allowed transitions.

Measurement of the lifetime width of a core hole state is complicated by the existence of other contributions to the width of the experimental line from (1) instrumental resolution, (2) phonon broadening, and (3) inhomogeneous broadening. The determination of the lifetime width requires a least-squares fitting of the experimental data with a line shape generated by the convolution of a Lorentzian with the other known broadening functions.

The instrumental resolution function can have a variety of shapes. In instruments using unmonochromatized Al or Mg $K\alpha$ radiation it contains the long-tailed Lorentzian shape due to the lifetime width of the K-shell hole in Al or Mg. This makes it difficult to measure the additional Lorentzian width due to the hole state under study. In instruments using an X-ray monochromator the resolution function is significantly narrower and tends towards a Gaussian profile.

Phonon broadening is due to the excitation of lattice vibrations by the photo-ionization process. In ionic crystals, it arises from the electrostatic interactions of the suddenly created core hole with its neighbouring ions. In covalent materials, the change in atomic radius due to the contraction of the outer orbitals in response to the core hole similarly causes a sudden relaxation of the neighbouring atoms. Theoretical analysis[26] is usually based on the Frank–Condon principle. The excitation energy is much greater than typical phonon energies, so that many phonons are created. A multiphonon process has a Poisson probability distribution, which in the many-phonon limit is well-approximated by a Gaussian shape.

Inhomogeneous broadening arises from a superposition of lines with different chemical shifts, and should be important mainly in disordered system, e.g. alloys, amorphous materials, or submonolayer adsorbates.

3.3.3 Intensities

Line intensities (areas) are of interest for quantitative chemical analysis. Since one is generally interested in relative concentrations only the ratio of areas of lines from the various elements in a given sample are required. The composition ratio is then calculated according to

$$\frac{[A]}{[B]} = \frac{\sigma_b \zeta_b \lambda_b \eta_b I_a}{\sigma_a \zeta_a \lambda_a \eta_a I_b} \tag{3.7}$$

where σ is the photoelectric cross-section, ζ the fraction of photoelectric events which take place without intrinsic plasmon excitation, λ the mean-free-path in the sample, η the kinetic energy dependent spectrometer transmission, and I the area of the line. (Constant X-ray flux and equal counting times are assumed.)

The cross-sections are advantageously taken from the calculations of Scofield[27] which have given satisfactory results. For theoretical details, see Manson.[28] The value of the mean-free-path λ may be estimated from the 'universal curve' shown in Fig. 3.3. In spectrometers employing retardation, η is proportional to the reciprocal of the kinetic energy. In solids ζ is difficult to estimate because intrinsic- and extrinsic-loss structures are superimposed and indistinguishable. As a result the entire loss tail is often arbitrarily assigned to extrinsic events, setting ζ equal to 1. A significant error is introduced by this assignment since ζ is not necessarily the same even for two elements in a single compound. For example, in an oxide with a largely oxygen $2p$ derived valence band, intrinsic valence band plasmons are more likely to be excited during emission of an oxygen $1s$ electron than during emission of an electron from the other element. As a result the oxygen concentration will be underestimated. The inability to estimate ζ places a serious limitation on the use of XPS for quantitative chemical analysis.

3.3.4 Time-scales

The time-scales relevant to photoemission experiments arise not from the photoelectric effect itself, but from the various forms of relaxation which follow it. The time-scale in which the relaxation energy E_R is converted into a particular electronic excitation is of the order of $\hbar/E_R$. For a typical E_R of 30 eV this corresponds to 2×10^{-17} second. This does not imply that all spectral features have the corresponding 30 eV width, but rather that the width of the entire spectrum is 30 eV. The width of the individual components is determined by the lifetime of the particular state which has been created. For the fully screened main line it is the lifetime of the core hole state which determines the width; for a plasmon it is the lifetime of the collective plasmon mode.

Another time which is relevant is the interval which the photoelectron spends in electromagnetic contact with its atom of origin after photoexcitation. For a 1 keV electron and a 2 Å distance, this yields a time of 10^{-17} s, a value comparable to the estimate of the inter-atomic relaxation time. Note that both intervals refer to the time *after* photoelectric effect, and are therefore not relevant to temporal properties of the atom in the initial state. (This issue arises below in the discussion of mixed-valence compounds.) If the measurement were performed with an interaction time short compared to the relaxation time one might expect to find only a broad line centred on the Koopmans' energy. If it is done with very long interaction time, so that the sudden approximation is invalid, the intensity of the shake-up structure will be reduced.

3.4 Satellites and multiplets

In the discussion of the intra-atomic relaxation which accompanies the ejection of a core electron we have alluded to the fact that all the electrons of an atom are affected when one is removed. The basic reason is that the electronic wave-functions of the atom in the final state must be eigen functions of the hole-state atom. The symmetry of the final state is determined by the coupling of the spin and orbital angular momentum of the hole-state to the spin and orbital angular momentum of the initial state atom. Outer shell monopole ($\Delta l = 0$) shake-up transitions do not change the net symmetry and are not forbidden by selection rules. They are responsible for much of the satellite structure which is observed in free atoms. Such structure can be recognized by the fact that it occurs for all core electron states. There are, however, other types of final state structure which are different for each core hole state. One type is due to final state configuration interaction, others occur when there is an incomplete outer subshell.

3.4.1 Final state configuration interaction

The final state relaxation which accompanies photoemission of a core electron generally results in outer shell excitations, which are usually referred to as 'shake-up' transitions. When the photoexcitation takes place within the outer shell itself, the concept of shake-up must be generalized to include all final states which have the same symmetry as that of the original hole state.

This process is most readily understood in terms of a specific example. Consider a K^+ ion with $[Ne]3s^23p^6$ configuration, in a ^1S state. Photoionization of the $3s$ shell produces a ^2S state ion with $[Ne]3s^13p^6$ configuration. The photoemission spectrum, however, shows not only this line but also a satellite 14 eV towards greater binding energy, Fig. 3.7. Other K^+ core

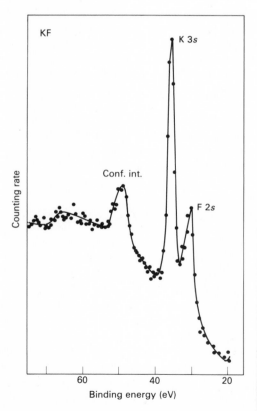

Fig. 3.7. Configuration interaction satellite of the K $3s$ shell in KF. (After Wertheim and Rosenwaig.[29])

electron photoemission lines do not have 14 eV satellites ruling out an interpretation based on the monopole excitation mechanism. In this case an examination of the atomic states of K^+, shows that there are other ^2S states of the form $[Ne]3s^23p^43d$ or $[Ne]3s^23p^44s$, which do indeed have energies 14 eV greater than the $[Ne]3s^13p^6$ state. Calculations have shown that only states which involve transitions within the M-shell are appreciably excited in this case, and thus the observed satellite is identified as the $[Ne]3s^23p^43d$ final state. A similar satellite has also been observed on the Rb 4s state where the corresponding transitions take place in the N-shell; there is no corresponding satellite in Na because the L-shell is filled, there being no d-states.[29,30]

A general principle can now be stated: *photoemission from states belonging to a shell with an empty or partially filled subshell will excite all states accessible by redistributing the electrons (within that shell) that have the same symmetry as the state produced by the one-electron transition.* This coupling between states with the same symmetry is an example of final state configuration interaction. It is possible whenever a two-hole state exists that has the same symmetry as the one-hole state, but the two-hole state is significantly populated only when its energy is not too different from that of the one-hole state. These conditions are satisfied by the $4p_{\frac{1}{2}}$ state of Ba where a $4p^54d^{10}$ final state can couple to a $4p^64d^84f^1$ state.[31] Configuration interaction will be met again in the discussions of multiplet structure in open-shell ions.

3.4.2 Core electron multiplet splitting

In the final state produced by photoemission all spins and orbital angular momenta are coupled. In an ion with an outer, partially occupied shell, such as a rare-earth or transition-metal ion, the coupling of the outer shell to the core-hole will result in a set of final states with distinct energies. In some cases the core state may be split into only a few well-resolved components, while in others a multitude of lines may be obtained. The spectrum of states is readily obtained by applying the rules of atomic physics, and the splitting by using the relevant overlap integrals.[32,33]

The simplest cases are found in core s-electron spectra of S-state ions like $Mn^{2+}3d^5$, ^6S or $Gd^{3+}4f^7$, ^8S. In the absence of orbital angular momentum only two states are produced by coupling the core hole spin of $\frac{1}{2}$ to the spin of the outer shell. For Mn they are ^5S and ^7S. The 2s spectrum of Mn^{2+} is indeed found to be split into two components with the expected 5:7 intensity ratio. The 3s spectrum, however, deviates significantly from this ratio because there are additional final states made available by the configuration interaction process. The one-electron final states have the configuration $[Ne]3s^13p^63d^5$, coupled into ^5S and ^7S. Configuration interaction makes available an additional ^5S state of the form $[Ne]3s^23p^43d^6$, which lies at higher binding energy and reduces the intensity of the corresponding one-electron state.[34]

The multiplet splitting is largest for the core state that have the same principal quantum number as the incomplete shell, e.g. for the 3s and 3p states of the 3d-group transition metals, and for the 4s, 4p, and 4d states of the rare earths. The multiplet spectra of shells with non-zero angular momentum tend to be very complex. Those of the 2p and 3p shells of the transition-metal ions, and of the 4d shell of some of the rare earths, have been calculated.

Core electron multiplet splitting can sometimes be used to identify outer shell properties. For example the incomplete d-shell of a diamagnetic, strong-crystal-field complex like $Fe(CN)_6^{4+}$ does not produce multiplet splitting, since these electrons are coupled into a ^1S state. The Fe 2p electron states of such a compound are distinctly sharper than those of weak-crystal-field compounds, which are broadened or split by multiplet coupling.

The systematics of the multiplet splitting in the rare earths and transition metals has been quite thoroughly explored. For rare-earth compounds it has been verified that the 4s splitting is proportional to $2S+1$, where S is the spin of the 4f shell. It peaks in the middle of the series at a value of 8 eV for Gd^{3+} with $4f^7$ configuration. For 3d-group transition-metal ions the dependence of the 3s splitting on valence and covalency has been demonstrated. For iron the largest splitting of 7.0 eV is found in the ionic $3d^5$ compound K_3FeF_6, and progressively smaller values for FeF_3, FeF_2, $FeCl_2$, $FeBr_2$, metallic iron, and FeB. For Cr (III) compounds, values range from 4.2 eV in CrF_3 to 3.1 eV in $CrBr_3$ (see Carver, Schweitzer, and Carlson[35] for details). The underlying theory is developed by Viinikka and Larsson.[36]

3.4.3 Photoionization of open-shells

The discussion here is restricted to open shells with localized electrons. The 4f shells of the rare-earth elements provide the prime examples. Open shells which form bands are discussed in the following section. The distinction between the photoionization of closed and open shells arises because closed shells yield spin-orbit doublets (or single lines if $l=0$) while open shells yield complex multiplet spectra. These can be analysed in terms of the optical excited states of the atom with next lower atomic number.[37] Consider, for example, photoemission from the $4f^8$ shell of Tb^{3+} which is coupled according to Hund's rule into a $^7F_6$ configuration with a 7 spin-up and 1 spin-down electrons. Two types of states are produced: ^8S if the spin-down electron is removed, or one of a set of sextet states if a spin-up electron is ejected, Fig. 3.8. The ^8S state corresponds to the ground-state of the adjacent element Gd, the sextet states to its excited states. One can therefore interpret the final state spectrum resulting from the photoionization of a rare-earth ion in terms of the optical spectrum of the adjacent element with next lower Z.[37] The energy scale of the spectrum requires an adjustment because the final states exist on a core with one extra charge, Tb rather than Gd. This produces an overall expansion of the energy scale which is typically of the order of 10–15 per cent.

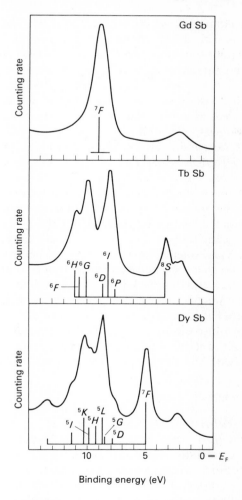

Fig. 3.8. Multiplet structure in photoemission from rare-earth ions. Photoionization of the incomplete $4f^n$ shell produces a set of final states corresponding to the ground and excited states of $4f^{n-1}$. The samples are cubic rare-earth antimonides. The Sb p valence band appears between 4 eV and the Fermi level. (After Campagna, Wertheim, and Bucher.[38])

The multiplet structure of the $4f$ shell of the rare earths is a final-state phenomenon. It belongs to the f^{n-1} ion and has no counterpart in the initial state. The final-state point of view is clearly essential to an understanding of these photoemission data. In order to obtain intensities of the final states, a fractional parentage calculation is required. Basically the initial ground state is projected on to the accessible final states plus an outgoing plane wave representing the photoelectron. Such calculations have been carried out in detail and are invaluable in the detailed interpretation of XPS data.[38]

The fact that photoemission from $4f^n$ shells yields distinct final-state multiplet structures for each value of n has been turned to good use in the study of rare-earth mixed-valence compounds, e.g. SmB_6, $EuCu_2Si_2$, TmSe, and $YbAl_3$.[39] In these materials the $4f$ level lies at the Fermi energy and is fractionally occupied. Mössbauer isomer shift measurements, especially on Sm and Eu compounds, give a clear indication of an intermediate average valence. X-ray photoemission always yields sharp multiplet spectra for the two adjacent integer valences, never an average or broadened result. This has sometimes been erroneously taken as an indication that the time-scale of photoemission is much smaller than the fluctuation time of these mixed-valence systems. The proper interpretation is that emission of a $4f$ electron necessarily destroys the fluctuating state, forcing the system into one of two final states with discrete valence. It is really the final-state nature of XPS that sets it apart from Mössbauer effect, which does not perturb the fluctuations.

When a *core* electron of a mixed-valence system is photoexcited only discrete valences are again obtained. From an equivalent-cores point of view it is clear that the fluctuations cannot persist in the final state, when the electronic structure has been converted into that of a $Z+1$ atom. Thus, the observation of two discrete valences in XPS of transition-metal binuclear complexes does not in itself distinguish between a system with two distinct valences and one with half-integral average valences.

3.4.4 Screening in metals

Photoemission lines from metals are broadened on the high binding energy side by the excitation of electron-hole pairs at the Fermi surface.[40, 41] The response of the conduction electrons to the creation of a core hole is collective. All electrons are affected, and excitations of all magnitudes from zero to the width of the conduction band are possible. The essential result is that the electron–hole pairs created by the screening response of the conduction electrons produce a line shape with the form of a one-sided singularity

$$I(\omega) \propto [\xi/\hbar(\omega - \omega_0)]^{1-a}, \; (\omega > \omega_0) \tag{3.8}$$

where ξ is an energy of the order of the bandwidth, and the singularity index a lies in the interval $0 < a < 0.5$. The optimally screened final state lies at ω_0, i.e. at the asymptote of the power law in Eqn 3.8. The power law itself represents the energy which has gone into creating electron–hole pairs and, in a sense, corresponds to the satellite structure of insulators. In metals it becomes an integral part of the main photoemission line because there is no gap in the density of states at the Fermi level.

A closed-form expression for the line shape of simple metals, combining Eqn 3.7 with the lifetime Lorentzian, has been obtained by Doniach and Sunjic.[40]

$$I(\omega) = \frac{\cos[\pi a/2 + (1-a)\arctan(\varepsilon/\gamma)]}{(\varepsilon^2 + \gamma^2)^{(1-a)/2}} \quad (3.9)$$

where $\varepsilon = \omega - \omega_0$, and γ is the half width at half height of the Lorentzian. For $a = 0$ the function is a Lorentzian. For finite values it has a long tail to higher binding energy. It has been found that eqn 3.9 can be used to fit experimental line shapes from a wide variety of metallic systems, provided an explicit representation of the plasmons is included.[42]

3.5 Valence band spectra

Photoemission spectroscopy of the outermost electronic states has long been the province of UPS. In recent years He and Ne resonance radiation, and synchrotron radiation, have extended the range over the entire valence shell and provided resolution better than 0.05 eV. The most significant accomplishment of UV photoemission is the mapping of the band structure using angle-resolved spectroscopy. This has been carried out for single crystals of a number of materials, and is now a widely used spectroscopic technique.

UPS is, however, not without problems. At UV energies the background of degraded electrons severely distorts the spectrum near the work-function cut-off. The cross-sections of the rare-earth $4f$ and actinide $5f$ states are small. The shape of the photoemission spectrum is significantly modulated by the unoccupied density of states. Although XPS offers lower resolution and lower rates of data-acquisition it is insensitive to the empty density of states and is consequently able to provide a clear view of the disposition of the occupied electronic orbitals.

As an example of the range of behaviour of an electronic orbital in different chemical states, consider the behaviour of the outer electrons of Au, [Xe]$4f^{14}5d^{10}6s^1$, in two chemically different systems (Fig. 3.9). In the elemental metal the $5d$ electrons form a broad band lying in the interval between 2 and 8 eV. The $6s$ electrons are seen between 2 eV and the Fermi cut-off, but the $6s$ band extends to much greater binding energy, and is strongly hybridized with the $5d$ band. The agreement between the details of the data and band structure calculations is generally satisfactory, especially if the progressive lifetime broadening of states at increasing binding energy is taken into account. Contrast this behaviour with that of CsAu, a red, transparent semiconductor with a 2.6 eV bandgap. Because of its large electronegativity, the Au becomes a negative ion with a filled $6s^2$ shell. The data show that the $5d^{10}$ shell of Au is a core-like spin-orbit doublet with 1.5 eV splitting, which corresponds closely to the free-atom value because the overlap between the

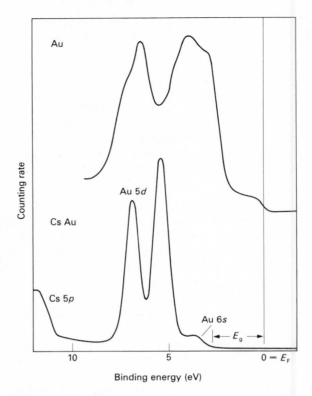

Fig. 3.9. Comparison of the valence band structure of metallic gold and the semiconductor CsAu. Note that the broad d-band of the metal is reduced to a spin-orbit doublet in the compound. (After Wertheim, Bates, and Buchanan.[43])

d-orbitals of neighbouring atoms is negligibly small. The $6s$ electrons constitute the valence band and lie about 3 eV below the Fermi energy, indicating strong n-type conductivity.[43]

Valence band spectra of successive elements, starting with the d-group metals, provide striking examples of the transition from valence-like to core-like behaviour with increasing atomic number (Fig. 3.10). As long as the d-band is partially filled, as in Ni, it intersects the Fermi level. In the noble metals, e.g. Cu, the complete d^{10} shell falls well below E_F, but clearly retains its band-like character. In Zn where the $3d$ binding energy reaches 10 eV little band-like character remains, and the spectrum has the appearance of a spin-orbit doublet.

The image of the occupied band structure obtained by XPS can thus provide valuable tests of theoretical models of bonding and of band structure calculations. It must be borne in mind, however, that the data that are

X-ray photoelectron spectroscopy and related methods 107

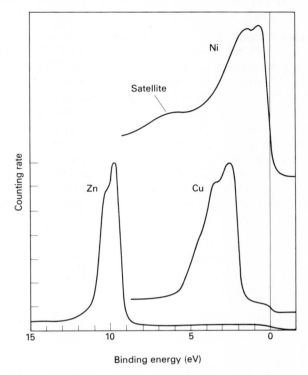

Fig. 3.10. The transition of the 3d electrons from band-like in Ni and Cu, to core-like in Zn.

obtained do not correspond in every detail to the total one-electron density of states. There are three main reasons: (a) lifetime broadening, (b) differences in cross-sections, and (c) multi-electron excitations.

(a) Lifetime broadening

The valence band spectrum is distorted by the hole-state lifetime width, which changes rapidly with binding energy. In metals at E_F, no lifetime-broadening is in evidence, but at the bottom of the conduction band it may be substantial. The resulting Lorentzian tailing of the band edge, coupled with a rising plasmon loss-tail often makes it difficult to define the bottom of the conduction band. The bottom of the 4s band of Cu is generally not detectable in photoemission data. In comparisons of band-structure calculations with XPS data it has been found useful to broaden the theoretical density of states by a Lorentzian ranging in width from a few eV at the bottom of the band, to zero at the Fermi level.

(b) Differences in cross-sections

The photoelectric cross-sections may be quite different in magnitude for the various electronic states appearing in the valence region. This is illustrated by the comparison of band-structure calculations and XPS data for ReO_3, a copper-coloured conductor, in Fig. 3.11. The electronic configuration of Re^{6+} is $[Xe]4f^{14}5d^1$. Theory and experiment agree in showing the narrow filled portion of the $5d$ conduction band below E_F, well-separated from the O $2p$ valence band. However, the XPS data for the valence band do not agree with the details of the calculated *total* density of states (DOS). This is a reflection of the fact that there is substantial d-admixture into the valence band and a disparity in cross-sections which favours Re $5d$ over O $2p$ by a factor of 30. As a result the valence band peak at 3.5 eV, which is largely of O $2p$ character is almost totally suppressed in the experimental data, while the

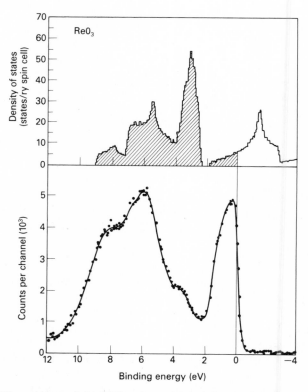

Fig. 3.11. Comparison of the XPS valence band of ReO_3 with the total density of states. The data reflect largely the Re $5d$ content of the band structure (see text). The O $2p$ band at 4 eV in the theoretical DOS is almost completely suppressed, and the conduction band enhanced, in the experimental data. (After Wertheim *et al.*[44])

lower part of the valence band and the conduction band at 1 eV are greatly enhanced.[44]

XPS data for valence bands of transition-metal oxides, especially towards the end of each series, are dominated by the d-admixture, and have a shape very different from the total density of states. Generally speaking, p-derived features are suppressed, and the more strongly d-hybridized features are enhanced. As one proceeds through the chalcogenide series to the sulfides, selenides, and tellurides this effect becomes weaker and may actually reverse, because the anion p cross-section increases markedly from oxygen to tellurium. Similarly, valence band spectra of rare-earth compounds, especially for elements beyond Nd, tend to be dominated by $4f$ emission. At Al $K\alpha$ energy the $4f$ cross-sections are an order of magnitude greater than those of the rare-earth $5d$, $6s$, or anion p-states. As a result the contribution from the latter may be largely obscured. At lower photon energy, in the UV, the situation is reversed: $4f$ cross-sections are small, so that complementary information is obtained. In general, then, a knowledge of the photoelectric cross-sections is essential for the interpretation of valence band spectra.

(c) Multi-electron excitations

The multi-electron character of the photoemission process may introduce peaks where none exist in the density of states. As a result one cannot guarantee that an observed feature in the valence band region corresponds to a feature in the one-electron density of states. Shake-up satellites are expected to be weak in the valence band region, because the relaxation energy associated with the creation of an outer shell hole is necessarily much smaller than that for a core hole. However, it is possible for the strong coupling of electrons within a shell to allow multiple hole states to be excited through configuration interaction or resonant processes. A feature 6 eV below E_F in Ni, a strongly correlated d-band metal, represents a two-hole state in the d-band created during the photoemission process[45] (see Fig. 3.10). This gives warning that even valence band spectra may contain satellites. However, since these effects are rare and usually weak, the first approach to the interpretation of valence band spectra should be an identification of the observed features with those of the one-electron band structure on the scheme of hybridized orbitals.

3.5.1. Some applications of valence band XPS

The sodium–tungsten bronzes with formula Na_xWO_3 are both structurally and electronically related to ReO_3. Both materials are made up of corner-sharing MO_6 octahedra so that the band structures are similar. Since tungsten has one electron less than rhenium, the conduction band of WO_3 is empty, and the compound is a transparent insulator. In the tungsten bronzes, the alkali metal donates its outer s-electron to the conduction band,

110 *Solid state chemistry: techniques*

producing an ReO$_3$-like d-band metal. The filling of the 5d band with increasing sodium content has been verified by both XPS and UPS. At low sodium concentration there is a chemically controlled metal–insulator transition.

Closely related to the crystalline tungsten bronzes are the electrochromic amorphous WO$_3$ films. They can be electrolytically charged with hydrogen producing a deep blue colour, which is due to the filling of localized W 5d states. The process is readily reversible, and may find application in display devices. A similar blue colour is also found in oxygen-deficient tungsten trioxide in which oxygen vacancies split off filled 5d-states from the bottom of the conduction band.

Temperature-driven metal–insulator transitions are found in a number of stoichiometric transition-metal oxides, e.g. VO$_2$, NbO$_2$. In VO$_2$ the high-temperature rutile phase is a metal, and the monoclinic low-temperature phase an insulator. VO$_2$, like ReO$_3$, has one electron per metal atom in a d-conduction band. In the distorted, insulating phase there is a pairing of vanadium atoms, and the opening of a gap in the density of states at the Fermi level. The transition is readily detected by XPS,[46] as shown in Fig. 3.12, and also by UPS.[47]

3.6 Spectroscopy of surface layers

It was emphasized above, that photoelectron spectroscopy is a surface-sensitive technique because the mean-free-path of energetic electrons in solids comprises only a few atomic layers. While this is a potential embarrassment for the study of bulk properties, it can be turned into an advantage for the study of surfaces.[48] On the simplest level it means that surface chemistry can be readily studied because even a single atomic layer contributes a significant part of the total signal. Furthermore, the surface sensitivity can be enhanced by accepting only electrons that leave the sample at a glancing angle. Take-off angles as large as 80°, measured from the surface normal, can be advantageously employed provided the surface is sufficiently flat.

3.6.1 Clean surfaces

It is known on theoretical grounds that the electronic structure of the first atomic layer of a clean surface is different from the bulk. In ionic crystals one expects a chemical shift at the surface simply because the Madelung potential at the surface must differ from that in the bulk.[49] Calculations have shown that this effect is not negligible. However, since photoemission lines from insulators are broadened by phonon excitation, the resulting shift may be difficult to resolve.

In metals, where the band structure reflects the lattice periodicity, the

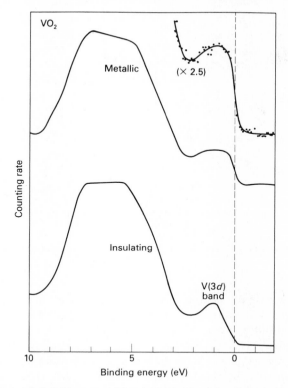

Fig. 3.12. The metal–insulator transition in VO_2. The data at the top, taken with the sample at 95°C, show the sharp Fermi cut-off typical of a metal; those below show the transfer of states away from E_F, opening a gap in the insulating phase. (After Wertheim.[46])

change at the surface arises in a different way. Simple theoretical considerations show that the width of bands is proportional to the square root of the coordination number.[50,51] Significantly narrower bands are consequently expected for the surface atoms. These are difficult to detect because they are superimposed on the bulk bands. The usual approach has been to look for band narrowing with increasing take-off angle.

Surface effects are actually more easily detected in *core electron* spectra. Surface atoms exhibit core-electron chemical shifts due to changes in the electronic structure at the surface. A simple rigid-band picture predicts core level shifts when a conduction band is narrowed at the surface (see Fig. 3.13). When a less-than-half-filled band is narrowed about its centroid, electrons will flow out of the band. Loss of valence charge increases the electron binding energies of all levels, allowing electrons to return to the conduction band. This tends to restore charge neutrality, so that the net charge flow is

112 *Solid state chemistry: techniques*

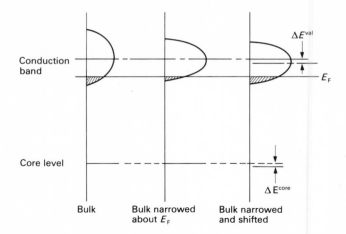

Fig. 3.13. Change in core electron binding energy due to surface band narrowing in a metal.

typically only a small fraction of one electron per atom. In the rigid-band model the core electron binding energies move correspondingly, resulting in an increased binding energy. For bands that are more than half-filled, a similar argument predicts decreased surface atom binding energies.

Following the original observation of a surface atom core level shift for gold by Citrin, Wertheim, and Baer,[52] such shifts have been reported for a large number of metals and semiconductors. Experimentally these shifts have been found to be of the order of 0.5 eV, i.e. detectable with well-monochromatized radiation. They were detected on tungsten by Duc et al.[53] and van der Veen et al.[54] (Fig. 3.14), and on iridium by van der Veen et al.[55] It has been shown that the shift is largely, but not entirely, confined to the first atomic layer, depends on surface orientation and reconstruction, and is sensitive to chemisorption. It is clear that a resolved surface atom signal opens up the possibility of doing sensitive surface chemistry by XPS and UPS. The magnitude of the shift is in reasonable agreement with estimates based on surface cohesive energies, using a Born-Haber cycle.[16,56]

3.6.2 Adsorbate systems

When a clean metal or semiconductor surface is exposed to foreign atoms the initial result is usually the formation of an adsorbed surface layer. In general, one describes a system as physisorbed if the bonding is largely by van der Waals interaction, and as chemisorbed if there has been some electronic interaction between the substrate and the adsorbate.

A rare gas adsorbed on graphite, or even on a metal, clearly falls into the physiosorption category. A number of interesting results have been obtained

X-ray photoelectron spectroscopy and related methods 113

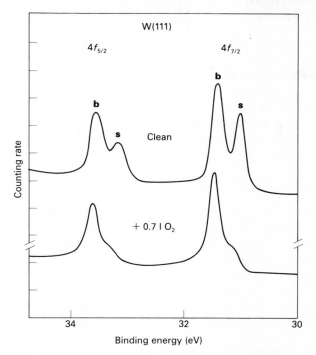

Fig. 3.14. Photoemission from the (111) surface of tungsten, with 70-eV photons. The surface signal 's' which is well resolved on the clean surface, is shifted after exposure to oxygen. (After van der Veen et al.[54])

with these adsorbate systems. With rare gases on metallic substrates it has been possible to measure the final-state screening response of the substrate conduction electrons when the adsorbate is photoionized. This is accomplished by measuring the core electron binding energy of successive monolayers; that of the first monolayer is smallest because the screening by charge in the conduction band of the substrate is more effective in lowering the energy of the final state, the closer the hole orbital is to the substrate.[57]

Halogens on noble metals might be expected to approach the chemisorption limit since electron transfer to the halogen would produce a filled shell. However, the adsorbate atom must not be visualized as a negative ion on top of a metallic substrate. The empty p-orbital is filled by sharing, rather than extracting an electron from the conduction band of the metal. As a result iodine adsorbed on Cu or Ag carries little dipole moment. This is known both from the small change in work-function when iodine is adsorbed on the noble metals, and from the small change with iodine coverage of the iodine core electron binding energy.[58] The fact that iodine forms incommensurate structures on Cu(001), and Cl on Ag(111), is a further indication that there are no adsorbate–substrate chemical bonds.

114 *Solid state chemistry: techniques*

Many studies have been carried out to delineate the reactions of clean surfaces with oxygen or water. Growth of oxide films, and corrosion, have been observed. Generally the chemical shift of a substrate core level provides the most direct information on the nature of the reactions, while the intensity of the adsorbate core level is more useful to monitor the extent of the reaction. The oxidation of silicon has been studied in great detail in order to define the nature of the technologically important interface between Si and SiO_2. The Si $2p$ core level shift for SiO_2 relative to Si is 4 eV, sufficiently large that intermediate valences of interfacial atoms can be detected.[59] Many studies of this kind, typically dealing with metals, have been reported.

3.6.3 Surface and bulk aspects of XPS

Once the signals from the first atomic layer have been identified, one can re-examine the validity of XPS as a technique for the study of bulk properties. The small shift associated with the surface atoms explains why surface effects remained elusive, even with the 0.5 eV resolution generally achieved with monochromatized Al $K\alpha$ radiation. It justifies ignoring the surface when working with the 1 eV resolution which is more generally available in XPS. For electrons with 1400 eV kinetic energy, an escape depth of about 15 Å makes the surface contribution a relatively small part of the total signal. In contrast, with synchrotron radiation in the 50–100 eV range, the escape depth is near its minimum of 5 Å, and a resolution of 0.1 eV can be achieved. Under these conditions the surface atom signal is readily resolved, and will contribute almost half of the total signal.

3.7 Related spectroscopies

Electron and X-ray spectroscopies can be used in a wide variety of ways to gain information about the chemical and electronic properties of matter. Photoelectron spectroscopy examines the occupied electronic states, bremsstrahlung isochromat spectroscopy the empty density of states. Auger spectroscopy has found its major application in the determination of (surface) chemical composition, extented X-ray absorption fine structure (EXAFS) spectroscopy in the measurement of bond lengths and the determination of coordination. The information obtained by these techniques is complementary, each one having its own major area of strength. The essential features of these spectroscopies are summarized below.

3.7.1 Auger electron spectroscopy (AES)

In AES the energy spectrum of electrons emitted from an atom due to the two-electron decay of a core-hole is measured. Since each element gives rise to its own characteristic spectrum, elemental analysis is readily performed.

Semi-quantitative analysis is possible as well. Analysis is limited to a layer with thickness defined by the mean-free-path of the Auger electrons, typically in the range 5–25 Å. The Auger decay is usually excited by bombardment with 3–5 keV electrons to produce a core hole which constitutes the initial state of the process. The Auger electron must consequently be detected in the presence of a large background of degraded electrons from the primary beam. Signal levels are typically sufficiently large so that current- rather than electron-detection can be used. Since the background is relatively smooth, it can be suppressed by taking the derivative of the total signal. This is usually accomplished by modulating the transmission energy of the analyser, and using phase-sensitive detection. The second derivative is generally recorded, and serves as the primary data of Auger spectroscopy. For further information, see Kirschner[60] and Wagner.[61]

Detailed interpretation of Auger spectra is complicated by the fact that the final state is a two-hole state. Moreover, each initial state hole gives rise to a variety of final states, because there are many allowed two-electron transitions. Interpretation of chemical shifts encounters all the problems familiar from photoemission, since the initial state of the Auger process is the final state of photoemission. In addition the analogous and more difficult problems of the two-hole final state must be solved. Auger events involving valence electrons are inherently more complex than those of photoemission. If one of the two decay electrons comes from the valence band, the spectrum reflects the local shape of the band in the presence of a core hole. If both come from the valence band, the shape may range from that of a self-convolution of the initial state band in the free electron limit, to a set of discrete, pseudo-atomic final states in the strongly localized limit.

The most widespread use of Auger spectroscopy lies in surface characterization, but both the shift and shape of Auger spectra are being used increasingly to obtain more detailed chemical information.

3.7.2 Bremsstrahlung isochromat spectroscopy (BIS)

Bremsstrahlung isochromat spectroscopy is based on the inverse of the photoelectric effect. A clean surface is bombarded with monochromatic electrons. When an electron is captured into an empty state of the band structure of the solid, the excess energy is radiated as a photon whose energy is measured. The energy of the electronic state is defined by the difference between the energy of the incident electron and that of the emitted photon. As the name indicates, the technique is usually implemented by detecting photons of a single energy while sweeping the kinetic energy of the incident electrons. It offers the possibility of mapping the empty band structure of solids. Since it depends on the kinetic energy of electrons propagating in a solid, it is limited to surface layers defined by the electron mean-free-path. An instrument for this spectroscopy is described by Lang and Baer.[62]

BIS has already produced interesting results, revealing the empty d-bands of the transition metals, and f-states of the rare earths. Its application is, of course, more limited than photoelectron spectroscopy because the empty states broaden rapidly with energy above the Fermi level. Many materials have only broad bands above E_F. Nevertheless BIS constitutes a valuable technique, complementary to photoemission by virtue of making the *empty* density of states accessible.

3.7.3 X-ray absorption edge spectroscopy and extended X-ray absorption fine structure (EXAFS)

X-ray absorption edge spectroscopy also holds promise for the study of the unoccupied band structure. By exciting dipole transitions from a core level to the empty DOS, the latter can, in principle, be mapped, and even the s-, p-, and d-character determined. Two additional phenomena modulate the absorption: the extended X-ray absorption fine structure (EXAFS), and, in metals, the absorption edge threshold anomaly.

The absorption edge threshold anomaly has its origin in the dynamical screening of the core hole by the conduction electrons. The details of this process were elucidated by Mahan[63] who showed that the shape of the edge could be profoundly altered by the excitation of electron–hole pairs at the Fermi level. The edge shapes of the simple metals have been studied in considerable detail.

Edge spectroscopy in insulators is of interest because the details of the empty states are less distorted by the extra-atomic screening. However, it must be borne in mind that the observed edge structure is that of a hole-state atom.

The second phenomenon, EXAFS, has recently commanded by far the greater interest. This is due in large measure to the new synchrotron radiation sources, which make these experiments easy to perform by providing a broad-band, tuneable source of radiation with uniform intensity in the X-ray region. The EXAFS appears as a modulation of the X-ray absorption which extends hundreds of electron volts beyond the edge towards higher energy. Its origin lies in the back-scattering of the photoelectron from neighbouring atoms to the one which has emitted the photoelectron. The result is an interference which modulates the photoelectric cross-section and hence the absorption. Since the interference depends on the phase of the back-scattered wave-function, the effect is sensitive to the phase-shift associated with the back-scattering, and to the distances to the neighbouring atoms. A change in the kinetic energy of the photoelectron changes the phase associated with the round trip to the neighbouring atom. As a result the absorption exhibits a periodic modulation when plotted against the electron momentum, from which the distance to the neighbouring atom can be precisely determined. The intensity of the modulation contains information about the coordination number, but its determination is much less reliable. See Lee et al.[64] for more detail.

In practice, EXAFS experiments are easy to perform. They are not inherently surface-sensitive so that bulk information can be obtained from powders. UHV techniques are not required. A typical experiment is performed in transmission, using a monochromatized beam from a synchrotron which is scanned over the energy interval of interest using a double-crystal monochromator. Two X-ray detectors are used, one as a monitor of the intensity of the incident beam, the other to measure the transmitted beam. The data consist of the ratio of the two intensities.

Data are typically more complex than suggested by the above discussion which considered only the near-neighbour environment in a crystal of high symmetry. In a real case, there will be reflections from second neighbours and possibly multiple reflections adding their own periodic modulation. In a crystal of low symmetry, there may be a range of distances to the atoms in the near-neighbour coordination sphere, with each distance producing its own frequency. Data analysis therefore usually requires harmonic analysis to unravel a possibly complicated set of modulations. Finally the energy-dependence of the back-scattering phase-shift must be considered since it is included in the calculation of the neighbour-distances. It may be obtained theoretically or, typically, from a calibration with a compound of known interatomic spacing.

EXAFS spectroscopy has a number of useful features. Because absorption edges of different elements are well separated it is possible to look selectively at a particular chemical constituent of a complex system, even when present in great dilution. For example, one can examine the environment of a particular atom in a complex biological molecule, such as iron in the porphyrin macrocycle of haemoglobin. Crystal structure is not required; the distribution of bond lengths in amorphous materials thus can be obtained. The sensitivity can be enhanced by detecting the Auger electrons from the decay of the core hole produced by the X-ray absorption. In this way even monolayer adsorbates on clean surfaces can be studied. EXAFS is consequently a powerful technique for the determination of bond lengths and bonding sites, providing valuable information especially in systems where standard X-ray crystallographic techniques are of limited utility.

3.7.4 Electron energy loss spectroscopy (EELS)

The basic purpose of this spectroscopy is to determine the excitation spectrum of a solid or surface by measuring the energy loss spectrum of electrons scattered by it. Such experiments are done with electron kinetic energies ranging from a few electron-volts to hundreds of thousands of electron-volts. At the highest energies the spectroscopy is closely related to X-ray absorption edge spectroscopy, in that it is used to study excitation of deep core electrons to the Fermi level of solids. Electrons provide more information than X-rays because both energy loss and momentum transfer can be measured.[65]

High-energy experiments are done in transmission. In a typical set-up the electron gun and monochromator are at ground potential and provide an electron beam with resolution better than 0.1 eV. The electrons enter the spectrometer with finite energy at ground potential, are then accelerated to the sample chamber, penetrate the sample, and are retarded back to ground potential where the energy distribution can then be conveniently measured. The energy loss spectrum is closely related to the dielectric function of the scattering medium, being proportional to $Im(-1/\varepsilon)$. Careful comparisons between optical and electron measurements have confirmed this theoretical prediction.[66,67,68] In the 10–50 keV range the technique has been used to study the plasmon response of solids, mapping out the plasmon dispersion by measuring the loss at specified values of momentum transfer.

At very low kinetic energy, EELS serves a different purpose; and the requisite instrumentation assumes quite different forms. The electron beam is reflected from a single-crystal surface and may be used to study the vibrational spectra of adsorbates. The best resolution obtained by EELS is 0.003 eV. For more detail, see Froitzheim.[69] The information obtained is very similar to that normally associated with infrared and Raman spectroscopy, but those techniques have only recently achieved the sensitivity to deal with monolayer adsorbates on solids like silicon, which are transparent in the infrared.

3.8 References

1. Einstein, A. *Ann. Physik* **17**, 132 (1905).
2. Siegbahn, K., Nordling, C., Fahlman, A., Nordberg, R., Hamrin, K., Hedman, J., Johansson, C., Bergmark, T., Karlson, S.-E., Lindgren, I., and Lindberg, B. ESCA atomic, molecular, and solid-state structure studied by means of electron spectroscopy. In *Nova Acta Regia Soc. Sci.* Upsaliensis, Ser. IV, Vol. 20 (1967).
3. Siegbahn, K., Nordling, C., Johansson, G., Hedman, J., Heden, P. F., Hamrin, K., Gelius, U., Bergmark, T., Werme, L. O., Manne, R., and Baer, Y. *ESCA applied to free molecules.* North Holland, Amsterdam (1969).
4. Barrie, A. Instrumentation for electron spectroscopy. In *Handbook of X-ray and ultraviolet photoelectron spectroscopy* (ed. D. Briggs) p. 79. Heyden, London (1977).
5. Orchard, A. F. Basic principles of photoelectron spectroscopy. In *Handbook of X-ray and ultraviolet photoelectron spectroscopy* (ed. D. Briggs) p. 1. Heyden, London (1977).
6. William, P. M. Ultraviolet photoemission from solids. In *Handbook of X-ray and ultraviolet photoelectron spectroscopy* (ed. D. Briggs) p. 313. Heyden, London (1977).
7. Roy, D. and Carette, J. D. Design of electron spectrometers for surface analysis. In *Electron spectroscopy for surface analysis* (ed. H. Ibach) Vol. 4. Springer, Berlin (1977).

8. Evans, S. Energy calibration in photoelectron spectroscopy. In *Handbook of X-ray and ultraviolet photoelectron spectroscopy* (ed. D. Briggs) p. 121. Heyden, London (1977).
9. Lundquist, B. I. *Phys. Kondens, Mater.* **6**, 167 (1967).
10. Koopmans, T. *Physica* **1**, 104 (1933).
11. Shirley, D. A. Many-electron and final-state effects: beyond the one-electron picture. In *Photoemission in solids* (eds M. Cardona and L. Ley) Vol. I, pp. 165–95. Springer, Berlin (1978).
12. Bearden, J. A., and Burr, A. F. *Revs. mod. Phys.* **39**, 125 (1967).
13. Citrin, P. H. and Thomas, T. D. *J. chem. Phys.* **57**, 4446 (1972).
14. Gelatt, C. D., Ehrenreich, H., and Watson, R. E. *Phys. Rev.* **B15**, 1613 (1977).
15. Williams, A. R. and Lang, N. D. *Phys. Rev. Let.* **40**, 954 (1978).
16. Johansson, B. and Martensson, N. *Phys. Rev.* **B21**, 4427 (1980).
17. Auger, P. *C. r. Acad. Sci.* **180**, 65; *J. Phys. Radium* **6**, 205 (1925).
18. Auger, P. *Ann. Phys. (Paris)* **6**, 183 (1926).
19. Coster, D. and Kronig, R. de L. *Physica* **2**, 13 (1935).
20. McGuire, E. J. *Phys. Rev.* **A2**, 273 (1970).
21. McGuire, E. J. *Phys. Rev.* **A3**, 587 (1971).
22. McGuire, E. J. *Phys. Rev.* **A5**, 1043 (1972).
23. Walters, D. L. and Bhalla, C. P. *Phys. Rev.* **A3**, 1919; **A4**, 2164 (1971).
24. Kostroun, V. O., Chen, M. H., and Crasemann, B. *Phys. Rev.* **A3**, 533 (1971).
25. Crasemann, B., Chen, M. H., and Kostroun, V. O. *Phys. Rev.* **A4**, 1 and 2161 (1971).
26. Citrin, P. H., Eisenberger, P., and Hamann, D. R. *Phys. Rev. Let.* **33**, 965 (1974).
27. Scofield, J. H. *J. Electron Spectrosc.* **8**, 129 (1976).
28. Manson, S. T. The calculations of photoionization cross-sections: an atomic view. In *Photoemission in solids* (eds M. Cardona and L. Ley) Vol. 1, p. 135. Springer, Berlin (1978).
29. Wertheim, G. K. and Rosencwaig, A. *Phys. Rev. Let.* **26**, 1179 (1971).
30. Reader, J. *Phys. Rev.* **A7**, 1431 (1973).
31. Wendin, G. and Ohno, M. *Physica Scripta* **14**, 148 (1976).
32. Fadley, C. S., Shirley, D. A., Freeman, A. J., Bagus, P. S., and Mallow, V. J. *Phys. Rev. Let.* **23**, 2397 (1969).
33. Freeman, A. J., Bagus, P. S., and Mallow, V. J. *Int. J. Magnetism* **4**, 35 (1973).
34. Kowalczyk, S. P., Ley, L., Pollak, R. A., McFeely, F. R., and Shirley, D. A. *Phys. Rev.* **B7**, 4409 (1973).
35. Carver, J. C., Schweitzer, G. K., and Carlson, T. A. *J. chem. Phys.* **57**, 973 (1972).
36. Viinikka, E.-K. and Larsson, S. *J. Electron Spectrosc.* **7**, 163 (1975).
37. Wertheim, G. K., Rosencwaig, A., Cohen, R. L., and Guggenheim, H. J. *Phys. Rev. Let.* **27**, 505 (1971).
38. Cox, P. A. *Struct. Bonding* **24**, 59 (1975).
39. Campagna, M., Wertheim, G. K., and Bucher, E. *Struct. Bonding* **30**, 99 (1976).
40. Doniach, S. and Sunjic, M. *J. Phys.* **C3**, 285 (1970).
41. Hüfner, S., Wertheim, G. K., Buchanan, D. N. E., and West, K. W. *Phys. Lett.* **46A**, 420 (1974).
42. Citrin, P. H., Wertheim, G. K., and Baer, Y. *Phys. Rev.* **B16**, 4256–82 (1977).
43. Wertheim, G. K., Bates, C. W., Jr., and Buchanan, D. N. E. *Solid State Commun.* **30**, 473 (1979).
44. Wertheim, G. K., Mattheiss, L. F., Campagna, M., and Pearsall, T. P. *Phys. Rev. Let.* **32**, 997 (1974).
45. Hüfner, S. and Wertheim, G. K. *Phys. Lett.* **51A**, 299 (1975).

46. Wertheim, G. K. *J. Franklin Inst.* **298**, 289 (1974).
47. Beatham, N., Fragala, I. L., Orchard, A. F., and Thornton, G. *J. chem. Soc.* **FII, 76,** 929 (1980).
48. Ibach, H. (ed.) *Electron spectroscopy for surface analysis.* Springer, Berlin (1977).
49. Watson, R. E., Davenport, J. W., Perlman, M. L., and Sham, T. K. *Phys. Rev.* **B24,** 1791 (1981).
50. Cyrot-Lackman, F. *Adv. Phys.* **16,** 393 (1967).
51. Cyrot-Lackman, F. *J. Phys. Chem. Solids* **29,** 1235 (1968).
52. Citrin, P. H., Wertheim, G. K., and Baer, Y. *Phys. Rev. Lett.* **41,** 1425 (1978).
53. Duc, T. M., Guillot, C., Lassailly, Y., Lecante, J., Jugnet, Y., and Vedrine, J. C. *Phys. Rev. Lett.* **43,** 789 (1979).
54. van der Veen, J. F., Heiman, P., Himpsel, F. J., and Eastman, D. E. *Solid State Commun.* **37,** 555–9 (1981).
55. van der Veen, J. F., Himpsel, F. J., and Eastman, D. E. *Phys. Rev. Lett.* **44,** 189 (1980).
56. Rosengren, A. and Johansson, B. *Phys. Rev.* **B22,** 3706 (1980).
57. Kaindl, G., Chiang, T. C., Eastman, D. E., and Himpsel, F. J. *Phys. Rev. Lett.* **45,** 1808 (1980).
58. DiCenzo, S. B., Wertheim, G. K., and Buchanan, D. N. E. *Surf. Sci.* **121,** 411 (1982).
59. Grunthaner, F. J., Grunthaner, P. J., Vasquez, R. P., Lewis, B. F., Maserjian, J., and Madhukar, A. *Phys. Rev. Lett.* **43,** 1683 (1979).
60. Kirschner, J. Electron-excited core level spectroscopies. In *Electron spectroscopy for surface analysis* (ed. H. Ibach). Springer, Berlin (1977).
61. Wagner, C. D. The role of auger lines in photoelectron spectroscopy. In *Handbook of X-ray and Ultraviolet photoelectron spectroscopy* (ed. D. Briggs) p. 249. Heyden, London (1977).
62. Lang, J. K. and Baer, Y. (1979). *Rev. sci. Instrum.* **50,** 221 (1979).
63. Mahan, G. D. *Phys. Rev.* **163,** 612 (1967).
64. Lee, P. A., Citrin, P. H., Eisenberger, P., and Kincaid, B. M. *Revs mod. Phys.* **53,** 769 (1981).
65. Schnatterly, S. E. Inelastic electron scattering spectroscopy. In *Solid-state physics* (eds H. Ehrenveich, F. Seitz, and D. Turnbull) Vol. 34, pp. 275–358. Academic Press, New York (1979).
66. Daniels, J., von Festenberg, C., Raether, H., and Zeppenfeld, K. Optical constants of solids by electron spectroscopy. In *Springer tracts in modern physics* (ed. G. Höhler) Vol. 54. Springer, Berlin (1970).
67. Raether, H. Solid state excitations by electrons. In *Springer tracts in modern physics* (ed. G. Höhler) Vol. 38. Springer, Berlin (1965).
68. Raether, H. Excitations of plasmons and interband transitions in solids. In *Springer tracts in modern physics,* Vol. 88, pp. 1–196. Springer, Berlin (1980).
69. Froitzheim, H. Electron energy loss spectroscopy. In *Electron spectroscopy for surface analysis* (ed. H. Ibach). Springer, Berlin (1977).

3.9 Bibliography

Brundle, C. R. and Baker, A. D. (eds), *Electron spectroscopy, theory, techniques, and applications,* Vols I and II. Academic Press, London (1977 and 1978).

Cardona, M. and Ley, L. (eds), *Photoemission in solids I, general principles.* Springer, Berlin (1978).

──── ──── *Photoemission in solids II, case studies.* Springer, Berlin (1979).

Carlson, T. A., *Photoelectron and Auger spectroscopy.* Plenum press, New York (1975).

Day, P. (ed.), *Electronic states of inorganic compounds: new experimental techniques.* Reidel, Dortrecht (1975).

Fiermans, L., Vennik, J., and DeKeyser, W. (eds), *Electron and ion spectroscopy of solids.* Plenum, New York (1978).

Hannay, N. B. (ed.), *Treatise on solid state chemistry, Vol. 6A: Surfaces I.* Plenum, New York (1976).

Ibach, M. (ed.), *Electron spectroscopy for surface analysis*, Topics in current physics, Vol. 4. Springer, Berlin (1977).

4 Magnetic measurements

W. E. Hatfield

4.1 Introduction

In this chapter we will be interested in obtaining chemical information from the response of materials to magnetic fields. Substances will be broadly classified as to their response to an inhomogeneous magnetic field. When placed in an inhomogeneous magnetic field, a *diamagnetic* substance will tend to move towards the weakest region of the inhomogeneous magnetic field, while a *paramagnetic* substance will tend to move to the strongest region of the magnetic field.

Magnetic properties arise from the spin and orbital motions of electrons. If all of the electrons in a substance are paired, then the substance will usually be *diamagnetic*. However, if unpaired electrons are present, then the substance may be paramagnetic. Co-operative interionic interactions may lead to long-range magnetic ordering resulting in ferromagnetism, antiferromagnetism, or ferrimagnetism, properties which are discussed in Volume 2 of this work.

4.2 Basic concepts

4.2.1 Substances in magnetic fields

The intensity of the magnetic field (density of magnetic lines of force) within a body differs from that surrounding the body in the following manner:

$$B = H + \Delta H. \tag{4.1}$$

The magnetic induction is given by

$$B = H + 4\pi M \tag{4.2}$$

and along a specific direction, say i

$$B_i = H_i + 4\pi M_i. \tag{4.3}$$

Upon dividing through by H_i, there results

$$B_i/H_i = 1 + 4\pi M_i/H_i. \tag{4.4}$$

The ratio (M_i/H_i) is the susceptibility of a body towards induction in a field of strength H_i. This ratio is the volume magnetic susceptibility and is denoted by κ, that is

$$\kappa_i = (M_i/H_i) \text{ cm}^{-3} \tag{4.5}$$

or, in isotropic cases,

$$\kappa = (M/H) \text{ cm}^{-3}. \tag{4.6}$$

It is usually more convenient to manipulate magnetic susceptibilities in units of weight. The gram magnetic susceptibility χ_g is obtained by dividing the volume magnetic susceptibility κ by the density and the molar magnetic susceptibility χ_M is obtained by multiplying χ_g by the molecular weight.

Table 4.1 lists some conversion factors.

4.3 Experimental techniques

4.3.1 The Gouy method

The general experimental arrangement for the Gouy method is shown in Fig. 4.1, where it may be seen that a cylindrical sample is suspended from the beam of an analytical balance. One end of the cylindrical sample is positioned between the plane parallel pole pieces of a magnet, with the sample being of such a length that the other end of the sample is in a region of zero field. The theory which relates the force exerted on a sample in such a magnetic field gradient to magnetic susceptibility will now be developed.

The magnetic potential energy per unit volume of a body in a magnetic field is given by

$$E = (HB)/8\pi \tag{4.7}$$

where H is the magnetic field intensity and B is the magnetic field induction. For anisotropic cases

$$B = (1 + 4\pi\kappa)H \tag{4.8}$$

where κ is the volume magnetic susceptibility tensor. Thus the magnetic potential energy becomes

$$E = H(1 + 4\pi\kappa)H/8\pi$$
$$= (H^2/8\pi) + H\kappa H/2. \tag{4.9}$$

Or, for isotropic cases which will be utilized in the subsequent development,

$$E = H^2/8\pi + \kappa H^2/2. \tag{4.10}$$

Table 4.1 Conversion factors for magnetic quantities

Multiply the number for		By	To obtain the number for	
Gaussian quantity	Unit		SI quantity	Unit
Flux density B	G	10^{-4}	flux density B	$T(\equiv Wb\,m^{-2}$ $\equiv VS\,m^{-2})$
Magnetic field strength H	Oe	$10^3/4\pi$	magnetic field strength H	$A\,m^{-1}$
Volume magnetic susceptibility κ	emu cm^{-3} (dimensionless)	4π	rationalized volume susceptibility κ	dimensionless
Gram magnetic susceptibility χ_g	emu g$^{-1}(\equiv cm^3\,g^{-1})$	$4\pi \times 10^{-3}$	rationalized gram susceptibility κ_g	$m^3\,kg^{-1}$
Molar magnetic susceptibility χ_M	emu mol$^{-1}(\equiv cm^3\,mol^{-1})$	$4\pi \times 10^{-6}$	rationalized molar susceptibility κ_{mol}	$m^3(mol)^{-1}$
Magnetization M	G or Oe	10^3	magnetization M	$A\,m^{-1}$
		$4\pi \times 10^{-4}$	magnetic polarization J	T
Magnetization $4\pi M$	G or Oe	$10^3/4\pi$	magnetization M	$A\,m^{-1}$
		10^{-4}	magnetic polarization J	T
Magnetization M	μ_B per atom or μ_B per form. unit, etc.	1	magnetization M	μ_B per atom or μ_B per form. unit etc.
Magnetic moment of a dipole m	erg G^{-1}	10^{-3}	magnetic moment of a dipole m	$J/T(\equiv Am^2)$
Demagnetizing factor N	dimensionless	$1/4\pi$	rationalized demagnetizing factor N	dimensionless

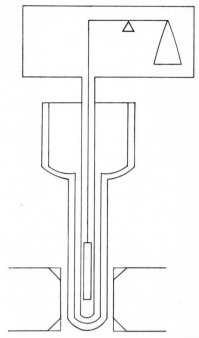

Fig. 4.1. Schematic diagram of the apparatus for the determination of magnetic susceptibilities by the Gouy method.

The difference in magnetic potential energy per unit volume between a sample with volume magnetic susceptibility κ and the displaced medium with volume magnetic susceptibility κ_0 is

$$E = (H^2/2)(\kappa - \kappa_0). \tag{4.11}$$

Since energy is given by force acting through a distance,

$$dE = F_s \, ds$$

it follows that

$$dF_s/dV = (d/ds)\{(H^2/2)(\kappa - \kappa_0)\}. \tag{4.12}$$

Then

$$dF_s = (\kappa - \kappa_0)\{H_x(dH_x/ds) + H_y(dH_y/ds) + H_z(dH_z/ds)\}dV. \tag{4.13}$$

Now, let the direction $s \equiv z$, and position the sample in the magnetic field so that

$$dH_x/dz = 0 \text{ and } dH_y/dz = 0. \tag{4.14}$$

Thus,

$$dF_z = (\kappa - \kappa_0)H(dH/dz)dV. \tag{4.15}$$

To integrate the force equation, assume a cylindrical sample of cross-section A and length l, and suspend the sample such that the axis of the cylinder is parallel to z with one end in the centre of the magnetic field (field strength H_l). The other end of the cylinder is not in the applied magnetic field (field strength H_0). Furthermore, the volume element is Adz. We may now write

$$dF_z = (\kappa - \kappa_0)A\int_{H_0}^{H_1} HdH \qquad (4.16)$$

and

$$F_z = (\kappa - \kappa_0)(A/2)(H_l^2 - H_0^2). \qquad (4.17)$$

If the cylindrical sample is weighed in the presence and absence of a magnetic field, the change in weight is $(g\delta m)$, and

$$(g\delta m) = \tfrac{1}{2}(\kappa - \kappa_0)A(H_l^2 - H_0^2). \qquad (4.18)$$

Attention to some physical characteristics of the experimental apparatus simplifies the practical implementation of the Gouy technique. First, if the magnet pole piece diameter is large with respect to the magnet pole gap, the position of the bottom of the sample in the magnetic field is not critical. Second, if the length of the sample is large with respect to the diameter of the pole pieces, the magnetic field H_0 at the top of the sample may be neglected with respect to H_l. Also, it may be that $\kappa \gg \kappa_0$, and that κ_0 may be neglected. Assuming that these details have been properly arranged, then

$$g\delta m = \tfrac{1}{2}\kappa A H_l^2. \qquad (4.19)$$

The density of the cylindrical sample, ρ, is given by m/Al, where m is the mass of the sample, so

$$g\delta m = \tfrac{1}{2}\kappa(m/\rho l)H_l^2 \qquad (4.20)$$

and

$$\kappa = 2(g\delta m)\rho l/mH_l^2. \qquad (4.21)$$

Gram magnetic susceptibility is given by

$$\chi_g = 2l(g\delta m)/mH_l^2.$$

If l, $g\delta m$, H_j^2, and m are known, then χ_g can be calculated. However, in practice, a magnetic susceptibility standard with an accurately known χ_g is used.

There are some serious disadvantages of the Gouy method. The long length of the samples requires the use of relatively large samples which may not be available. Perhaps a more important disadvantage arises from the requirement in the theory that the sample be of uniform density. It is difficult to pack powdered samples in a quartz tube such that the density is uniform over the length l of the sample. With practice, the packing error reduces to an uncertainty of 2–5 per cent.

4.3.2 The Faraday method

Some of the disadvantages of the Gouy method are eliminated in the Faraday method. Consider the differential equation for force

$$dF_z = (\kappa - \kappa_0)H(dH/dz)dV. \tag{4.22}$$

If the experimental conditions are chosen so that $H(dH/dz)$ is a constant for the volume element dV, we have

$$F_z = (\kappa - \kappa_0)H(dH/dz)V \tag{4.23}$$

and if $\kappa \gg \kappa_0$

$$F_z = m\chi H(dH/dz) \tag{4.24}$$

since $\kappa = \rho\chi = (m/V)\chi$, and $\kappa V = m\chi$.

The Faraday method requires a magnet with pole pieces shaped to provide the constant $H(dH/dz)$ over the volume of the sample. The profile of such pole pieces[1] are shown in the diagram of the experimental apparatus (Fig. 4.2). Alternatively, the field gradient may be provided by induction coils[2]

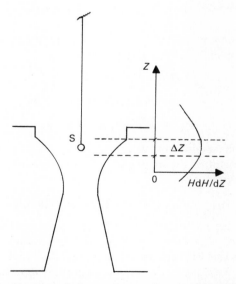

Fig. 4.2. Schematic diagram of the apparatus for the determination of magnetic susceptibilities by the Faraday method.

mounted on parallel planar magnet pole pieces. Since the sign of the field gradient can be changed by reversing the direction of current flow in the coils, it is possible to increase the sensitivity of the technique by alternating the direction of the current flow and using phase-sensitive detection techniques.[3]

The Faraday method offers several advantages, including small sample size (of the order of a few milligrams), the absence of packing errors, and only a small volume needing to be thermostatted for temperature variation measurements. However, it is not readily applicable to solutions or for measurement of magnetic anisotropies of single crystals. In the latter case the magnetic anisotropy may lead the single crystal to re-orient in the magnetic field. Reorientation can be prevented if a rigid suspension system is used, but this is not generally possible with microbalances commonly used for the force measurements. The sensitivity of the Faraday method is about 10^{-8} emu, and the experimental uncertainties can be less than ± 1 per cent.

4.3.3 Change-in-flux methods

The methods to be discussed here are based on changing the magnetic flux of a solenoid and involve measuring that change when a sample is introduced into the solenoid. Changes in flux are measured by direct current, alternating current, and induction methods in the presence of a static magnetic field, an oscillating magnetic field, or both. Oscillating magnetic fields of low frequency are used to measure static magnetic susceptibilities, while high frequencies are used in experiments involving relaxation times of paramagnetic states.

The applied magnetic field may consist of a static component H_0 and an oscillating component H_1 and is given at time t by

$$H(t) = H_0 + H_1 \cos \omega t. \tag{4.25}$$

The magnetization of a substance in this oscillating field is given by

$$M(t) = M_0 + M_1 \cos (\omega t - \varphi) \tag{4.26}$$

where φ is the phase shift by which the magnetization lags behind the oscillating component of the magnetic field. In terms of magnetic susceptibilities, we may write

$$M(t) = \chi_o H_0 + \chi' H_1 \cos (\omega t) + \chi'' H_1 \sin (\omega t) \tag{4.27}$$

where χ' is the high-frequency (in-phase) component of the magnetic susceptibility, and χ'' is the paramagnetic dispersion or out of phase component. The value of measurements at both high and low frequencies is apparent. At low frequencies the magnetization M_1 does not lag behind the applied field, φ is zero, and in the low-field limit $\chi' = \chi_0$, while $\chi'' = 0$. Experiments at high frequencies using phase-sensitive detection yield information about the additional component of the magnetic susceptibility.

If a sample is placed inside a coil, the inductance is given by

$$L = L_0(1 + 4\pi\zeta\kappa) \qquad (4.28)$$

where L is the observed inductance, L_0 is the inductance of the coil in a vacuum, ζ is a filling factor, and κ is the magnetic susceptibility. The inductance of the displaced medium is small and can be neglected. The inductance L can be measured with an inductance bridge.

If the coil containing the sample is made to oscillate in an LC-circuit, the oscillation frequencies f (with) and f_0 (without the sample) are related by

$$f_0 = f(1 + 4\pi\zeta\kappa)^{\frac{1}{2}}. \qquad (4.29)$$

Since $4\pi\zeta\kappa$ is small with respect to unity, we can approximate f as

$$f = f_0(1 - 2\pi\zeta\kappa). \qquad (4.30)$$

A measurement of the frequency shift yields the volume magnetic susceptibility

$$\kappa = (f_0 - f)/(2\pi\zeta f_0). \qquad (4.31)$$

Induced voltages may also be measured. The voltage induced in a coil when a sample is introduced is given by

$$V = \mu N^2 A f / S \qquad (4.32)$$

where N is the number of turns of wire in the coil, A is the cross-sectional area, f is the frequency of current oscillation, S is the length of the coil, and μ is the magnetic permeability of the substance within the coil.

In practice, mutual inductances are usually measured with a bridge circuit, usually consisting of a primary coil and two secondary coils wound in opposition and connected in series. If the two secondary coils were perfectly balanced, the induced voltage would cancel. Introducing a sample in one of the secondary coils unbalances the bridge, and the unbalance is a measure of the magnetic susceptibility.

Inductance techniques frequently are not very sensitive. A d.c. inductance method utilizing the Meissner effect of superconductors provides a highly sensitive technique for measuring magnetic susceptibilities. The Meissner effect is the quantized expulsion of magnetic fields from a superconducting substance when the substance is cooled below its superconducting transition temperature. The only possible values of magnetic flux which may be trapped within a thick superconducting ring are integral multiples of $hc/2e = 2 \times 10^{-7}$ Oe cm^2. Using this property, it is possible to produce a region of zero magnetic field in the following way. If a cylinder made of a superconducting substance is cooled below its transition temperature in an external magnetic field that produces a flux of less than $hc/2e$, a current is introduced in the superconducting cylinder to exactly cancel the flux. This results in the production of the lowest quantized state, that being zero magnetic field. The current in the cylinder is a direct measure of the flux that existed in it initially. The device is called a SQUID (superconducting quantum interference device).

A 'SQUID' is a superconducting ring with a weak link and is capable of amplifying small changes in magnetic fields into large electrical signals. A typical superconducting magnetometer consists of two superconducting coils L_1 and L_2 wound in opposition with one of them coupled inductively to a third superconducting coil L_3. If a sample is passed through L_1, the persistent current induced is proportional to the magnetization of the sample. Although SQUIDs are expensive to purchase, and expensive to operate, their very high sensitivity (changes in magnetic susceptibility of 10^{-12} emu) ensures wide application for important chemical problems.

4.3.4 Vibrating sample magnetometer

A vibrating sample magnetometer is a device in which a sample is vibrated in a uniform magnetizing field, and the magnetization of the sample is detected. The instrument allows precise magnetization measurements to be made as a function of temperature, magnetic field strength, and crystallographic orientation. However, the sensitivity of the technique is limited.

When a specimen is placed in a uniform magnetic field, a magnetization M is induced in the sample equal to the product of the susceptibility and the applied magnetic field. If the sample is vibrated, say in a sinusoidal motion, an electrical signal due to the magnetization of the sample can be induced in stationary pick-up coils placed in a suitable position. The electrical signal is proportional to the magnetic moment, the vibrational amplitude, and the vibrational frequency.

In the commercial system available from Princeton Applied Research, a transducer unit with beryllium–copper springs is used to vibrate the sample. The transducer assembly is mounted on a rotating platform, and the sample is mounted on a rigid, but readily demountable rod.

The electronic system is designed so that variations in the amplitude and frequency of the vibration are accounted for, and a signal proportional only to the magnetization of the sample is monitored and displayed as sample magnetization. The magnetization is converted to molar magnetic susceptibility by the expression

$$\chi_M = M_s(\text{MW})_s / H_a w_s \qquad (4.33)$$

where M_s is the magnetization of the sample, $(\text{MW})_s$ is the molecular or formula weight, H_a is the applied magnetic field, and w_s is the weight of the sample.

4.3.5 Evans method for determination of magnetic susceptibilities in solution by nuclear magnetic resonance

Nuclear magnetic resonance provides a convenient method for determining paramagnetic susceptibilities of substances in dilute solutions.[4] Specific advantages of the technique are that only about 0.2 ml of solution is required, and contributions to the magnetic susceptibility by other diamagnetic substances, including ligands, can be eliminated.

The resonance position of a line in the nuclear resonance spectrum, say *tert*-butyl alcohol in aqueous solution, depends on the bulk magnetic susceptibility of the medium, and the shift in the resonance position of the proton caused by the presence of paramagnetic ions is given by

$$\Delta H/H = (2\pi/3)\Delta \kappa \qquad (4.34)$$

where $\Delta \kappa$ is the difference in volume susceptibilities of the liquids with, and without, the paramagnetic ions.

In practice, a solution of the paramagnetic substance containing an internal standard is placed in the inner of two concentric NMR tubes, and a solution of the internal standard in the same solvent is placed in the outer tube. Two separate NMR lines arising from the internal standard are observed, with the line at higher frequency corresponding to the standard in the solution containing the paramagnetic substance. The mass susceptibility of the dissolved paramagnetic substance is given by

$$\chi = (3\Delta f/2\pi fm) + \chi_0 + \chi_0(d_0 - d_s)/m \qquad (4.35)$$

where Δf is the frequency difference of the two lines in hertz, f is the frequency at which the magnetic resonance lines are being recorded, m is the mass of the paramagnetic substance per millilitre of solution, χ_0 is the mass susceptibility of the solvent, d_0 is the density of the solvent, and d_s is the density of the solution.[5]

4.3.6 Calibration standards

Absolute values of magnetic susceptibilities are rarely measured in the laboratory. Instead, the apparatus is usually calibrated with standard substances that have well-known values. It is generally advisable to use a standard which has a gram magnetic susceptibility comparable in magnitude to that of the unknown. Thus, for diamagnetic materials, water ($\chi_g = -0.720 \times 10^{-6}$ cgs units at 293 K)[6] is frequently used. The compound $HgCo(NCS)_4$ ($\chi_g = 16.44 \times 10^{-6}$ cgs units at 293 K)[7] is widely used as a standard for paramagnetic substances, even though the gram magnetic susceptibility is rather large. $HgCo(NCS)_4$ can be used over a range of temperatures since the Curie–Weiss parameters are well known.[8,9] The manganese Tutton salt $(NH_4)_2Mn(SO_4)_2 \cdot 6H_2O$ is widely used since the magnetic susceptibility of this compound obeys the Curie law with $C=4.375$.[10] In fact, it is often used to calibrate thermometers used in the cryogenic range. The compound $(Me_2enH_2)CuCl_4$ has been suggested[11] as a calibration standard for work with less paramagnetic materials since the gram magnetic susceptibility is significantly smaller than that of $HgCo(NCS)_4$ or the manganese Tutton salt.

4.4 Formulae for data handling

4.4.1 The Van Vleck equation

The equation for calculating magnetic susceptibilities which was derived by Van Vleck in 1932 holds for paramagnetic systems in which the magnetic susceptibility is independent of the magnetic field, and is applicable for moderate magnetic fields and temperatures. The system described by the theory consists of a series of n states, each of which is m-fold degenerate. The energy of the state $E_{n,m}$ in a magnetic field H is expanded in terms of a power series in H:

$$E_{n,m} = E_{n,m}^{(0)} + HE_{n,m}^{(1)} + H^2 E_{n,m}^{(2)} + \ldots \quad (4.36)$$

The magnetic moment of the level $E_{n,m}$ in the magnetic field direction ($i = x, y, z$) is given by

$$\mu_{n,m} = -\partial E_{n,m}/\partial H_i \quad (4.37a)$$

$$= (-\partial/\partial H_i)(E_{n,m}^{(0)} + HE_{n,m}^{(1)} + H^2 E_{n,m}^{(2)} + \ldots) \quad (4.37b)$$

$$= -E_{n,m}^{(1)} - 2HE_{n,m}^{(2)} - \ldots \quad (4.37c)$$

The magnetic moment per gram atom for the array of states is obtained by taking a statistical mean over the populated states weighted by a Boltzmann distribution yielding

$$M_i = \frac{N_A \Sigma_{n,m} \mu_{n,m} \exp(-E_{n,m}/kT)}{\Sigma_{n,m} \exp(-E_{n,m}/kT)} \quad (4.38)$$

where N_A is the Avogadro number.

Now consider the exponential

$$\exp(-E_{n,m}/kT) = \exp\{(-E_{n,m}^{(0)} - HE_{n,m}^{(1)} - \ldots)/kT\}. \quad (4.39)$$

By definition

$$\exp(x) = 1 + x + x^2/2! + x^3/3! + \ldots \quad (4.40)$$

so we can write

$$\exp(-E_{n,m}/kT) = \exp\{(-E_{n,m}^{(0)})/kT\}$$
$$\times \exp\{(-HE_{n,m}^{(1)})/kT\} \exp\{(-H^2 E_{n,m}^{(2)} - \ldots)/kT\} \quad (4.41)$$

and approximate as follows:

$$\exp(-E_{n,m}/kT) = \exp(-E_{n,m}^{(0)}/kT)(1 - HE_{n,m}^{(1)}/kT). \quad (4.42)$$

(This approximation is rather good when x is small with respect to 1.)
Substitution into the expression for M_i yields

$$M_i = \frac{N_A \Sigma_{n,m}(-E_{n,m}^{(1)} - 2HE_{n,m}^{(2)} - \ldots)\exp(-E_{n,m}^{(0)}/kT)(1 - HE_{n,m}^{(1)}/kT)}{\Sigma_{n,m}\exp(-E_{n,m}^{(0)}/kT)(1 - HE_{n,m}^{(1)}/kT)}. \quad (4.43)$$

Since there is no magnetic moment (magnetization) in the absence of an applied magnetic field, then the term $\Sigma_{n,m} - E_{n,m}^{(1)}\exp(-E_{n,m}^{(0)}/kT)$ vanishes. By definition, magnetic susceptibility χ_i is M_i/H_i, and it follows that

$$\chi_i = \frac{N_A \Sigma_{n,m}\{(E_{n,m}^{(1)})^2/kT - 2E_{n,m}^{(2)} + HE_{n,m}^{(2)}E_{n,m}^{(1)}\}\exp(-E_{n,m}^{(0)}/kT)}{\Sigma_{n,m}\exp(-E_{n,m}^{(0)}/kT)(1 - HE_{n,m}^{(1)}/kT)}. \quad (4.44)$$

Furthermore, in view of the constraint that the expression is for systems in which the magnetic susceptibility is independent of magnetic field, the *Van Vleck equation* becomes

$$\chi_i = \frac{N\Sigma_{n,m}\{(E_{n,m}^{(1)})^2/kT - 2E_{n,m}^{(2)}\}\exp(-E_{n,m}^{(0)}/kT)}{\Sigma_{n,m}\exp(-E_{n,m}^{(0)}/kT)}. \quad (4.45)$$

4.4.2 The Curie law

We will now identify the terms $E_{n,m}^{(1)}$ and $E_{n,m}^{(2)}$ in the Van Vleck equation as the first- and second-order Zeeman coefficients. Consider an octahedral molecule with the direction of the applied field H taken parallel to z, so we can use the $\hat{L}_z + 2\hat{S}_z$ components of the Zeeman operator $\hat{L} + 2\hat{S}$. Thus,

$$E_{n,m}^{(1)} = <\Psi_{n,m}|\hat{L}_z + 2\hat{S}_z|\Psi_{n',m'}>\mu_B \quad (4.46)$$

and

$$E_{n,m}^{(2)} = \Sigma_n|<\Psi_n|\hat{L}_z + 2\hat{S}_z|\Psi_0>\mu_B|^2/(E_0 - E_n). \quad (4.47)$$

Ignoring the second-order term, the magnetic susceptibility of an orbitally nondegenerate ground state becomes

$$\chi_z = (N\mu_B^2/kT)|<\Psi_{n'm}|2\hat{S}_z|\Psi_{n',m'}>|^2/(2S+1) \quad (4.48a)$$

$$= \{N\mu_B^2 g^2/kT(2S+1)\}\Sigma_{n=-S}^{S} m_S^2 \quad (4.48b)$$

$$= \{N\mu_B^2 g^2/kT(2S+1)\}\{\tfrac{1}{3}S(S+1)(2S+1)\} \quad (4.48c)$$

$$= (N\mu_B^2 g^2/3kT)\{S(S+1)\}. \quad (4.48d)$$

This last equation is the spin-only formula where Landé g factor has been introduced to admit the possibility of spin-orbit coupling with excited states. The Landé g factor is given by

$$g = 1 + \{J(J+1) + S(S+1) - L(L+1)\}/\{2J(J+1)\}. \quad (4.49)$$

In the case of spin-only paramagnetism, the Landé g factor is 2 since $L=0$.
We frequently use the Curie law

$$\chi = C/T \quad (4.50)$$

where C is the Curie constant. Thus,

$$C = N\mu_B^2 g^2 S(S+1)/3k. \quad (4.51)$$

The Curie law was found experimentally in the 1890s.

4.4.3 Paramagnetism

In the general case the magnetic properties will depend on the total angular momentum J, the magnetic moment will be μ_J, and the Hamiltonian for a single magnetic ion will be

$$\mathcal{H} = -\mu_J H_0 \quad (4.52)$$

where H_0 is the applied magnetic field. Along one direction, say the z-direction, the Hamiltonian becomes

$$\mathcal{H} = -\mu_{J,z} H_z. \quad (4.53)$$

The eigenvalues are $-g\mu_B m_J H_z$ where m_J can take on the values $J, J-1, \ldots, -J$. The partition function is

$$Z_J = \Sigma_{m_J} \exp(g\mu_B H_z/kT) \quad (4.54a)$$

$$= (\sinh[\{(2J+1)/2J\}x])(\sinh(x/2J)) \quad (4.55b)$$

where $x = g\mu_B J H_z/kT$ is the ratio between magnetic and thermal energies. The magnetic moment becomes

$$M = Ng\mu_B \Sigma_{m_J} m_J \exp(m_J x/J)/Z_J(x) \quad (4.56)$$

which can be expressed in terms of the Brillouin function as

$$M = Ng\mu_B J B_J(x) \quad (4.57a)$$

where

$$B_J(x) = \{(2J+1)/2J\}\coth[\{2J+1)/2J\}x] - (1/2J)\coth(x/2J). \quad (4.57b)$$

Two limiting cases can be envisaged. In the first case $x \ll 1$. This would correspond to a typical paramagnetic substance at room temperature. For example, with $J=1$, $g=2$, and $H_0 = 10\,000$ Oe at room temperature, we can calculate $x = 5 \times 10^{-3}$. In such case, the Brillouin function reduces to

$$B_J(x) = \{(J+1)/3J\}x \tag{4.58}$$

and the magnetic moment is given by

$$M = Ng^2\mu_B^2 J(J+1)H_z/3kT. \tag{4.59}$$

Since the magnetic susceptibility is defined by $\chi = M/H$, then

$$\chi = C/T \tag{4.60}$$

where the Curie constant C is

$$C = \{Ng^2\mu_B^2 J(J+1)\}/3k \tag{4.61}$$

in agreement with the results obtained earlier considering the spin only.

An effective magnetic moment may be defined as

$$\mu_{\text{eff.}} = g\{J(J+1)\}^{\frac{1}{2}} \tag{4.62}$$

or, for spin-only paramagnetism, by

$$\mu_{\text{eff.}} = g\{S(S+1)\}^{\frac{1}{2}}. \tag{4.63}$$

For $g=2$, the spin-only formula may be expressed alternately as

$$\mu_{\text{eff.}} = \{n(n+2)\}^{\frac{1}{2}} \tag{4.64}$$

where n is the number of unpaired electrons. Magnetic moments calculated with these simple formulas are compared with experimental magnetic moments in Table 4.2.

The other limiting case occurs when the magnetic energy is large with respect to thermal energy, i.e. large values of x. This corresponds to paramagnetic saturation. We may write

$$M/Ng\mu_B J = B_J(x) = 1 - (1/J)\exp(-g\mu_B H_z/kT) \tag{4.65}$$

which has unity as a limiting value as $T \to 0$. Plots of $B_J(x)$ as a function of x are shown in Fig. 4.3 for several J values. It is clear that magnetization studies can be very useful in determining angular momentum quantum numbers since the shape of the magnetization curves are very sensitive to the value of J.

4.4.4 Data analysis

Measurements of magnetic properties at one temperature, say room temperature, or over a limited temperature range, yield data which may be useful; but generally measurements should be made over wide ranges of tempera-

Table 4.2 *Typical experimental magnetic moments compared with theoretical predictions*

Ion	O_h ground state	S	L	J	g	$\mu_{\text{eff}} = g\{J(J+1)\}$	$\mu_{\text{eff}} = 2\{S(S+1)\}$	μ_{eff} expt'l B.M.
Ti^{3+}	$^2T_2$	$\frac{1}{2}$	2†	$\frac{3}{2}$		1.55	1.73	1.8
V^{3+}	$^3T_1$	1	3†	2		1.63	2.83	2.8
Cr^{3+}	$^4A_2$	$\frac{3}{2}$	3†	$\frac{3}{2}$		0.77	3.87	3.8
Mn^{3+}	^5E	2	2†	0		0.00	4.90	4.9
Fe^{3+}	$^6A_1$	$\frac{5}{2}$	0†	$\frac{5}{2}$		5.92	5.92	5.9
Fe^{2+}	$^5T_2$	2	2†	4		6.70	4.90	5.4
Co^{2+}	$^4T_1$	$\frac{3}{2}$	3†	$\frac{9}{2}$		6.63	3.87	4.8
Ni^{2+}	$^3A_2$	1	3†	4		5.59	2.83	3.2
Cu^{2+}	^2D	$\frac{1}{2}$	2†	$\frac{5}{2}$		3.55	1.73	1.9
La^{3+}		0	0	0	—	0.00		0.0
Ce^{3+}		$\frac{1}{2}$	3	$\frac{5}{2}$	$\frac{6}{7}$	2.54		2.6
Pr^{3+}		1	5	4	$\frac{4}{5}$	3.58		3.5
Nd^{3+}		$\frac{3}{2}$	6	$\frac{9}{2}$	$\frac{8}{11}$	3.62		3.5
Pm^{3+}		2	6	4	$\frac{3}{5}$	2.68		
Sm^{3+}		$\frac{5}{2}$	5	$\frac{5}{2}$	$\frac{2}{7}$	0.85		1.5
Eu^{3+}		3	3	0	—	0.00		3.4
Gd^{3+}		$\frac{7}{2}$	0	$\frac{7}{2}$	2	7.94		8.0
Tb^{3+}		3	3	6	$\frac{3}{2}$	9.72		9.5
Dy^{3+}		$\frac{5}{2}$	5	$\frac{15}{2}$	$\frac{4}{3}$	10.65		10.5
Ho^{3+}		2	6	8	$\frac{5}{4}$	10.61		10.4
Er^{3+}		$\frac{3}{2}$	6	$\frac{15}{2}$	$\frac{6}{5}$	9.58		9.5
Tm^{3+}		1	5	6	$\frac{7}{6}$	7.56		7.3
Yb^{3+}		$\frac{1}{2}$	3	$\frac{7}{2}$	$\frac{8}{7}$	4.54		4.5
Lu^{3+}		0	0	0	—	0.00		0.0

† free ion

tures and magnetic field strengths. Such measurements may yield adequate information for detailed descriptions of electronic structure and bonding.

Experimental magnetic susceptibility data must be corrected for the diamagnetism of the constituent atoms, which is practically temperature independent. Diamagnetism may be estimated rather precisely by the following procedure developed by Pascal:

$$\chi_M(\text{dia}) = \sum_i n_i \chi(A)_i + \sum_j n_j \chi(B)_j \quad (4.66)$$

where n_i is the number of atoms of kind i and diamagnetic susceptibility χ_i, and χ_j is the susceptibility associated with n_j structural features (such as double or triple carbon–carbon bonds). Extensive self-consistent tables of magnetic susceptibilities of atoms, ions, and structural features have been tabulated and sources are listed in the bibliography.

The next step is to plot the molar magnetic susceptibility χ_M, corrected for

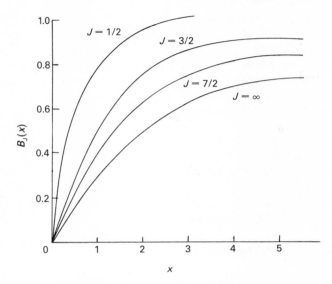

Fig. 4.3. The Brillouin function for selected J values.

the diamagnetism of the constituent atoms, versus T^{-1}. If a straight line is obtained, then the data follow the Curie law, $\chi = C/T$. The effective magnetic moment can be calculated from the expression

$$\mu_{\text{eff.}} = 2.828(\chi_M T)^{\frac{1}{2}} \tag{4.67}$$

and the Bohr magneton number can be used along with any complementary data to solve the problem at hand.

If a straight line which intercepts the positive χ_M axis at $T^{-1} = 0$ is obtained, then it is possible that temperature-independent paramagnetism is present, or alternatively, that the diamagnetic correction has been overestimated. If the straight line intercepts the negative χ_M axis at $T^{-1} = 0$, it is likely that the diamagnetic correction has been underestimated.

If the line in the plot of χ_M versus T^{-1} is not straight, then plot χ_M^{-1} versus T. A straight line indicates that the data obey the Curie–Weiss law

$$\chi_M = C/(T - \theta) \tag{4.68}$$

which, when recast, yields

$$T = C/\chi_M + \theta. \tag{4.69}$$

The slope of the line is the Curie constant, and the intercept is the Weiss constant.

It is usually instructive to plot $\chi_M T$ versus T. Curie law behaviour holds if a straight line parallel to the T-axis is obtained. For orbitally nondegenerate systems, deviations from straight-line behaviour towards larger $\chi_M T$ values with decreasing temperature indicate an array of populated electronic states with states of larger spin multiplicity lying lowest. The array of states may arise from ferromagnetic interaction. Deviations of $\chi_M T$ to smaller values with decreasing temperature indicate the presence of array of populated electronic states with lower spin multiplicities lying lowest. Such an array of states may arise from antiferromagnetic interactions. If a straight line is not obtained from the Curie–Weiss plot, then it is clear that more detailed theory is necessary for data analysis, and the Van Vleck equation is the starting point for such an analysis.

A serious complication in the interpretation of magnetic data may arise if ferromagnetic impurities are present even in small quantities. In favourable cases it is possible to account for such impurities by making measurements as a function of magnetic field strength.

4.5 Crystal field theory

The results of crystal field theory are very useful in discussions of magnetic properties of transition-metal compounds. Here, for the purposes of illustration, we will discuss crystal field theory from a weak-field approach. Those interested in more complete treatments beyond the scope of this chapter are referred to discussions in the appropriate monographs. Perturbation theory is used, assuming free-ion eigenfunctions and eigenvalues, and the Hamiltonian is

$$\mathcal{H} = -(h/8\pi^2 m)\sum_i \nabla_i^2 - \sum_i (Ze^2/r_i) + \tfrac{1}{2}\sum_{i \neq k}(e^2/r_{i,k}) + \sum_i (\zeta_i(r)l_i s_i) + V_{CF}. \tag{4.70}$$

The following cases may be envisaged:

1. $V_{CF} < l_i s_i$
2. $l_i s_i < V_{CF} < e^2/r_{i,j}$
3. $V_{CF} > e^2/r_{i,j}$

Complexes of the rare-earth elements provide examples of case 1, complexes of the first-transition-series metals provide cases of case 2, and complexes with covalent bonds provide examples of case 3. We shall be concerned with case 2 and discuss an octahedral transition-metal complex with a d^1 electronic configuration.

4.5.1 The octahedral crystal field

The octahedral crystal field removes the five-fold orbital degeneracy of the ^2D free-ion state yielding a three-fold degenerate state at $-4Dq$ and a two-fold degenerate state at $+6Dq$. The results of the crystal field calculation are collected in Table 4.3.

Table 4.3 *Crystal field splitting of ^2D free-ion state*

$E = -4Dq$	$E = 6Dq$
$\Psi_1 = (\lvert 2 \rangle - \lvert -2 \rangle)/\sqrt{2}$	$\Psi_4 = (\lvert 2 \rangle + \lvert -2 \rangle)/\sqrt{2}$
$\Psi_2 = \lvert 1 \rangle$	$\Psi_5 = \lvert 0 \rangle$
$\Psi_3 = \lvert -1 \rangle$	

The symmetries of the crystal field states are determined by examining the transformation properties of $\Psi_1, \ldots, \Psi_5$ in the point group O, since it is known that d wave-functions are even under inversion, and O_h is $O \times i$. The representations of the point group O to which the functions belong can be determined using the information in Table 4.4. The set of functions Ψ_1, Ψ_2, Ψ_3 transforms like T_2 in O, and the set Ψ_4, Ψ_5 transforms like E. In O_h, these become T_{2g} and E_g, respectively. Thus, the octahedral crystal field removes the orbital degeneracy of the ^2D free-ion state resulting in the formation of two crystal field states, $^2T_{2g}$ and ^2E_g.

4.5.2 Spin-orbit coupling in octahedral crystal field states

The degeneracies of the crystal field states may be further split by coupling of orbital angular momenta of the electrons to their spin angular momenta. In the present development, we are interested in spin-orbit coupling in $^2T_{2g}$ and

Table 4.4 *Transformation properties of $\lvert M \rangle$ functions in point group O*

$\lvert M \rangle$ [a]	E	C_3 [b]	C_2 [c]	C_4 [d]	C_2' [e]
$\lvert 2 \rangle$	$\lvert 2 \rangle$	$-(\sqrt{3/8})\lvert 0 \rangle - \tfrac{1}{4}\lvert 2 \rangle - \tfrac{1}{4}\lvert -2 \rangle - \tfrac{1}{2}\lvert 1 \rangle - \tfrac{1}{2}\lvert -1 \rangle$	$\lvert 2 \rangle$	$-\lvert 2 \rangle$	$-\lvert -2 \rangle$
$\lvert 1 \rangle$	$\lvert 1 \rangle$	$(i/2)\lvert 2 \rangle - (i/2)\lvert -2 \rangle + (i/2)\lvert 1 \rangle - (i/2)\lvert -1 \rangle$	$-\lvert 1 \rangle$	$-i\lvert 1 \rangle$	$i\lvert -1 \rangle$
$\lvert 0 \rangle$	$\lvert 0 \rangle$	$-\tfrac{1}{2}\lvert 0 \rangle + (\sqrt{3/8})\lvert 2 \rangle + (\sqrt{3/8})\lvert -2 \rangle$	$\lvert 0 \rangle$	$\lvert 0 \rangle$	$\lvert 0 \rangle$
$\lvert -1 \rangle$	$-\lvert 1 \rangle$	$-(i/2)\lvert 2 \rangle + (i/2)\lvert -2 \rangle + (i\lvert 2)\lvert 1 \rangle - (i/2)\lvert -1 \rangle$	$-\lvert -1 \rangle$	$i\lvert -1 \rangle$	$-i\lvert 1 \rangle$
$\lvert -2 \rangle$	$\lvert -2 \rangle$	$-(\sqrt{3/8})\lvert 0 \rangle - \tfrac{1}{4}\lvert 2 \rangle - \tfrac{1}{4}\lvert -2 \rangle + \tfrac{1}{2}\lvert 1 \rangle + \lvert -1 \rangle$	$\lvert -2 \rangle$	$-\lvert -2 \rangle$	$-\lvert -2 \rangle$

[a] The functions are $\lvert 2 \rangle = (\sqrt{3/8})(x+iy)^2$
$\lvert 1 \rangle = -(\sqrt{3/2})(x+iy)z$
$\lvert 0 \rangle = \tfrac{1}{2}(3z^2 - r^2)$
$\lvert -1 \rangle = (\sqrt{3/2})(x-iy)z$
$\lvert -2 \rangle = (\sqrt{3/8})(x-iy)^2$

[b] for the operation $(x \to y, y \to z, z \to x)$
[c] for the operation $(z \to z, x \to -x, y \to -y)$
[d] for the operation $(z \to z, x \to y, y \to x)$
[e] for the operation $(z \to -z, x \to y, y \to x)$

^2E_g states. Again, we use group theory, and in the case of odd numbers of electrons, double groups must be used.

We proceed by determining the representations spanned by the spin function and the orbital function in the double group, take the direct product of these representations, and reduce the direct product representation to a sum of irreducible representations.

In this example, we will determine the representations spanned by the doublet spin function and by the orbital functions (E and T_2) in the double group O' (Table 4.5) using the following expressions:

$$\chi(E) = 2l + 1 \tag{4.71a}$$

$$\chi(R) = -(2l + 1) \tag{4.71b}$$

$$\chi(C_a) = \sin(2l+1)a/\sin(a/2) \tag{4.71c}$$

Table 4.5 *Character table for double group O'*

O'	E	R	$4C_3^2R$ $4C_3$	$4C_3R$ $4C_3^2$	$3C_2R$ $3C_2$	$3C_4^3R$ $3C_4$	$3C_4R$ $3C_4^3$	$6C_2'R$ $6C_2'$
Γ_1	1	1	1	1	1	1	1	1
Γ_2	1	1	1	1	1	-1	-1	-1
Γ_3	2	2	-1	-1	2	0	0	0
Γ_4	3	3	0	0	-1	1	1	-1
Γ_5	3	3	0	0	-1	-1	-1	1
Γ_6	2	-2	1	-1	0	$\sqrt{2}$	$-\sqrt{2}$	0
Γ_7	2	-2	1	-1	0	$-\sqrt{2}$	$\sqrt{2}$	0
Γ_8	4	-4	-1	1	0	0	0	0

The results are summarized in Table 4.6. Reduction of the reducible representations $\Gamma(E \times S)$ and $\Gamma(T_2 \times S)$ yield

$$\Gamma(E \times S) = \Gamma_8 \tag{4.72a}$$

$$\Gamma(T_2 \times S) = \Gamma_7 + \Gamma_8. \tag{4.72b}$$

Thus, in an octahedral crystal field the ^2E_g state is not split by spin-orbit coupling, but transforms as the irreducible representation Γ_8 in the double group O', while the $^2T_{2g}$ state is split by spin-orbit coupling giving new states Γ_7 and Γ_8.

In subsequent sections, the influence of the ^2E_g state will be ignored, even though the symmetry arguments developed above show that it can mix with the ground state by spin-orbit coupling. This neglect is for pedagogical reasons; careful research work will require the addition of the ^2E_g state.

Table 4.6 *Spin-orbit interaction in d^1 O_h crystal field states*

O'	E	R	$4C_3^2R$ $4C_3$	$4C_3R$ $4C_3^2$	$3C_2R$ $3C_2$	$3C_4^3R$ $3C_4$	$3C_4R$ $3C_4^3$	$6C_2'R$ $6C_2'$
$\Gamma(S=\tfrac{1}{2})$	2	−2	1	−1	0	$\sqrt{2}$	$-\sqrt{2}$	0
$\Gamma(E)$	2	2	−1	−1	2	0	0	0
$\Gamma(T_2)$	3	3	0	0	−1	−1	−1	1
$\Gamma(E \times S)$	4	−4	−1	1	0	0	0	0
$\Gamma(T_2 \times S)$	6	−6	0	0	0	$-\sqrt{2}$	$\sqrt{2}$	0

The term that is added to the Hamiltonian to account for spin-orbit coupling is

$$\mathcal{H}_{so} = \sum_i (2m^2c^2r_i)^{-1}(\partial V/\partial r)\hat{l}_i \cdot \hat{s}_i \quad (4.73a)$$

$$= \sum_i \zeta_i \hat{l}_i \cdot \hat{s}_i. \quad (4.73b)$$

It may be noted that this term is valid for spherical symmetry only, but the error introduced in lower symmetry cases is not great. The spin-orbit coupling constant could be taken from the spectra of gaseous ions, but it is usually treated as a parameter with the fitted parameter in the complex being less than that of the free ion.

If the interactions between different Russell–Saunders states are not large, the spin-orbit coupling operator may be written

$$\mathcal{H}_{so} = \lambda_{LS}\hat{L} \cdot \hat{S}. \quad (4.74)$$

For a shell less than half-filled

$$\lambda_{LS} = \zeta/2S \quad (4.75)$$

while a shell that is more than half-filled has

$$\lambda_{LS} = -\zeta/2S. \quad (4.76)$$

$\hat{L} \cdot \hat{S}$ is expanded in terms of raising and lowering operators in the following manner:

$$\hat{L} \cdot \hat{S} = \hat{L}_x \hat{S}_x + \hat{L}_y \hat{S}_y + \hat{L}_z \hat{S}_z. \quad (4.77)$$

By defining

$$\hat{L}_\pm = \hat{L}_x \pm i\hat{L}_y \quad (4.78a)$$

$$\hat{S}_\pm = \hat{S}_x \pm i\hat{S}_y \quad (4.78b)$$

we can write

$$\hat{L}_x = \tfrac{1}{2}(\hat{L}_+ + \hat{L}_-) \quad (4.79a)$$

$$\hat{L}_y = (1/2i)(\hat{L}_+ - \hat{L}_-) \qquad (4.79b)$$
$$\hat{S}_x = \tfrac{1}{2}(\hat{S}_+ + \hat{S}_-) \qquad (4.80a)$$
$$\hat{S}_y = (1/2i)(\hat{S}_+ - \hat{S}_-) \qquad (4.80b)$$

and

$$\hat{L}\cdot\hat{S} = \tfrac{1}{4}(\hat{L}_+ + \hat{L}_-)(\hat{S}_+ + \hat{S}_-) + \tfrac{1}{4}(\hat{L}_+ - \hat{L}_-)(\hat{S}_+ - \hat{S}_-) + \hat{L}_z\hat{S}_z \qquad (4.81a)$$
$$= \tfrac{1}{2}\hat{L}_+\hat{S}_- + \tfrac{1}{2}\hat{L}_-\hat{S}_+ + \hat{L}_z\hat{S}_z. \qquad (4.81b)$$

The raising and lowering operators have the following important properties:

$$\hat{L}_+|\Psi(M_L,M_S)\rangle = \{(L+M_L+1)(L-M_L)\}^{\frac{1}{2}}|\Psi(M_L+1,M_S)\rangle \qquad (4.82a)$$
$$\hat{L}_-|\Psi(M_L,M_S)\rangle = \{(L-M_L+1)(L+M_L)\}^{\frac{1}{2}}|\Psi(M_L-1,M_S)\rangle. \qquad (4.82b)$$

That is, application of a raising (or lowering) operator to a function characterized by the quantum number M_L generates a new function characterized by the quantum number (M_L+1), or (M_L-1). Comparable raising and lowering operators for spin are obtained by substituting S and M_S for L and M_L, respectively, in the expressions given above.

The crystal field corrected wave-functions may be written as

$$\varphi_1 = (1/\sqrt{2})(d_2 - d_{-2})\alpha \qquad (4.83a)$$
$$\varphi_2 = (1/\sqrt{2})(d_2 - d_{-2})\beta \qquad (4.83b)$$
$$\varphi_3 = d_1\alpha \qquad (4.83c)$$
$$\varphi_4 = d_1\beta \qquad (4.83d)$$
$$\varphi_5 = d_{-1}\alpha \qquad (4.83e)$$
$$\varphi_g = d_{-1}\beta \qquad (4.83g)$$

where the subscripts on d are the magnetic quantum numbers, and α denotes spin $+\tfrac{1}{2}$ and β denotes spin $-\tfrac{1}{2}$.

The secular determinantal equation which results from the action of $\hat{L}\cdot\hat{S}$ on the set of functions $\varphi_1, \ldots, \varphi_6$ is:

$$\begin{vmatrix} |\varphi_1\rangle & |\varphi_2\rangle & |\varphi_3\rangle & |\varphi_4\rangle & |\varphi_5\rangle & |\varphi_6\rangle \\ -E & 0 & 0 & 0 & 0 & -\lambda/\sqrt{2} \\ 0 & -E & \lambda/\sqrt{2} & 0 & 0 & 0 \\ 0 & \lambda/\sqrt{2} & \lambda/2 - E & 0 & 0 & 0 \\ 0 & 0 & 0 & \lambda/2 - E & 0 & 0 \\ 0 & 0 & 0 & 0 & \lambda/2 - E & 0 \\ -\lambda/\sqrt{2} & 0 & 0 & 0 & 0 & \lambda/2 - E \end{vmatrix} = 0. \qquad (4.84)$$

The solutions reveal that there are four roots with $E = -\lambda/2$ and two roots with $E = +\lambda$, and we deduce that Γ_8 corresponds to the four-fold degenerate level with Γ_7 corresponding to the two-fold degenerate level.

The wave-functions corrected for spin-orbit coupling may now be obtained. We will designate these new functions as $\chi_1 \ldots, \chi_6$. The effect of spin-orbit coupling does not mix φ_4 or φ_5 with the others, and for

$$E = -\lambda/2: \quad \chi_1 = \varphi_4; \quad \chi_2 = \varphi_5. \tag{4.85a}$$

The secular equations are used to generate the remaining wave-functions:

$$E = -\lambda/2: \quad \chi_3 = (\sqrt{2}\varphi_1 + \varphi_6)/\sqrt{3}; \quad \chi_4 = (\sqrt{2}\varphi_2 - \varphi_3)/\sqrt{3} \tag{4.85b}$$

$$E = +\lambda: \quad \chi_5 = (\varphi_1 - \sqrt{2}\,\varphi_6)/\sqrt{3}; \quad \chi_6 = (\varphi_2 + \sqrt{2}\,\varphi_3)/\sqrt{3}. \tag{4.85c}$$

The new corrected wave-functions will be used in the next section to calculate the Zeeman effect, that is the effect of an applied magnetic field.

4.5.3 The Zeeman effect

The operator which represents the interaction with the magnetic field is

$$\mathcal{H} = \mu_B \sum_{i=1}^{N} H(\hat{l}_i + 2\hat{s}_i) \tag{4.86}$$

and integrals of the following sort arise:

$$\langle \chi_i | \hat{l}_q + 2\hat{s}_q | \chi_j \rangle \mu_B H_q \quad (q = x, y, z). \tag{4.87}$$

Since the directions x, y, z are equivalent in O_h, it is necessary to calculate only one directional property. The wave-functions that have been used here are quantized with respect to z and are eigenfunctions of $\hat{l}_z$ and $\hat{s}_z$. Therefore, the following integrals will be evaluated:

$$\langle \chi_i | \hat{l}_z + 2\hat{s}_z | \chi_j \rangle \mu_B H_z \tag{4.88}$$

where $\chi_1, \ldots, \chi_6$ are listed in the previous section.

The secular determinantal equation which results is

$$\begin{vmatrix} |\chi_1\rangle & |\chi_2\rangle & |\chi_3\rangle & |\chi_4\rangle & |\chi_5\rangle & |\chi_6\rangle \\ -\lambda/2-E & 0 & 0 & 0 & 0 & 0 \\ 0 & -\lambda/2-E & 0 & 0 & 0 & 0 \\ 0 & 0 & -\lambda/2-E & 0 & \sqrt{2}\mu_B H_z & 0 \\ 0 & 0 & 0 & -\lambda/2-E & 0 & \sqrt{2}\mu_B H_z \\ 0 & 0 & \sqrt{2}\mu_B H_z & 0 & \lambda-\mu_B H_z-E & 0 \\ 0 & 0 & 0 & \sqrt{2}\mu_B H_z & 0 & \lambda+\mu_B H_z-E \end{vmatrix} = 0.$$

It is appropriate to use perturbation theory to approximate the energies since $\lambda \gg \mu_B H$. The first-order Zeeman energy is the first-order perturbation energy, and it may be seen that there is no first-order Zeeman effect in the four states at $-\lambda/2$. However, there is a first-order Zeeman effect splitting the two states at $+\lambda$, with one state being destabilized by $\mu_B H$ and the other state being stabilized by $\mu_B H$.

The second-order Zeeman energy is the second-order perturbation energy, i.e.

$$E_i'' = \sum_j' <\Psi_i|(\hat{l}_z+2\hat{s}_z)\mu_B H_z|\Psi_j> <\Psi_j|(\hat{l}+2\hat{s}_z)\mu_B H_z|\Psi_i>/(E_i^0-E_j^0) \quad (4.90)$$

where the prime on the summation means that the term with $i=j$ is omitted. The second-order Zeeman energies are

$$E''(\chi_1)=0; \; E''(\chi_3)=-4\mu_B^2 H_z^2/3\lambda; \; E''(\chi_5)=+4\mu_B^2 H_z^2/3\lambda;$$
$$E''(\chi_2)=0; \; E''(\chi_4)=-4\mu_B^2 H_z^2/3\lambda; \; E''(\chi_6)=+4\mu_B^2 H_z^2/3\lambda. \quad (4.91)$$

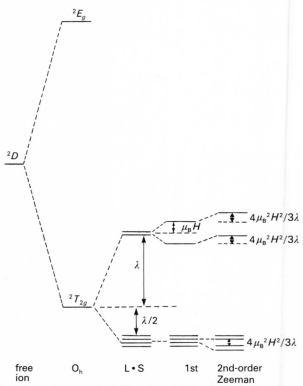

Fig. 4.4. The effect of an octahedral crystal field, spin-orbit coupling, and first- and second-order Zeeman interactions on the ^2D state.

An energy level diagram which shows the effects of the octahedral crystal field, spin-orbit coupling, and an external magnetic field on the ^2D free-ion term is shown in Fig. 4.4. The following expression for magnetic susceptibility may be generated by substituting the first- and second-order Zeeman coefficients and the zero-field energies into the Van Vleck equation:

$$\chi_M = (N\mu_B^2/3kT)\{8+(3\lambda/kT-8)\exp(-3\lambda/2kT)\}/(\lambda/kT)\{2+\exp(-3\lambda/2kT)\}. \quad (4.92)$$

Since, by definition,

$$\mu_{\text{eff}}^2 = (3kT/N\mu_B^2)\chi_M \quad (4.93)$$

then

$$\mu_{\text{eff}} = [\{8 + (3\lambda/kT - 8)\exp(-3\lambda/2kT)\}/(\lambda/kT)\{2 + \exp(-3\lambda/2kT)\}]^{\frac{1}{2}}. \quad (4.94)$$

A plot of this expression for μ_{eff} as a function of temperature is given in Fig. 4.5 where it may be seen that μ_{eff} tends towards zero as T approaches zero. We can understand this in terms of a typical wave-function for Γ_8, one being $\chi_1 = |d_1\beta\rangle$. The moments arising from the orbital and spin motions cancel.

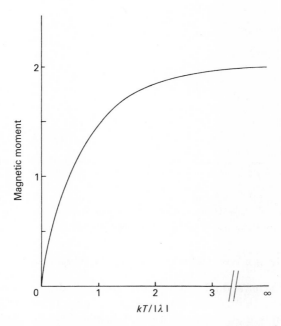

Fig. 4.5. The temperature-dependence of the magnetic moment of a hypothetical $TiL_6^{q\pm}$ complex.

4.5.4 Estimation of the second-order Zeeman effect in titanium (III)

The spin-orbit coupling constant is 155 cm^{-1}, and at 0 K the expression for the magnetic susceptibility of the d^1 O_h case reduces to

$$\chi_M = 4N\mu_B^2/3\lambda. \quad (4.95)$$

It is of interest to examine the plot of χ_M^{-1} as a function of temperature shown in Fig. 4.6. Above about 200 K the data obey a Curie–Weiss law with a large θ value, but below about 200 K the susceptibility deviates from the Curie–Weiss law and becomes constant. This behaviour may be understood in terms of the calculation. At low temperatures only Γ_8 is populated, and Γ_8 exhibits only a second-order Zeeman effect. This example points out an important result, large θ values in the Curie–Weiss law are not always associated with long- or short-range cooperative magnetic effects.

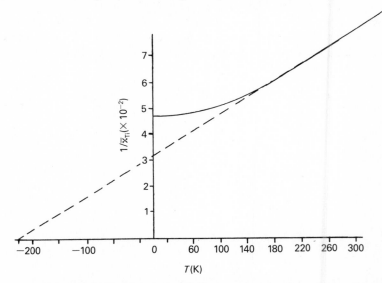

Fig. 4.6. A plot of the inverse susceptibility versus temperature for a hypothetical $TiL_6^{q\pm}$ which shows a large deviation from the Curie–Weiss law.

4.5.5 Example of a low-symmetry crystal field component

The effect of a tetragonal distortion of an octahedron will be investigated here. Consider an elongation or contraction of an octahedron along a C_4 axis such that there are two equivalent metal–ligand bond distances along the z-axis with length b, and four equivalent metal–ligand bond distances perpendicular to the z-axis with length a.

Since D_4 is a subgroup of the point group O, we can determine, from symmetry arguments, the effect of the lower symmetry on the ^2E_g and $^2T_{2g}$ states as follows:

$$^2E_g \rightarrow {}^2A_{1g} + {}^2B_{1g}$$
$$^2T_{2g} \rightarrow {}^2B_{2g} + {}^2E_g.$$

Now the energies of the states must be calculated. For example, for $^2A_{1g}$ we have to evaluate

$$<d_0|V(D_{4h})|d_0>. \tag{4.96}$$

$V(D_{4h})$ has additional terms as a result of the lower symmetry, and it is necessary to define new constants in addition to Dq. Two popular constants are

$$Ds = C_2^0 \int_0^\infty R_{n,2}^2(r) r^2 r^2 dr \qquad (4.97a)$$

$$Dt = C_2^0 \int_0^\infty R_{n,2}^2(r) r^4 r^2 dr. \qquad (4.97b)$$

Other constants appear in the literature, but Ds and Dt predominate. The energies are

$$A_{1g} = 6Dq - 2Ds - 6Dt \qquad (4.98a)$$

$$B_{1g} = 6Dq + 2Ds - Dt \qquad (4.98b)$$

$$B_{2g} = -4Dq + 2Ds - Dt \qquad (4.98c)$$

$$E_g = -4Dq - Ds + 4Dt \qquad (4.98d)$$

Since the low-symmetry component of the crystal field and spin-orbit coupling are of comparable magnitude, these perturbations to the octahedral results must be applied simultaneously, i.e. $H' = \lambda \hat{l} \cdot \hat{s} + V'(D_{4h})$. We will restrict our attention to the components which arise from the $^2T_{2g}$ octahedral state. Using symmetry arguments and the double group D_4', it may be seen that the six-fold degenerate $^2T_{2g}$ state splits into three doubly degenerate levels.

The secular determinantal equation which results from the action of H' on the set of functions $\varphi_1, \ldots, \varphi_6$ is

$$\begin{vmatrix} |\varphi_1\rangle & |\varphi_2\rangle & |\varphi_3\rangle & |\varphi_4\rangle & |\varphi_5\rangle & |\varphi_6\rangle \\ -E & 0 & 0 & 0 & 0 & -\lambda/\sqrt{2} \\ 0 & -E & \lambda/\sqrt{2} & 0 & 0 & 0 \\ 0 & \lambda/\sqrt{2} & \lambda/2 + \Delta - E & 0 & 0 & 0 \\ 0 & 0 & 0 & -\lambda/2 + \Delta - E & 0 & 0 \\ 0 & 0 & 0 & 0 & -\lambda/2 + \Delta - E & 0 \\ -\lambda/\sqrt{2} & 0 & 0 & 0 & 0 & \lambda/2 + \Delta - E \end{vmatrix} = 0$$

(4.99)

where $\Delta = E(E_g) - E(B_{2g})$, which is $5Dt - 3Ds$. As defined, a positive tetragonal distortion Δ puts the B_{2g} level lowest in energy. Solution of the Hamiltonian matrix yields the following values for the energy states:

$$E_1 = \Delta - \lambda/2 \qquad (4.100a)$$

$$E_2 = \Delta - \lambda/2 \qquad (4.100b)$$

$$E_{3,4} = \tfrac{1}{2}\{(\lambda/2 + \Delta) \pm (\Delta^2 + \lambda\Delta + 9\lambda^2/4)^{\frac{1}{2}}\} \qquad (4.100c)$$

$$E_{5,6} = \tfrac{1}{2}\{(\lambda/2 + \Delta) \pm (\Delta^2 + \lambda\Delta + 9\lambda^2/4)^{\frac{1}{2}}\}. \qquad (4.100d)$$

Corrections to the energies and corrected wave-functions may be generated by substitution of the energies into the appropriate secular equations. For the details and solutions, the paper by Figgis[12] should be consulted. It is sufficient to note that the details of the low-lying energy levels, which determine the magnetic properties, are dependent on the spin-orbit coupling constant and the crystal field component.

Anisotropic magnetic properties are anticipated in the axial system under consideration, where the Zeeman effect in the direction parallel to the C_4 axis may be calculated with $(\hat{l}_z + 2\hat{s}_z)\mu_B H_z$ and, in the perpendicular direction, by use of $(\hat{l}_x + 2\hat{s}_x)\mu_B H_x$.

The steps to be followed may be summarized as follows:

(1) Secular determinants must be constructed for each of the three doubly degenerate levels in both directions.
(2) First-order Zeeman coefficients must be evaluated for each direction.
(3) Matrix elements connecting the three secular determinants must be evaluated to yield second-order Zeeman coefficients.
(4) The first- and second-order Zeeman coefficients must be substituted into the Van Vleck equation to yield the anisotropic magnetic susceptibilities $\chi_\perp$ and $\chi_\parallel$. Generally anisotropic magnetic properties are discussed in terms of $\mu_\parallel$ and $\mu_\perp$ since the variation of the anisotropic components are much more strongly dependent on temperature than that of $\bar{\mu}$.

4.6 Exchange coupling

Up to now we have considered the magnetic properties of assemblies of single ions, where the properties of a given single ion in an assembly is independent of the presence of other ions. However, the possibility of interactions between ions was admitted in the discussion of the Curie–Weiss Law. Many interesting problems arise when interactions between paramagnetic ions are considered, and such interactions give rise to materials with properties that are theoretically and technologically important.

It is convenient to define two types of dimensionality: namely spin (or magnetic) dimensionality, and structural dimensionality. There are three classes of spin dimensionality: the highly anisotropic one-dimensional Ising case, the two-dimensional XY-case, and the fully isotropic Heisenberg case. In general, these classes arise from structural features of materials which govern interactions between paramagnetic ions, or from the electronic structure of the single ions.

In the next section we will show that the interaction between paramagnetic ions corresponds roughly to the exchange of electrons in the orbitals of the ions, and so they are called *exchange interactions*. If the paramagnetic ions are linked together by diamagnetic ligand bridges, and if the orbitals and electrons of the ligand bridge are involved in the exchange interaction, then the process is called a *superexchange interaction*.

There are four classes of structural dimensionality. The simplest class is that in which the structure is made up of isolated clusters containing two or more exchange-coupled paramagnetic ions. Such systems are defined to be zero-dimensional, 0-D. The exchange interactions between the paramagnetic ions in the cluster may have spin dimensionalities of one (Ising case), two (XY-case), or three (Heisenberg). If the number of paramagnetic ions in the cluster is small, exact expressions for magnetic properties can be derived for any of the three spin dimensionalities. As we shall see later, small clusters of exchange-coupled paramagnetic ions are frequently used as models for more complicated extended systems.

The second structural class is that of a linear chain of exchange-coupled ions. In the ideal example of a one-dimensional system, 1-D, each paramagnetic ion has two nearest neighbours linked by suitable ligand bridges, and the next-nearest neighbours are much farther away and are not directly bridged to the chain. As above, the exchange interactions may have spin dimensionalities of one, two, or three. Important phenomena may appear. In real systems there will be finite interactions between chains, and in careful studies on selected systems, an increase in structural dimensionality can be detected. The one-dimensional exchange-coupled chain is inherently unstable towards distortion to an alternatingly spaced chain, where additional electronic energy is gained by strong exchange coupling between pairs of paramagnetic ions. This is called the *spin-Peierls distortion*. The spin-Peierls distortion is rarely seen in one-dimensional chains of transition-metal ions linked by ligand bridges since increases in structural dimensionality usually occur at temperatures higher than that predicted for the spin-Peierls distortion.

The last structural class of low-dimensional magnetic systems is an isolated layer or sheet containing exchange-coupled paramagnetic ions. The role that spin dimensionality plays in determining the nature of long-range cooperative behaviour in two-dimensional arrays (2-D) will be discussed in a subsequent section.

In three-dimensional magnetic systems (3-D), exchange interactions may occur in all directions. These may range from simple cubic systems to very complex arrays which reflect structural details of superexchange pathways and the electronic structures of the paramagnetic ions.

4.6.1 Exchange, an orbital effect

Consider a two-electron system

in which there are two nuclei **a** and **b** each with charge Z^+. The Schroedinger equation for this system may be written as

$$(\mathcal{H}^0 + \mathcal{H}')\Psi = E\Psi \tag{4.101}$$

where

$$\mathcal{H}^0 = (-h^2/8\pi^2 m)[\nabla_1^2 + \nabla_2^2] - Ze^2/r_a(1) - Ze^2/r_b(2) \tag{4.102a}$$

$$\mathcal{H}' = Z^2 e^2/r_{a,b} + e^2/r_{1,2} - Ze^2/r_b(1) - Ze^2/r_a(2). \tag{4.102b}$$

For the purpose of this discussion, the nuclear repulsion term $Z^2 e^2/r_{a,b}$ and the electron repulsion term $e^2/r_{1,2}$ will be ignored. $\mathcal{H}'$ now contains only Coulomb attraction terms, namely the attraction of nucleus **a** for electron **2** and the attraction of nucleus **b** for electron **1**.

In view of the indistinguishability principle, the wave-functions in the zero-overlap approximation are

$$\Psi_\pm = (1/\sqrt{2})\{\varphi_a(1)\varphi_b(2) \pm \varphi_a(2)\varphi_b(1)\}. \tag{4.103}$$

The φ_j are eigenfunctions of the Hamiltonian

$$\mathcal{H} = -(h^2/8\pi^2 m)\nabla^2 - Ze^2/r \tag{4.104}$$

with the eigenvalues E^0. If $Z = 1$, then $E^0 = E_H$. We may write

$$E_+ = E^0 + \tfrac{1}{2}\langle \Psi_+ | -Ze^2/r_b(1) - Ze^2/r_a(2)|\Psi_+\rangle. \tag{4.105}$$

The electrostatic or Coulomb integrals are

$$K = \langle \varphi_a(1)\varphi_b(2)| -Ze^2/r_b(1) - Ze^2/r_a(2)|\varphi_a(1)\varphi_b(2)\rangle \tag{4.106a}$$

$$= \langle \varphi_a(2)\varphi_b(1)| -Ze^2/r_b(1) - Ze^2/r_a(2)|\varphi_a(2)\varphi_b(1)\rangle \tag{4.106b}$$

and the exchange integrals are

$$J = \langle \varphi_a(1)\varphi_b(2)| -Ze^2/r_b(1) - Ze^2/r_a(2)|\varphi_a(2)\varphi_b(1)\rangle \tag{4.107a}$$

$$= \langle \varphi_a(2)\varphi_b(1)| -Ze^2/r_b(1) - Ze^2/r_a(2)|\varphi_a(1)\varphi_b(2)\rangle \tag{4.107b}$$

Upon collection of terms we find that

$$E_+ = E^0 + \tfrac{1}{2}(2K + 2J) \tag{4.108a}$$

$$E_- = E^0 + (K - J). \tag{4.108b}$$

The energy separation $\Delta E = E_+ - E_- = 2J$ within the zero-overlap approximation and the neglect of electron–electron and nuclear–nuclear repulsion terms.

It is important to recognize that this interaction is independent of electron spin, and that it is electrostatic in origin. Spin dependence arises as a result of the Pauli exclusion principle when spin properties are added to the wave functions $\Psi_\pm$. Let the electron spin be designated by

$$\alpha \text{ for } m_s = +\tfrac{1}{2} \tag{4.109a}$$

$$\beta \text{ for } m_s = -\tfrac{1}{2} \tag{4.109b}$$

and write possible permutations of spin coordinates. The symmetrical spin wave-functions are

$$a(1)a(2); \qquad \sum m_s = +1 \qquad (4.110a)$$

$$\beta(1)\beta(2); \qquad \sum m_s = -1 \qquad (4.110b)$$

$$\{a(1)\beta(2) + \beta(1)a(2)\}/\sqrt{2}; \qquad \sum m_s = 0 \qquad (4.110c)$$

and the antisymmetric wave function is

$$\{a(1)\beta(2) - a(2)\beta(1)\}/\sqrt{2}; \qquad \sum m_s = 0. \qquad (4.111)$$

According to the Pauli principle, the symmetric spin-functions must combine with Ψ_-, and the antisymmetric spin-functions with Ψ_+. When these combinations are made, we find that Ψ_+ corresponds to a singlet spin state

$$\Psi_S = \tfrac{1}{2}\{\varphi_a(1)\varphi_b(2) + \varphi_a(2)\varphi_b(1)\}\{a(1)\beta(2) - \beta(1)a(2)\} \qquad (4.112)$$

and that Ψ_- corresponds to a triplet spin state

$$\Psi_{T1} = 1/\sqrt{2}\{\varphi_a(1)\varphi_b(2) - \varphi_a(2)\varphi_b(1)\}\{a(1)a(2)\} \qquad (4.113a)$$

$$\Psi_{T2} = \tfrac{1}{2}\{\varphi_a(1)\varphi_b(2) - \varphi_a(2)\varphi_b(1)\}\{a(1)\beta(2) + \beta(1)a(2)\} \qquad (4.113b)$$

$$\Psi_{T3} = 1/\sqrt{2}\{\varphi_a(1)\varphi_b(2) - \varphi_a(2)\varphi_b(1)\}\{\beta(1)\beta(2)\}. \qquad (4.113c)$$

4.6.2 Vector model for exchange

It is convenient to develop a vector model for exchange in terms of spin. In the vector model, we take $\hat{S}_1$ and $\hat{S}_2$ to be spin angular momentum operators in units of $h/2\pi$ with the vector sum $S_{1,2} = S_1 + S_2$, and

$$(S_{1,2})^2 = (S_1 + S_2)^2 = (S_1)^2 + (S_2)^2 + 2S_1 \cdot S_2. \qquad (4.114)$$

The eigenvalues of the operators $(\hat{S}_{1,2})^2$, $(\hat{S}_1)^2$, and $(\hat{S}_2)^2$ are $S_{1,2}(S_{1,2}+1)$, $S_1(S_1+1)$, and $S_2(S_2+1)$, respectively, and we may write

$$\hat{S}_1 \cdot \hat{S}_2 = \tfrac{1}{2}\{S_{1,2}(S_{1,2}+1) - S_1(S_1+1) - S_2(S_2+1)\}. \qquad (4.115)$$

We deduce the exchange Hamiltonian to be

$$\mathcal{H} = -2J_{1,2}\hat{S}_1 \cdot \hat{S}_2 \qquad (4.116)$$

since we had previously found that $E_{S,T} = 2J$.

In the multielectron case where atom **a** has total spin S_a and atom **b** has total spin S_b, the exchange Hamiltonian is written as

$$\mathcal{H} = -2J_{a,b}\hat{S}_a \cdot \hat{S}_b. \qquad (4.117)$$

In a cluster of exchange-coupled ions the Hamiltonian is generalized to

$$\mathcal{H} = -2\sum_{n<m} J_{n,m}\hat{S}_n \cdot \hat{S}_m. \qquad (4.118)$$

Van Vleck developed a convenient formula for the energies of the spin states which result from exchange coupling in a cluster with n magnetically equivalent magnetic ions. The expression is

$$E(S') = -J\{S'(S'+1) - nS_i(S_i+1)\} \tag{4.119}$$

where S' can take the values allowed by the vector summation rule. For $S = \tfrac{1}{2}$ and $n=2$, $S'=0, 1$; for $n=3$, $S'=\tfrac{1}{2}, \tfrac{3}{2}$; for $n=4$, $S'=0, 1, 2$; etc.

It is clear that more than one spin state with a given S' may arise. Van Vleck showed that the number of states W with a given S' is given by

$$W(S') = \Omega(S') - \Omega(S'+1) \tag{4.120a}$$

where $\Omega(S')$ is the coefficient of $x^{S'}$ in the expansion of

$$(x^S + x^{S-1} + \ldots + x^{-S})^n \tag{4.120b}$$

As an example, consider an assembly of three paramagnetic ions with $S_i = \tfrac{1}{2}$. The expression to be expanded is $(x^{\frac{1}{2}} + x^{-\frac{1}{2}})^3$ which gives $x^{3/2} + 3x^{\frac{1}{2}} + \ldots$ Thus, there is one state with $S' = +\tfrac{3}{2}$ and two states with $S' = \tfrac{1}{2}$. If all interacting ions are magnetically equivalent, then all states of a given S' are degenerate.

For the specific case of $S = \tfrac{1}{2}$, Van Vleck's cluster equation and degeneracy treatment yields the data given in Table 4.7 for clusters with two, three, and four exchange-coupled ions. The cluster formula may also be used for ions with $S_i > \tfrac{1}{2}$. Results for exchange-coupled pairs of ions with S_i of 1 and $\tfrac{3}{2}$ are also given in Table 4.7. Magnetic susceptibility expressions may be readily obtained by substitution of the zero-field energies and first-order Zeeman coefficients into the Van Vleck equation.

Table 4.7 *Zero-field energy levels and degeneracies for representative exchange-coupled clusters*

No. of spins	S_i	S'	$E(S')$	Degeneracy
2	$\tfrac{1}{2}$	0	$+3J/2$	1
		1	$-J/2$	1
3	$\tfrac{1}{2}$	$\tfrac{1}{2}$	$+3J/2$	2
		$\tfrac{3}{2}$	$-3J/2$	1
4	$\tfrac{1}{2}$	0	$+3J$	2
		1	$+J$	3
		2	$-3J$	1
2	1	0	$+4J$	1
		1	$+2J$	1
		2	$-2J$	1
2	$\tfrac{3}{2}$	0	$+15J/2$	1
		1	$+11J/2$	1
		2	$+3J/2$	1
		3	$-9J/2$	1

4.6.3 Anisotropic and antisymmetric exchange

Frequently, the exchange Hamiltonian

$$\mathcal{H} = -2\sum_{n<m} J_{n,m} \hat{S}_n \cdot \hat{S}_m \qquad (4.121)$$

cannot adequately explain magnetic properties and it is necessary to take additional kinds of interactions into consideration. The Hamiltonian as written above implies that exchange interactions are isotropic and symmetric, but we have already seen that exchange interactions may be anisotropic. If we rewrite the Hamiltonian as

$$\mathcal{H} = -2\sum_{n<m} J_{n,m} \{\alpha \hat{S}_n^z \hat{S}_m^z + \beta \hat{S}_n^x \hat{S}_m^x + \delta \hat{S}_n^y \hat{S}_m^y\} \qquad (4.122)$$

anisotropic exchange interactions can be treated.

Consider the Ising case, i.e. $\alpha = 1$ and $\beta = \delta = 0$. For our example we will use the pair of exchange-coupled $S = \tfrac{1}{2}$ ions for which spin wave-functions were given above. The exchange coupling operator is

$$\mathcal{H} = -2J_{1,2}{}^z S_1^z S_2^z. \qquad (4.123)$$

Applying the operator to the basis set yields the determinantal equation

$$\begin{vmatrix} |aa\rangle & |a\beta + \beta a\rangle/\sqrt{2} & |\beta\beta\rangle & |a\beta - \beta a\rangle/\sqrt{2} \\ -\tfrac{1}{2}J_{1,2}{}^z - E & 0 & 0 & 0 \\ 0 & \tfrac{1}{2}J_{1,2}{}^z - E & 0 & 0 \\ 0 & 0 & -\tfrac{1}{2}J_{1,2}{}^z - E & 0 \\ 0 & 0 & 0 & \tfrac{1}{2}J_{1,2}{}^z - E \end{vmatrix} = 0.$$

It may be seen that there are two roots at $E = +\tfrac{1}{2}J_{1,2}{}^z$ and two at $E = -\tfrac{1}{2}J_{1,2}{}^z$, with an energy separation of $|J_{1,2}{}^z|$.

Frequently anisotropic exchange is taken into account by adding the term $\hat{S}_n \Gamma_{n,m} \hat{S}_m$ to the Hamiltonian. In this term, $\Gamma_{n,m}$ is a symmetric tensor, and its principal value is approximately $(\Delta g/g)^2 J_{n,m}$ where $\Delta g = |g-2|$. The effect of anisotropic exchange is to cause a zero-field splitting of the triplet state. In axial cases, a singlet state and a doublet state result, while all degeneracy is lifted in rhombic cases.

Antisymmetric exchange interactions may arise between pairs of exchange-coupled ions when the single ion g values are different, a condition which implies that the exchange-coupled ions are not related by a centre of inversion. The term that is added to the Hamiltonian to account for antisymmetric exchange is $D_{n,m}[\hat{S}_n \times \hat{S}_m]$, being theoretically established by Moriya following a phenomenological development by Dzialoshinski. Essentially, this interaction tends to align the spins perpendicular to each other, and in this manner tends to oppose the isotropic exchange interaction which, depending on the sign of $J_{n,m}$, tends to align the spins either parallel, or antiparallel to one another.

The magnitude of the antisymmetric coupling constant may be estimated from the expression $|D_{n,m}| = (\Delta g/g) J_{n,m}$. Consider a case with $\Delta g = 0.1$ and

$J_{n,m} = 100$ cm^{-1}; we estimate $|D_{n,m}|$ to be about 5 cm^{-1}. This effect can be seen readily in EPR measurements, and in magnetic studies especially if $J_{n,m}$ is positive.

4.6.4 Dirac's permutation operator

Dirac has shown that the exchange Hamiltonian

$$\mathcal{H} = -2\sum_{i<j} J_{i,j} \hat{S}_i \cdot \hat{S}_j \tag{4.125}$$

can be represented by

$$\mathcal{H} = -\tfrac{1}{2}\sum_{i<j} J_{i,j}\{2P_{i,j} - 1\} \tag{4.126}$$

where $P_{i,j}$ is a permutation operator which exchanges the spin coordinates of electrons i and j.

As an example we will examine the two-electron case, for which we may write

$$\mathcal{H}_{1,2} = (-J_{1,2}/2)\{2P_{1,2} - 1\}. \tag{4.127}$$

Using the spin wave-functions for the exchange-coupled two-electron case, it is necessary to calculate the matrix elements of the secular determinantal equation. Since the operator only permutes spin coordinates, it is clear that there will be no off-diagonal matrix elements between states with different M_S. Furthermore, since the Hamiltonian must be invariant under all operations of the point group to which the exchange-coupled system belongs, there will be no off-diagonal matrix elements between states of different symmetry. With these results in mind, we see that it is necessary to calculate only two matrix elements to define the energies of this simple exchange coupled system.

4.6.5 Exchange in linear chains

The Hamiltonian for exchange in linear chains may be written

$$\mathcal{H} = \mathcal{H}_{ex} + \mathcal{H}_z + \mathcal{H}_a \tag{4.128}$$

where the first term is for nearest-neighbour exchange-coupling

$$\mathcal{H}_{ex} = -2J_{i,j} \sum_{i<j} (\alpha \hat{S}_i^z \hat{S}_j^z + \beta \hat{S}_i^x \hat{S}_j^x + \gamma \hat{S}_i^y \hat{S}_j^y) \quad . \tag{4.129}$$

The second term is the Zeeman effect, and the third term is the single-ion anisotropy, which can be written

$$\mathcal{H}_a = D(\hat{S}_z^2 - \tfrac{1}{3}S(S+1)) + E(\hat{S}_x^2 - \hat{S}_y^2). \tag{4.130}$$

As defined above, the exchange-coupling constant J can be either positive or negative. A negative J value corresponds to anti-parallel coupled spins, and in such cases the interaction is designated antiferromagnetic. Alternately, a positive coupling constant indicates ferromagnetic interactions where the spins are coupled parallel to each other.

The isotropic Heisenberg Hamiltonian

The isotropic ($\alpha = \beta = \gamma$) Heisenberg Hamiltonian has been extensively examined with $\mathcal{H}_z$ and H_a taken to be zero. No exact solutions are presently known, but results of many approximate methods exist. The paper by Bonner and Fisher[13] on short chains and rings of ($S = \frac{1}{2}$) ions is a key work in this area. For antiferromagnetic interactions, the maximum magnetic susceptibility, and the temperature at which it occurs, are uniquely defined by

$$kT_{max}/|J| = 1.282 \tag{4.131}$$

and

$$\chi_{max}|J|/Ng^2\mu_B = 0.07346. \tag{4.132}$$

The reduced-coordinate magnetic susceptibility versus temperature curve given by Bonner[13] may be fitted by the function

$$f(y) = (Ay^2 + By + C)/(y^3 + Dy^2 + Ey + F) \tag{4.133}$$

with the best-fit parameters being $A = 0.25$, $B = 0.14995$, $C = 0.30094$, $D = 1.9862$, $E = 0.68854$, and $F = 6.0626$. Consequently, the expression for the magnetic susceptibility of ($S = \frac{1}{2}$) ions isotropically coupled in antiferromagnetic linear chains becomes

$$\chi_M = (Ng^2\mu_B^2/kT)\{(A + Bx + Cx^2)/(1 + Dx + Ex^2 + Fx^3)\} \tag{4.134}$$

where $x = |J|/kT$. For ferromagnetic systems with $S = \frac{1}{2}$, the Padé approximation technique has been used to derive the following expression for the reduced magnetic susceptibility:

$$\chi_M(K) = \{(1.0 + 5.7979916K + 16.902653K^2 + 29.376885K^3 + 29.832959K^4$$
$$+ 14.036918K^5)/(1.0 + 2.7979916K + 7.0086780K^2 + 8.6538644K^3$$
$$+ 4.5743114K^4)\}^{\frac{2}{3}} \tag{4.135}$$

where $K = J/2kT$. The ferromagnetic case has also been described by a high-temperature series-expansion.[14]

Other results exist for the general case of spin S greater than $\frac{1}{2}$ for both ferromagnetic and antiferromagnetic exchange, including a solution for the classical infinite spin case for the infinite chain. This solution is applicable to both antiferromagnetic and ferromagnetic exchange coupling. The infinite spin results may be scaled to the leading terms of a high-temperature series-expansion to yield

$$\chi_M = \{Ng^2\mu_B^2 S(S+1)/3kT\}\{(1+\mu)/(1-\mu)\} \quad (4.136a)$$

where

$$\mu = \coth\{2JS(S+1)/kT - 1/2JS(S+1)\}. \quad (4.136b)$$

For the antiferromagnetic case the susceptibility curves will exhibit a maximum, approaching $Ng^2\mu_B^2/12|J|$ as the zero-temperature susceptibility.

The following high-temperature series-expansions provide solutions useful for ferromagnetic and antiferromagnetic cases, respectively, as designated:

$$\chi_F = (Ng^2\mu_B^2/3kT)\sum_{n=0}^{\infty} a_n(J/kT)^n \quad (4.137a)$$

$$\chi_{AF} = (Ng^2\mu_B^2/3kT)\sum_{n=0}^{\infty} (-1)^n a_n(J/kT)^n. \quad (4.137b)$$

The coefficients a_n are dependent on the spin value of the system and are tabulated up to a_6 for eight lattice types including the linear chain.[14] When using the results of this theory, the condition that $kT/|J|$ is greater than 1.5 should be observed.

An interpolation scheme has been developed for the analysis of exchange in linear chains of spin $S = 1, \frac{3}{2}, 2, \frac{5}{2}$, and 3.[15] The function

$$\chi_M = (Ng^2\mu_B^2/kT)\{A + Bx^2\}\{1 + Cx + Dx^3\}^{-1} \quad (4.138)$$

where $x = |J|/kT$, has been fitted to the numerical magnetic susceptibility results from the interpolation model, and the coefficients generated for the spin cases $S = \frac{1}{2}, 1, \frac{3}{2}, 2, \frac{5}{2}$ and 3 are available.

Anisotropic Heisenberg interaction

When $\beta = \gamma$, an exchange parameter $\eta = \beta = \gamma$ can be defined. In the limits of $\eta = 0$ and $\eta = 1$, the Ising and isotropic Heisenberg exchange Hamiltonians are obtained, respectively. Within these limits the following zero-field magnetic susceptibility expression has been derived:

$$\chi_M = (Ng^2\mu_B^2/kT)(X/3S^2)\{1 + \sum_{r=0}^{\infty} b_r(\eta,X)K^r\} \quad (4.139)$$

where $K = J/2S^2kT$ and $X = S(S+1)$. The coefficients $b_r(\eta,X)$ is a double polynomial in η and X of degree r. Expansions of the coefficients have been determined for $r = 1$ to $r = 6$.

The XY (or planar Heisenberg) interaction

The applications of the XY model Hamiltonian ($\alpha = 0$, $\beta = \gamma = 1$) have been very limited, although the specific heats and magnetic susceptibilities have been determined for several XY cases. With the field applied perpendicular to the chain direction, the susceptibility of a ($S=\frac{1}{2}$) system with XY anisotropy is given by

$$\chi = (Ng^2\mu_B^2/kT)[(1/\pi)\int dw/\{\cosh^2(2K\cos w)\}] \quad (4.140)$$

where $w=2\pi k/N$ and $K=J/2kT$. There are no results for the parallel susceptibility.

The Ising interaction

When $a=1$ and $\beta=\gamma=0$, the Ising model Hamiltonian is described. Closed-form solutions exist both for the parallel[16] and perpendicular[16,17] magnetic susceptibilities of ($S=\frac{1}{2}$) chains. These expressions are

$$\chi_\| = (Ng^2\mu_B^2/4kT)\exp(2|J|/kT) \quad (4.141a)$$

$$\chi_\perp = (Ng^2\mu_B^2/8|J|)\{\tanh(|J|/kT) + (|J|/kT)\text{sech}^2(|J|/kT)\}. \quad (4.141b)$$

Various exact solutions also exist for the parallel susceptibility of $S \geqslant 1$.[18–20] The general results for the Heisenberg, XY, and Ising models which relate χ, J, and T_{max} are tabulated in Table 4.8. Detailed information required for application of these results for the analyses of experimental data are available in the original works referenced here.

Table 4.8 *Relationships between exchange-coupling constants and the maximum in the magnetic susceptibility of antiferromagnetically coupled linear chains*

Exchange model	Spin	Model abbreviation	$(kT_{max}/|J|)$		$(\chi_{max}|J|/Ng^2\mu_B^2)$	
Heisenberg	$\frac{1}{2}$	H-BF[a]	1.282		0.07346	
Heisenberg	1	H-W[b]	2.70		0.088	
Heisenberg	$\frac{3}{2}$	H-W[b]	4.75		0.091	
Heisenberg	2	H-W[b]	7.1		0.094	
Heisenberg	$\frac{5}{2}$	H-W[b]	10.6		0.095	
Heisenberg	3	H-W[b]	13.1		0.096	
XY	$\frac{1}{2}$	XY[c]	0.64	$\chi_\perp$	0.174	$\chi_\perp$
Ising	$\frac{1}{2}$	I-F[d]	1.0	$\chi_\|$	0.09197	$\chi_\|$
			0.4168	$\chi_\perp$	0.2999	$\chi_\perp$
Ising	1		2.55[e]		0.098[e]	
Ising	$\frac{3}{2}$		4.70[e]		0.10[e]	
Ising	2		7.46[e]		0.101[e]	
Ising	$\frac{5}{2}$		10.8[e]		0.105[e]	
Ising	3		14.8[e]		0.102[e]	

[a] Bonner–Fisher results
[b] Weng results
[c] XY model
[d] Fisher's Ising result
[e] Parallel susceptibility only

Effects of single-ion anisotropy

Low-symmetry crystal fields can produce single-ion effects which are implied by the general Hamiltonian given above and which are non-negligible in linear chain compounds when the single-ion spin value is 1 or greater. In axial fields, the zero-field splitting parameter D describes the removal of spin degeneracy, while crystal fields of orthorhombic or lower symmetry require that the rhombic term E be non-zero; the E term consequently removes any remaining degeneracy. It must be pointed out that $\mathcal{H}_a$ is written in many forms in the literature, and care must be taken to note if the energy baricentre has been preserved.

Although the observed magnitudes, signs, and experimental anisotropies are spin- and metal-ion-dependent, some general features are common to all linear chains in which single-ion effects are important. Experimentally, the observed anisotropies in the principal magnetic susceptibilities, ($\Delta\chi = x_\perp - \chi_\parallel$), are positive for positive D. The reverse situation results when $D<0$ and $\Delta\chi$ becomes negative. For very small values of $D/|J|$, the theoretical analyses can be carried out by treating $\mathcal{H}_a$ as a perturbation to the exchange interaction.[21] Generally, a positive D value tends to force the exchange-coupled spins to lie in the plane perpendicular to the unique z-axis, while a negative D value tends to favour the spins lying along the z-direction. Larger values of either sign of $D/|J|$ present much more difficult theoretical problems, and they are dependent on whether the total spin of the ground state is even or odd. It is sufficient to state here that large zero-field splittings can induce anisotropies in the exchange interaction itself and make it difficult to choose an appropriate model.

4.6.6 Exchange in two-dimensional layers

The Hamiltonian appropriate for a two-dimensional magnetic system may be given by

$$\mathcal{H} = -2\sum_{i<j} J_{i,j}(a\hat{S}_{i,z}\hat{S}_{j,z} + \zeta\hat{S}_{i,x}\hat{S}_{j,x} + \eta\hat{S}_{i,y}\hat{S}_{j,y}) - 2J'\sum_{k<1}\hat{S}_k \cdot \hat{S}_1 \quad (4.142)$$

where the $S_{i,w}$ ($w = x, y, z$) are the components of the spin operator along the Cartesian coordinates, J and J' are the intra- and inter-layer exchange-coupling constants, and the parameters ζ and η represent the anisotropy in the intralayer exchange. If the z-axis is taken as the easy axis of magnetization, then the exchange is Ising-like when $a = 1$ and $\zeta = \eta = 0$, XY-like when $a = 0$ and $\zeta = \eta = 1$, and Heisenberg-like when $a = \zeta = \eta$. The anisotropy parameters can be represented by effective fields by the use of the following molecular field equations with $D = 1 - \zeta$, and $E = 1 - \eta$:

$$H_{ex} = 2zJ/g\mu_B \quad (4.143a)$$

$$H_a^{in} = DH_{ex} \quad (4.143b)$$

$$H_a^{out} = EH_{ex} \quad (4.143c)$$

where H_{ex} is the effective exchange field arising from the intralayer exchange, and z is the number of nearest neighbours within the layer. The contribution to the anisotropy arising from the anisotropic exchange interaction which favours parallel alignment within a layer can be estimated by the expressions

$$H_{A,E} = 2z\Delta JS/g\mu_B \qquad (4.144a)$$

$$\Delta J = J_{a,c} - J_b \sim (\Delta g/g)^2 J \qquad (4.144b)$$

where $\Delta g = g_{a,c} - g_b$ and $g = (2g_{a,c} + g_b)/3$. In some cases it has been reported that the experimentally observed H_a^{out} is substantially smaller than the sum of the theoretical components which are expected to be most important. This immediately suggests that there must be an additional contribution to the anisotropy arising from exchange interactions either within the layer or between layers. One such additional term which may be important is the antisymmetric (Dzialoshinki–Moriya) exchange which was discussed in Section 4.6.3.

4.6.7 Exchange in three-dimensional systems

There are no exact solutions of the exchange Hamiltonian for three-dimensional systems, but various approximations are available. By far the simplest, developed by Pierre Weiss in 1907, considers only one magnetic ion, the interaction of which with the rest of the lattice is replaced with an effective field. Furthermore, it is assumed that the effective field is proportional to the average net magnetic moment of the macroscopic crystal. With this assumption the effective field H_e is given by

$$H_e = \gamma M \qquad (4.145)$$

where M is the magnetic moment of the system and γ is the Weiss molecular field coefficient. This approximation is called molecular field or mean field theory.

The first step is to rewrite the exchange Hamiltonian as

$$\mathcal{H}_{ex} = -2J\hat{S}_i \sum_j \hat{S}_j \qquad (4.146)$$

where the angular momentum operator for one magnetic ion has been extracted from the summation, which runs over the nearest neighbours of that ith ion. The exchange interactions are now replaced by an effective magnetic field and

$$\mathcal{H}_{ex} = -g\mu_B \hat{S}_i H_e. \qquad (4.147)$$

When the two expressions for $\mathcal{H}_{ex}$ are equated, there results, for the effective field, the expression

$$H_e = (2J/g\mu_B) \sum_j \hat{S}_j. \qquad (4.148)$$

If it is now assumed that each $\hat{S}_j$ can be replaced by its average value S_j, that there are z identical neighbours, then the total magnetic moment of the system is given by

$$M = Ng\mu_B <S_j> \tag{4.149}$$

and the effective field is given by

$$H_e = (2zJ/g\mu_B)<S_j>. \tag{4.150}$$

Substitution for $<S_j>$ yields

$$H_e = (2zJ/Ng^2\mu_B^2)M \tag{4.151}$$

with the molecular field coefficient being

$$\gamma = 2zJ/Ng^2\mu_B^2. \tag{4.152}$$

The effect of an applied field is obtained by taking the sum of the effective field and the applied field

$$H_T = H_o + H_e \tag{4.153}$$

where H_o is the applied field. In the absence of anisotropy, M, H_o, and H_e can be taken to lie along z, and the single-ion Hamiltonian is

$$\mathcal{H} = -g\mu_B \hat{S}_z H_{T_z} \tag{4.154}$$

with the eigenvalues

$$E = -g\mu_B m_s H_{T_z} \text{ with } m_s = S, S-1, \ldots, -S. \tag{4.155}$$

The magnetic moment is given by

$$M = Ng\mu_B S B_s(x) \tag{4.156}$$

where x is now $g\mu_B H_{T_z} S/kT$. At high temperatures, where $x \ll 1$, then

$$M = Ng^2\mu_B^2 S(S+1)H_{T_z}/3kT. \tag{4.157}$$

By substitution of $2zJM/Ng^2\mu_B^2$ for H_e, C for $Ng^2\mu_B^2 S(S+1)/3k$, and setting $\theta = 2zJS(S+1)/3k$, the Curie–Weiss law is obtained.

$$\chi = C/(T-\theta) \tag{4.68}$$

As mentioned above, many approximate treatments for the magnetic susceptibility of three-dimensional systems exist, but the level of sophistication is beyond that of this work. However, as an example, we mention that the magnetic susceptibility of a simple cubic lattice has been developed in terms of the series expansion

$$\chi_o J = (1/3\theta)\{1 - \sum_i B_i/\theta^i\} \tag{4.158}$$

where $\theta = kT JS(S+1)$, χ_o is the reduced magnetic susceptibility $\chi/Ng^2\mu_B^2$, and the coefficients have been tabulated for $i=1$ to 10. Similar results exist for other lattice types.

4.7 References

1. Heyding, R. D., Taylor, J. D., and Hair, M. L. *Rev. sci. Instrum.* **32,** 161 (1960).
2. Lewis, R. T. *Rev. sci. Instrum.* **42,** 31, 1971.
3. Reeves, R. *J. Phys.* **E5,** 547 (1972).
4. Evans, D. F. *J. chem. Soc.,* 2003 (1959).
5. Ostfeld, D., Cohen, I. A. *J. Chem. Educ.* **49,** 829 (1972).
6. Mulay, L. N. *Magnetic susceptibility.* Interscience, New York (1963).
7. Figgis, B. N. and Nyholm, R. S. *J. chem. Soc.* 4190 (1958).
8. Bünzli, J.-C. G. *Inorg. chim. Acta* **36,** L413 (1979).
9. O'Connor, C. J., Cucauskas, E. J., Deaver (Jr.), B. S., and Sinn, E. *Inorg. chim. Acta* **32,** 29 (1979).
10. Cooke, A. H. *Prog. low temp. Phys.* **1,** 328 (1955).
11. Brown, D. B., Crawford, V. H., Hall, J. W., and Hatfield, W. E. *J. phys. Chem.* **81,** 1303 (1977).
12. Figgis, B. N. *Trans. Faraday Soc.* **57,** 198 (1961). See also; Figgis, B. N., Gerloch, M., Lewis, J., Mabbs, F. E., and Webb, G. A. *J. chem. Soc.* **(A),** 2086 (1968) and references therein.
13. Bonner, J. C. and Fisher, M. E. *Phys. Rev.* **A135,** 640 (1964). See also Bonner, J. C. PhD dissertation. University of London (1968).
14. Rushbrooke, G. S. and Wood, P. J. *Mol. Phys.* **1,** 257 (1958).
15. Weng, C. H. Dissertation. Carnegie-Mellon University (1968).
16. Fisher, M. E. *J. math. Phys.* **4,** 124 (1963).
17. Katsura, S. *Phys. Rev.* **127,** 1508 (1962).
18. Suziki, M., Tsuijiyama, B., and Katsura, S. *J. math. Phys.* **8,** 124 (1967).
19. Obokata, T. and Oguchi, T. *J. phys. Soc., Japan* **25,** 322 (1968).
20. Smith, J., Gerstein, B. C., Liu, S. H. and Stucky, G. *J. chem. Phys.* **53,** 418 (1970).
21. Smith, T. and Friedberg, S. A. *Phys. Rev.* **176,** 660 (1968).

4.8 Bibliography

There are numerous monographs which treat magnetic properties and which should be consulted for additional information to supplement the introductory material presented in this chapter. A selected list of monographs is given here.

Boudreaux, E. A. and Mulay, L. N. (eds) *Theory and applications of molecular paramagnetism.* John Wiley & Sons, New York (1976).

Carlin, R. L. and van Duyneveldt, A. J. *Magnetic properties of transition-metal compounds.* Springer-Verlag, New York (1977).

Chakravarty, A. S. *Introduction to the magnetic properties of solids.* John Wiley & Sons, New York (1980).

De Jongh, L. J. and Miedema, A. R. Experiments on simple magnetic model systems. *Adv. Phys.* **23,** 1–260 (1974).

Day, P. (Senior Reporter) *Electronic structure and magnetism of inorganic compounds.* A Specialist Periodical Report, The Chemical Society, London, Vol. 1—continuing series (1972–. . .).

Earnshaw, A. *Introduction to magnetochemistry.* Academic Press, New York (1968).

Goodenough, J. B. *Magnetism and the chemical bond.* Interscience Publishers, New York (1963).

König, E. and König, G. *Magnetic properties of coordination and organometallic transition-metal compounds.* Landolt-Börnstein, New Series, Vol. II/11 (1981). [Supplement to earlier volumes.]

Mabbs, F. E. and Machin, D. J. *Magnetism and transition-metal complexes.* Chapman and Hall, London (1973).

Mattis, D. C. *The theory of magnetism.* Harper & Row, New York (1965).

McMillan, J. A. *Electron paramagnetism.* Reinhold, New York (1968).

Morrish, A. H. *The physical principles of magnetism.* Robert E. Krieger Publishing Company, Inc., Huntington, New York (1980).

Mulay, L. N. and Boudreaux, E. A. (eds) *Theory and applications of molecular diamagnetism.* John Wiley & Sons, New York (1976).

O'Connor, C. J. "Magnetochemistry—advances in theory and experimentation", in *Prog. Inorg. Chem.,* Lippard, S. J. (ed.), **29**, 203 (1982).

Schieber, M. M. *Experimental magnetochemistry.* North-Holland Publishing Co., Amsterdam (1967).

Smart, J. S. *Effective field theories of magnetism.* W. B. Saunders Co., Philadelphia (1966).

Van Vleck, J. H. *The theory of electric and magnetic susceptibilities.* Oxford University Press, London (1932).

Zijlstra, H. *Experimental methods in magnetism.* North-Holland Publishing Co., Amsterdam (1967).

5 Optical techniques

R. G. Denning

5.1 Introduction

There are two principal ways in which the optical properties of a solid differ from those of a solution of a gas. First, the orientation of the molecules, or the environment of the absorbing centre, is fixed in space, apart from minor thermal oscillations. This property enables us to arrange the directions of the incident light and, if necessary, perturbing electric and magnetic fields in a defined manner with respect to the atomic framework—a feature which greatly simplifies the correct identification of the absorption process and is an invaluable technical aid to the spectroscopist. Second, the assembly of atoms in the solid is qualitatively different from a single gas molecule fixed in space. This statement is quite obvious in the case of sodium chloride where the gas-phase molecule cannot be recognized in the solid but is less obvious in, for example, solid naphthalene, where diffraction techniques tell us that the gas-phase molecule exists with almost unchanged dimensions, at a distance from its neighbours which clearly characterize it as a separate molecule.

Between these examples there are many intermediate cases where the alterations to the molecule in the solid state are more or less drastic. The electronic effects which accompany these alterations form one of the central interests of this chapter. They can be assessed by a large number of techniques, but here we are only concerned with the absorption, reflection, or emission of electromagnetic radiation associated with the excitation of valence electrons to states in which they are still bound to the solid. Usually this will restrict us to energies in the range 5000–50 000 cm^{-1}. The upper limit is a practical one set by the absorption of air, while the lower limit defines the province of vibrational spectroscopy (see Chapter 9). This chapter only deals with the properties of insulators.

There are several generalizations that can be made about the electronic spectroscopy of solids. First, the characteristic absorption length, the distance in which the intensity of the incident light falls to $(1/e)$ of its initial value, is generally more than 1 μm and can be much greater. This means that the measurements reflect the nature of the bulk solid rather than its surface, unlike those kinds of spectroscopy which involve incident, emitted, or scattered electrons. Second, the optical photon carries very little momentum compared to heavy particles of equivalent energy, so that its absorption or emission from a crystal causes only a small change in the crystal momentum. In the language of Chapter 9 the wave-vector $k = 2\pi l/\lambda$, describing the crystalline wave-function, is effectively unchanged in an optical transition thereby placing a limitation on the excited states which can be explored optically. Third, a great technical advantage available in the spectroscopy of solids is the ability to cool the sample to low temperature. At liquid-helium temperature, 4.2 K, where $kT = 2.5\ \text{cm}^{-1}$, most energy levels corresponding to quanta of vibrational motion are essentially unpopulated. The residual thermal energy lies in the very-low-frequency elastic waves of the lattice—the acoustic phonons—or in electronic states which lie very close to the ground state, as a result of the slight resolution of an intrinsic degeneracy. In practice the optical spectrum is then determined by the availability of excited-state energy levels, without the confusion associated with transitions from thermally excited states.

A related advantage arises from the reduction in the phonon population at these temperatures. The interaction of the electronic states with the fluctuating lattice alters the phase of the time-dependent wave-functions describing those states. These phase fluctuations create a distribution of frequencies in the electronic states which is reflected in the width of the spectral lines. As a result the spectral lines can sharpen dramatically at low temperature, particularly in lattices containing heavy atoms where the effect of low-frequency phonons can be very significant. The accompanying increase in spectral resolution is helpful in the detection of small separations between electronic states, which can reflect weak electronic interactions in the lattice.

5.2 Experiments and their interpretation

In this section we shall examine the techniques of absorption, reflection, and emission in turn. The practical aspects of these experiments are, in principle, rather straightforward. For example, at its simplest, the absorption technique involves a measurement of the attenuation of the intensity of a monochromatic light beam as it traverses the sample. There are a great many variations using polarized light, external applied fields, or stress. The light source may be derived from a tungsten lamp, from a discharge lamp, from a highly monochromatic continuous wave laser, or from very intense pulsed lasers. Some experiments probe the change in the absorption following the absorption of a laser pulse. Absorption may be detected by measuring the intensity of the transmitted beam photoelectrically or by measuring the absorbed energy in the form of the heat deposited in the sample. In the latter case it is usual to chop the light beam such that the heat deposition generates a sound wave within the sample at the frequency of the chopper. The signal is detected by a microphone, this particular technique being described as photoacoustic spectroscopy. Rather than exploring every variation in experimental technique this chapter describes, through the use of specific examples, the type of electronic information which can be obtained from the more widely used methods.

5.2.1 Linearly polarized absorption

The primary mechanism for absorption involves the resonant response of electrons to the electric field of the radiation. The latter is a vector quantity orthogonal to the propagation direction. In a single crystal, which for the time being we assume to contain 'molecules', the orientation of the molecular axes is fixed, so that it is possible to manipulate the relative orientation of the electric vector of the radiation with respect to these axes. In the sense that the electronic transition creates a fluctuating electric dipole with directional properties within the molecule it will be possible to observe the polarization of the transition as a function of the polarization of the radiation field.

The intensity of a transition can be described by a dimensionless quantity f, the oscillator strength, which is related to the band area in absorption by;

$$f = (10^3 mc^2 \ln 10 / N_A \pi e^2) \int \varepsilon(v) dv \qquad (5.1)$$

where $\varepsilon(v)$ is the molar extinction coefficient, N_A is the Avogadro number, and v is the wavenumber of the spectrum. The oscillator strength is proportional to the square of the transition moment through:

$$f = (8\pi^2 m v_0 c / 3he^2)\{|<j|er|a>|^2 + |<j|(e/2mc)(l+2s)|a>|^2\} \qquad (5.2)$$

where v_0 is the mean wavenumber of the transition, $|a>$ and $|j>$ represent the ground and excited state wavefunctions, er is the electric dipole operator, and $(e/2mc)(l+2s)$ is the magnetic dipole operator. The electric dipole intensity is then controlled by $(r.E)^2$ where E is the electric field vector of the radiation.

In systems with symmetry the quantity $<j|er|a>$ must be invariant under all symmetry operations. Group theory can then be used to work out the symmetry constraints on $|a>$ and $|j>$. An equivalent procedure can be applied to magnetic dipole transitions.

As an example let us consider the polarized absorption spectrum of $Cs_2UO_2Cl_4$.[1] This material will be used in a number of examples because of the sharpness of its spectrum and its structural simplicity. $Cs_2UO_2Cl_4$ has only one molecule in each primitive unit cell, a feature which removes the possibility of structurally distinct sites and also the splitting of spectroscopic features caused by the interaction of structurally identical, symmetry-related molecules within the same unit cell, known as Davydov splitting. The site symmetry of the uranium atom is C_{2h} which means that it is at a centre of inversion symmetry. The two-fold axis is perpendicular to the axis of the triatomic UO_2 uranyl group, which is required to be linear by the site symmetry. All the UO_2^{2+} ions have the same orientation in the crystal and are surrounded by four chloride ions which are almost at the corners of a square.

The crystals grow with a natural face parallel to both the two-fold axis and the O–U–O direction. Radiation incident on this face can therefore be chosen with its electric field vector either parallel or perpendicular to the O–U–O axis (Fig. 5.1). There are six distinct polarization experiments of this type. The radiation can propagate along any one of the three orthogonal molecular axes with the electric field directed in each case along either of the other two axes. The notation $X(z)$ implies propagation in the x-direction with the electric vector in the z-direction. It follows that the magnetic field of the radiation, in this example, is in the y-direction.

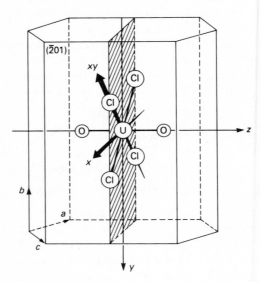

Fig. 5.1. Crystallographic axes, crystal habit, and molecular axes of $Cs_2UO_2Cl_4$.

Figure 5.2 shows the results of the six experiments. When the spectra labelled $X(y)$ and $Z(y)$ are compared much of their structure is seen to be identical, and is due to the electric field in the y-direction. However, strong, sharp features near $20\,100\,\text{cm}^{-1}$ and $20\,800\,\text{cm}^{-1}$ are present in the $Z(y)$ experiment but absent in the $X(y)$ experiment, while the converse is true of a weaker feature, labelled III, near $20\,405\,\text{cm}^{-1}$. The former absorptions are then apparently due to the magnetic field of the radiation in the x-direction while the $20\,405\,\text{cm}^{-1}$ absorption seems to be due to a magnetic field in the z-direction. Actually the latter band has a unique type of polarization which we shall discuss shortly, but the magnetic-dipole character of the feature labelled II is very informative. The band at $20\,810\,\text{cm}^{-1}$ is due to the excitation of one

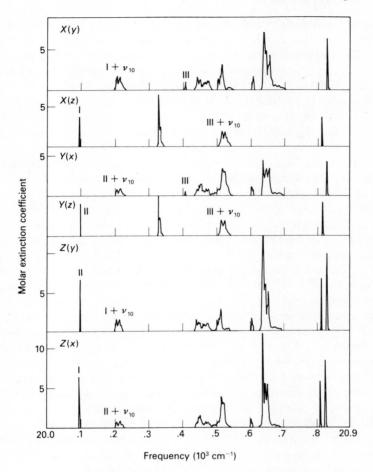

Fig. 5.2. Absorption spectra of single crystals of $Cs_2UO_2Cl_4$ in six linear polarizations; notation explained in the text.

quantum of the O–U–O symmetric stretching frequency in the electronic state responsible for II.

Because the molecular environment is centrosymmetric, the wave-functions can be classified as either *gerade* or *ungerade* upon coordinate inversion. The magnetic-dipole transition moment, induced by the magnetic field of the radiation, is a *gerade* function. Since the ground state is known to be *gerade*, as expected for a closed shell of electrons, it follows that the excited state is also *gerade* for the transition in question. Furthermore, conservation of angular momentum in the absorption process tells us that the transition is derived from one which in a spherical atom would have $\Delta M_l = 0, \pm 1$. In a cylindrical molecular environment, to which the O–U–O triatomic unit approximates, this angular momentum is constrained about the cylindrical axis. The orientation of the magnetic-dipole transition moment perpendicular to this axis requires that $\Delta M_l = \pm 1$. In the terminology of a linear triatomic molecule, the transition can then be described as $\Sigma_g^+ \to \Pi_g$, the Greek symbol describing the total angular momentum around the internuclear axis. In the $D_{\infty h}$ group, $(l_x + 2s_x)$ and $(l_y + 2s_y)$ form a basis for Π_g. With $|a\rangle$ forming a basis for Σ_g^+, invariance requires $|j\rangle$ to form a basis for Π_g.

A close look at Fig. 5.2 shows that there are actually two magnetic-dipole bands near $20\,100\,\text{cm}^{-1}$ with orthogonal polarizations, labelled I and II, separated by $1.6\,\text{cm}^{-1}$. These arise from a splitting of the degenerate Π_g excited state in the local crystalline field of the ion and have the expected polarization in the x- and y-directions. This experiment completely defines the symmetry of the first electronic excited states and it is now rather easy to deduce the symmetry of the vibrations that provide electric-dipole intensity when excited in conjunction with these origin bands. To analyse the vibronic structure associated with the Π_g electronic state we must consider the symmetry of the vibronic wave-function which is given by the direct product of the electronic and vibrational representations. For example, the degenerate Π_u bending mode of the O–U–O group yields vibronic excited state symmetries Σ_u^+, Σ_u^-, and Δ_u. Only the first is electric-dipole-allowed: the transition dipole has σ_u^+ symmetry, so that this vibration appears, near $20\,235\,\text{cm}^{-1}$, in spectra in which the electric field of the radiation is polarized parallel to the principal axis. On the other hand the σ_u^+ asymmetric stretching mode gives Π_u vibronic symmetry and is responsible for a feature near $20\,850\,\text{cm}^{-1}$ polarized perpendicular to the principal axis.

We saw that the band at $20\,405\,\text{cm}^{-1}$, which was labelled III in Fig. 5.2, appeared to be magnetic-dipole-allowed. The magnetic vector lies in the z-direction for any orientation of the propagation and electric field vectors in the xy-plane, so that this mechanism predicts that the intensity should not be a function of this orientation. Figure 5.3 shows three spectra all taken with the propagation and electric vectors constrained to the xy-plane, the propagation direction differing by 45° in each case. The intensity of the feature III varies cyclically with a period of $\pi/2$. The projection of a dipole on a dipolar field varies with a period of π, so that the $\pi/2$ period is interpreted as the result of projecting a quadrupole on to a quadrupolar field.[2]

Comparing the wavelength of the radiation, say 5000 Å, with the dimensions of the absorbing atom (1 Å), there will be a small gradient in the radiation field (or curvature in the potential) over the atomic dimensions which can be expressed by including a quadrupolar component in the radiation field. The induced molecular quadrupole moment can be pictured as analagous to the redistribution of charge which occurs when an electron is excited from a s-orbital to a d_{xy}-orbital. Group theory is again helpful in describing the situation. The transition moment is given by $\langle a|eQ_{xy}|j\rangle$. For

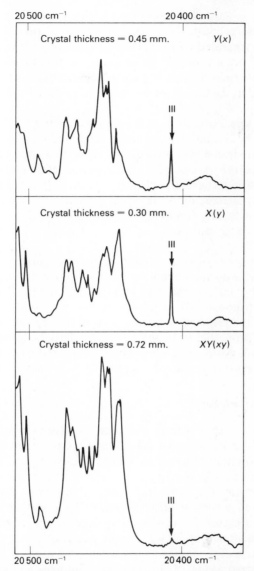

Fig. 5.3. The angular dependence of the feature labelled III in the 4.2 K absorption spectrum of $Cs_2UO_2Cl_4$, illustrating quadrupolar polarization. Axes are defined in Fig. 5.1.

the D_{4h} case where $|a\rangle$ has A_{1g} symmetry, both Q_{xy} and $|j\rangle$ have B_{2g} symmetry, whereas in the $D_{\infty h}$ symmetry of the isolated oxy-cation, Q_{xy} and $|j\rangle$ form a basis for the Δ_g representation. Here the transition can be described as $\Sigma_g^+ \rightarrow \Delta_g$ or as $\Delta M_l = +2$ by analogy with the spherical atom.

Actually these first excited states are approximately described as the components of a Russell–Saunders multiplet, labelled by $^3\varDelta_{1,2,3g}$ in the terminology of the cylindrical field, the different total angular momentum, or $\varOmega$, values being separated by the spin-orbit interaction. The linear polarization experiments are a vital part of the argument that establishes this aspect of the electronic structure, permitting a good description of the one-electron orbitals involved in the transition and identifying the role of these orbitals in the bonding.[3]

Although the linear polarization properties are particularly striking in this example, the same technique is generally applicable to absorbing centres in less than cubic environments, providing that the structural complexity is not so great as to obscure the orientational properties of an individual molecule. While the experiment usually identifies electric-dipole polarizations the ability to detect magnetic-dipole intensity is particularly valuable in the case of the intrinsically parity-forbidden transitions associated with the d–d and f–f transitions of transition-metal ions many of which are magnetic-dipole-allowed.

The effects are not, of course, confined to molecular systems. In continuous lattices of less than cubic symmetry the same arguments apply. For example the Cr^{3+} ions in the corundum lattice in ruby are in sites of C_3 symmetry, the trigonal axis of an octahedron of oxide neighbours being coincident with the principal symmetry axis of the crystal. In a later section we examine the properties of these ions in an external electric field. Similarly the Mn^{2+} ion in K_2MnF_4 is in a tetragonal lattice and we shall see later how the linear polarizations can be used to identify the pure electronic transitions.

5.2.2 Circular polarization

Circularly polarized light is easily prepared by passing plane-polarized light through a transparent material which has a refractive index which varies with the orientation of the polarization vector relative to the crystallographic axes. Such a material in the form of a plate, for example mica, has two principal values for the refractive index along axes that are mutually perpendicular and is said to be birefringent. If the thickness of the plate is such that the propagation time of a wave plane-polarized along one of these axes differs from an equivalent wave plane-polarized along the other axis by a quarter of a cycle of the radiation frequency (or a quarter of a wave-length) then light polarized along the bisector of these axes will emerge from the plate with a polarization vector which, as a function of time or position, describes either a right- or a left-handed helical path.

An elegant technique for studying the effect of circular polarization on absorption uses a plate of fused quartz which, in isolation, has no birefringence. This plate can be made to develop the necessary quarter-wavelength difference along two axes by means of mechanical stress. A convenient way of doing this is to establish a standing sound-wave in the plate. As a function of time, the element of the plate through which the light passes is first stretched and then compressed changing the sign of the induced difference in the refractive indices. The emergent beam therefore oscillates between left- and right-circular polarization.[4] If the sample has a different absorption coefficient for the two polarizations—a circular dichroism—the intensity of the transmitted beam will vary cyclically with the same frequency as the sound wave in the quartz plate. Such a modulation of the intensity is particularly easy to detect by selectively amplifying this frequency in phase with the oscillation.

There are two types of experiment which exploit circular dichroism. The first applies to naturally optically active solids. Here, in the region of each absorption band, the band area in the circular dichroism spectrum is proportional to a quantity called the rotational strength, which represents the differential transition probability between the two circular polarizations, and is given by

$$R_{aj} = \mathbf{Im} <a|er|j> \ <j|(e/2mc)l + 2s|a> \qquad (5.3)$$

where '**Im**' indicates that the imaginary part should be taken, and $|a>$ and $|j>$ are the ground- and excited-state wave-functions. It is easy to show that R_{aj} vanishes for all space groups that have planes or centres of symmetry.[5]

Natural circular dichroism (CD) has two types of spectroscopic application. Because R_{aj} is only appreciable, compared to the isotropic absorption intensity, for magnetic-dipole-allowed transitions, the experiment may be used to identify transitions allowed by this mechanism. Figure 5.4 shows the absorption and circular dichroism spectrum of a single crystal of sodium

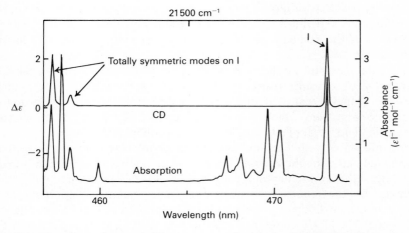

Fig. 5.4. The absorption and CD spectrum of a single crystal of sodium uranyl acetate at 4.2 K.

uranyl acetate at 4.2 K.[6] Very few of the features which appear in the absorption spectrum also appear in the circular dichroism spectrum. The CD picks out the electronic transitions with magnetic-dipole intensity, together with a progression in the symmetric O–U–O stretching mode, which is allowed by the same mechanism. Notice that the vibronic sidebands associated with this electronic state do not exhibit CD because the electric-dipole transition moment induced by the vibration oscillates in direction during the course of the vibration, its projection on to the static magnetic dipole transition moment averaging to zero.[7] Note that this crystal is cubic so that the linear polarization techniques for identifying magnetic-dipole transitions are not applicable in this case.

The second important application of CD is the correlation of the sign of the dichroism with the absolute configuration or chirality of the atomic framework surrounding the chromophore. A central example is the $Co(en)_3^{3+}$ ion whose visible CD spectrum in solution consists of two overlapping components of opposite sign. These components lie in the region where octahedral Co(III) complexes have a triply degenerate excited state of $^1T_{1g}$ symmetry. The two CD components are identified with the lowering of the local symmetry to D_3 by the chelating ethylenediamine ligands. In the Λ absolute configuration[5] the lower-energy component has positive CD. However, before any theoretical model of the mechanism inducing the dichroism can be tested it is vital to know which of the low symmetry components has this sign. This problem was resolved by the CD measurements of Mason[5] on single crystals. When the light propagates along the three-fold axis of the molecules oriented in the crystal the CD is much stronger than in solution and is entirely positive (Fig. 5.5). In the D_3 symmetry of the complex ion the parent $^1T_{1g}$ octahedral state is resolved into A_2 and E components. In this symmetry the electric dipole operator has A_2 symmetry in the direction of the threefold axis and E symmetry in the directions orthogonal to that axis. It follows that the excited state of E symmetry exhibits positive CD in this enantiomer.

A more powerful application of circular polarization exploits the CD induced in all optically absorbing solids by a magnetic field. This experiment uses the property that a circularly polarized photon carries a single unit of angular momentum which is conserved upon absorption. If an external magnetic field is applied parallel to the propagation axis of the light, electronic states with angular momentum about that axis will have their degeneracy lifted by the Zeeman interaction, in a sense which depends upon the sign of that momentum with respect to the field direction. As a consequence the frequency at which left-circularly polarized light is absorbed will generally be different from that at which right-circularly polarized light is absorbed. The CD is therefore simply related to the Zeeman splitting of the electronic states involved in the transition. The technique can be applied in

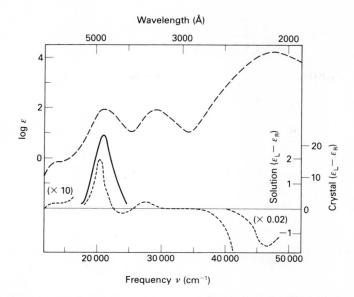

Fig. 5.5. The CD (....) and absorption spectrum (----) of (+)Co(en)$_3^{3+}$ ions in aqueous solution, and the single-crystal axial CD spectrum (———).

any medium in which circularly polarized light can propagate. In the solid state this implies cubic or uniaxial crystal systems or glasses.

The magnetic circular dichroism (MCD) is composed of three qualitatively different contributions. The differential molar extinction coefficient $\Delta\varepsilon$ can be described[8] by the expression

$$\Delta\varepsilon(a \to j) = (N_A \pi^2 \ln 10 / 250 \hbar c)\{A \cdot f_2 + (B + C/kT) \cdot f_1\} \cdot H\mu_B \qquad (5.4)$$

where H is the magnetic field parallel to the propagation direction, N_A is the Avogadro number, μ_B is the Bohr magneton, and f_1 and f_2 are functions of frequency defining the lineshape. A, B, and C are functions of the electric and magnetic dipole moments operating on the states $|a\rangle$ and $|j\rangle$ and, to a higher order, additional states $|k\rangle$. Such an expression only applies for the case where both the linewidth and kT are large compared with the Zeeman energy.

If the excited state is degenerate then the MCD spectrum resembles the derivative of the absorption spectrum (Fig. 5.6) and is described by the A-term of eqn (5.4). The sensitivity of the measurement is such that the magnetic moment of the state can be determined even for absorption bands with a width of 1000 cm^{-1} or more, even though the Zeeman splitting is of the order of 1 cm^{-1}.

If the ground state is degenerate the process is identical (Fig. 5.6) with the

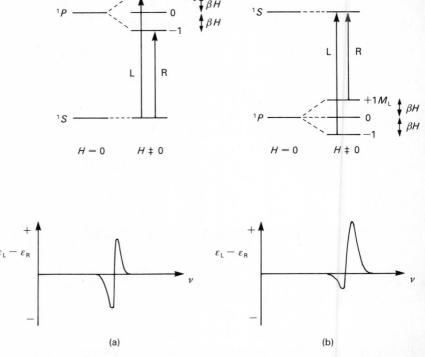

Fig. 5.6. The circular polarizations and form of the MCD spectrum arising from the Zeeman splitting of (a) a degenerate excited state, and (b) a degenerate ground state.

important exception that the circularly polarized transitions now have unequal probability on account of the different thermal populations of the Zeeman components. If the field is high enough, and the temperature is low enough, the transition can become completely circularly polarized, in which case the MCD is said to be saturated, in parallel with the magnetization. A temperature-dependent MCD is, consequently, a characteristic of a paramagnetic ground state. At low fields the MCD increases linearly with the field as reflected in the C-term of eqn (5.4), the approach to saturation occurring at field strengths determined by the magnetic moment. A system with a large moment will saturate at low fields. The magnetic moment can therefore be determined by fitting the saturation curve of MCD versus field. Such a technique is valuable where other methods of determining the moment may fail. For example the relaxation-time constraint in the ESR experiment is avoided and providing that the optical spectrum of the species of interest can be identified it is possible to determine its magnetic moment even in the presence of other paramagnetic species.

The third contribution to the MCD, which is the least important, arises from the tendency of the magnetic field to restore orbital angular momentum which has been quenched as a result of the lifting of a degeneracy by the local crystalline field, or by the perturbation of neighboring atoms. It corresponds to the second-order Zeeman contribution to magnetization and leads to rather weak MCD having the same spectral dispersion as the absorption band and is described by the *B*-term of eqn (5.4). There is a straightforward review of the MCD technique and some of its applications.[8]

In favourable cases the MCD measurement provides both ground and excited state magnetic moments together with constraints on the wavefunctions and their symmetry types. As an illustration consider part of the MCD spectrum of $CsNpO_2(NO_3)_3$ shown together with the absorption spectrum in Fig. 5.7. It is possible to interpret this spectrum in terms of the states of the linear NpO_2^{2+} ion, slightly perturbed by the nitrate ligands.[9] The linear ion states are characterized by a magnetic quantum number M_Ω describing the angular momentum around the unique axis. In the ground state this takes the values $\pm \frac{5}{2}$, the single unpaired electron residing in an *f*-orbital. It is then quite easy to show that electric-dipole-allowed transitions are brought about by the influence of the trigonal field of the nitrate ions. Using the methods of group theory, it can also be shown that predominantly positive MCD is expected in transitions to excited states having $M_\Omega = \pm \frac{3}{2}$, $\pm \frac{9}{2}$, but that negative MCD characterizes excited states with $M_\Omega = \pm \frac{1}{2}$, $\pm \frac{11}{2}$, $\pm \frac{13}{2}$. In Fig. 5.7 the features labelled V and VII fall in the former category, while those labelled IV and VI fall in the latter. The derivative shape arising from transitions with opposite circular polarization from the Zeeman components of the ground state is also clearly visible. The feature labelled V has only a single sign because the two circularly polarized transitions occur at identical wavelengths but are weighted by different thermal population factors. The superimposition implies that the ground- and excited-state magnetic moments are identical. In this compound the combination of the magnetic moments of the excited states and the sign of the MCD makes the identification of each excited state unambiguous.[9]

It is important to realize that the MCD experiment provides an extension of the study of electronic angular momentum in solids beyond what is possible by straightforward measurements of the Zeeman effect in absorption because it can be applied in systems with very broad bands. An example is the study of the F-centre absorption in alkali halides. The absorption of the electron located at, for example, a chlorine vacancy is broad and structureless, and corresponds to a transition analogous to the *s*–*p* transition in an alkali-metal atom. The MCD is temperature-dependent as expected for the unpaired ground-state electron, and resembles the derivative of the absorption spectrum. The analysis of the spectrum allows a determination of the magnitude of the spin-orbit coupling in the excited state, and the deduction

176   *Solid state chemistry: techniques*

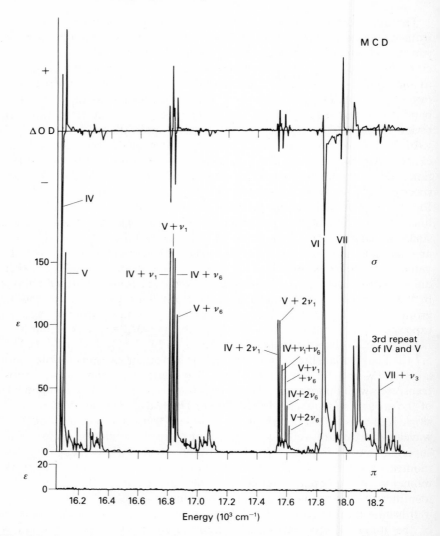

Fig. 5.7. A portion of the absorption and MCD spectrum of a single crystal containing $CsNpO_2(NO_3)_3$ at 4.2 K. The magnetic field is 5.0 T.

that there is a strong Jahn–Teller distortion in that state, as might have been expected from its orbital degeneracy.[10]

Finally the MCD spectrum can be particularly helpful in identifying the nature of the optical transitions in magnetic solids. As an example, consider the spectrum of the two-dimensional antiferromagnet, K_2MnF_4 (Fig. 5.8).[11] In the antiferromagnetic phase, this material consists of two magnetic

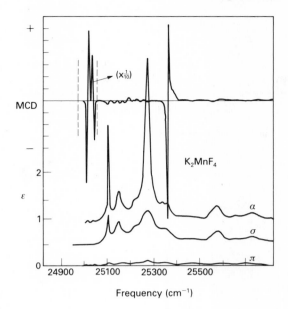

Fig. 5.8. Absorption and MCD spectra of K_2MnF_4 at 4.2 K. In the α spectrum propagation is along the tetragonal axis; in the σ and π polarization it is perpendicular to the axis with the electric vector respectively perpendicular and parallel to this axis.

sublattices on which the spins, through the agency of the spin-orbit interaction, are spatially interlocked, the spins on one sublattice being anti-parallel to those on the other. The interaction between the two sublattices is due to electron exchange between them. The net effect of this exchange (see Chapter 4) resembles a Zeeman interaction in the way that it lifts degeneracies, but is not a consequence of any real magnetic field. So, for example, in the isolated Mn^{2+} ion the ground state is a sextet with six degenerate levels having $M_s = \pm\frac{1}{2}$, $\pm\frac{3}{2}$, $\pm\frac{5}{2}$. In the presence of the exchange-interaction this degeneracy is completely lifted, so that the ground state on lattice A has $M_s = -\frac{5}{2}$, while that on lattice B has $M_s = +\frac{5}{2}$; the remaining M_s components lie above these levels at energies in excess of kT for T below the magnetic-ordering temperature.

Now consider one of the spin forbidden d–d transitions on a single ion to a spin quartet state. The situation is shown in Fig. 5.9. In the presence of a magnetic field the $(-\frac{5}{2} \to -\frac{3}{2})$ transition suffers a Zeeman shift of $2\mu_B H$ on sublattice A while the $(+\frac{5}{2} \to +\frac{3}{2})$ transition on the sublattice B suffers a shift of $-2\mu_B H$. These two transitions are oppositely circularly polarized so that a very strong MCD spectrum of derivative shape is observed. Unlike the MCD of a paramagnet, the amplitude of this MCD is not temperature-dependent,

178   Solid state chemistry: techniques

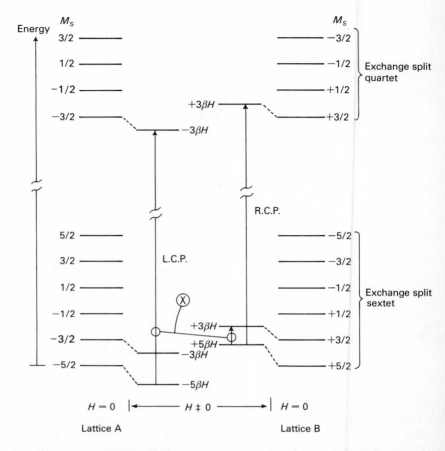

Fig. 5.9. Source of the MCD in the sextet to quartet single-ion transitions of antiferromagnetic Mn^{2+} compounds. 'X' represents the coupled exciton–magnon transition.

below the ordering temperature, because the magnetization of each sublattice is saturated.

Figure 5.8 shows that the transitions with very large MCD correspond to very weak sharp features in absorption which, from their linear polarizations, can be seen to be magnetic-dipole-allowed—the mechanism which would be anticipated for a single-ion pure electronic d–d transition in a centrosymmetric site. In the same spectrum there are broad intense electric-dipole transitions. These are co-operative transitions in which the excitation is not confined to a single site or sublattice. They occur by virtue of the exchange interaction between sublattices. Such a mechanism has a drastic effect on the spectral intensities because it can break both the spin and parity selection rules operating at a single site.[12] In the present case the sextet-to-quartet transition ($-\frac{5}{2} \to -\frac{3}{2}$) on lattice A is accompanied by a spin-inversion on lattice B; the energy of the latter process is related to the magnitude of the exchange-splitting separating the $+\frac{5}{2}$ and $+\frac{3}{2}$ components on the B lattice (see Fig. 5.9). In an external magnetic field, such a transition has equal and opposite Zeeman contributions from the two sublattices and suffers no Zeeman shift; there is no MCD, as Fig. 5.7 makes clear. The power of MCD in distinguishing the co-operative transitions is obvious.

The additional energy associated with the spin-inversion on the B sublattice should actually be considered as a spin-wave or magnon extending over the whole lattice. The energy will vary with the wave-vector k_m and has a characteristic dispersion. Remembering that an optical transition conserves the total crystal momentum, the observed transitions will be such that

$$k_e + k_m = 0 \tag{5.5}$$

where k_e is the wave-vector describing the pure electronic excitation. Frequently the dispersion of the energy with respect to k_e is much less than that with respect to k_m, in which case the optical spectrum primarily reflects the dispersion of the magnon, the intensity depending on the density of magnon states.[12] It is this dispersion which is responsible for the broadness of the electric-dipole bands in Fig. 5.8.

5.2.3 Other absorption techniques

In the previous sections we have shown how to exploit the linear and circular dichroism which arises from the disposition of atoms in the lattice or from the application of an external magnetic field. There are however a variety of other techniques measuring the linear dichroism or spectroscopic splitting due to externally applied stress, electric fields, and magnetic fields tranverse to the axis of propagation. Usually these techniques convey information on the intrinsic degeneracy of the electronic states by contriving to remove it. The experimental aspects are the subject of a book.[13]

As an example, Fig. 5.10 shows the influence of an electric field on the spectrum of one of the spin-forbidden transitions of ruby.[14] The pair of lines in the absence of the field is due to a small, 0.38 cm^{-1}, intrinsic splitting of the ground state. The further splitting is induced by an electric field applied parallel to the C_3 axis of each Cr^{3+} ion site. There are two such sites in the corundum lattice that, in the absence of the extern. l field, are related by inversion symmetry. Each of the sites can be thought of as possessing a dipole moment parallel to the C_3 axis, its symmetry-related counterpart

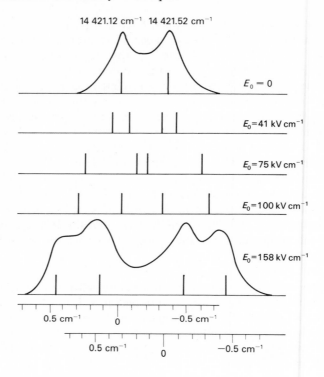

Fig. 5.10. Splitting of the $^2E \rightarrow {}^4A_2$ ruby emission as a function of the electric field applied parallel to the C_3 axis.

having a moment with the opposite sign. Since the magnitude of the dipole moment in the excited state differs from that in the ground state, the application of the field leads to an increase in the energy of the transition in one set of sites, and a decrease in the other set of sites. This pseudo-Stark effect has been used as the basis for a very elegant high-resolution study of the absorption where, in effect, the spectrum is recorded by traversing the absorption band past a fixed-frequency laser by sweeping the electric field applied to the sample.[15]

5.2.4 Reflectivity

The reflectivity of a single crystalline sample in air is given by

$$R = \{(n-1)^2 + k^2\}/\{(n+1)^2 + k^2\} \tag{5.6}$$

where n is the refractive index and k is the index of extinction. The dispersion of the reflectivity in the region of a single absorption band is not therefore simply related to the absorption spectrum, particularly as the dispersion of n follows the approximate form

$$n(v)^2 = n_0^2 + D(v_0^2 - v^2)/(v_0^2 - v^2)^2 + \gamma^2 v^2\} \tag{5.7}$$

where v_0 is the centre frequency of the transition, n_0 is the background contribution to the refractive index from other transitions, D is a measure of the transition probability, and γ relates to the width of the transition.

The main feature of $n(v)$ is that the dispersion is significant at frequencies well away from the region of absorption. In general it is hard to separate the contribution of a single transition to the composite dispersion. However, by comparing the reflectivity of the sample in contact with air with the reflectivity in contact with, say, quartz, it is possible to make an experimental determination of $n(\gamma)$. It then becomes possible to use eqn (5.6) to derive the absorption parameters. This technique is not widely used, but becomes essential for intense transitions in pure crystals where it is not possible to prepare samples thin enough to transmit light.

As an example, Fig. 5.11 shows the reflectivity of $BaPt(CN)_4$, polarized both parallel and perpendicular to the axis along which the $Pt(CN)_4^{2-}$ ions are stacked in chains.[15] Note that the reflectivity is highly polarized, and that the absorption spectrum derived from it can be seen to contain a single polarized transition centred at $23\,000\,\text{cm}^{-1}$. We shall return to the significance of this transition in due course.

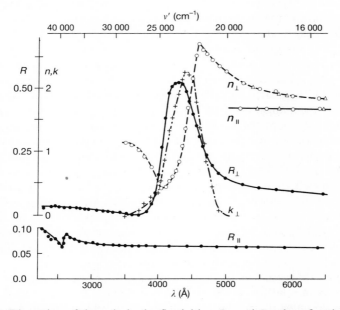

Fig. 5.11. Dispersion of the polarized reflectivities, $R_\|$ and $R_\perp$, the refractive indices, $n_\|$ and $n_\perp$, and the derived absorption coefficient $k_\|$.

5.2.5 Luminescence

Luminescence contains three distinct types of information: a spectral distribution, a polarization, and a time-dependence. The first two of these provide data which are qualitatively similar to that available in absorption. The main difference is that electronic states lying near the ground state, which are not thermally occupied under the low-temperature conditions suitable for high-resolution spectroscopy, can often be observed in luminescence. Observation of these states by absorption in the infrared is often obscured by pure vibrational transitions, or by unfavourable selection rules from the ground state.

More important information comes from the time-dependence and efficiency of the luminescence. Once an excited state has been created it can either emit or decay non-radiatively into thermal energy, or, alternatively, the excitation may migrate to a different site within the lattice, which may itself either emit or decay non-radiatively. The migration process, known as energy transfer, tells us about the mechanism by which the excitation is coupled from the donor to the acceptor site.[17]

An impurity or defect site may act as a trap for the mobile excitation. If the energy of the excited state at the trap site lies below that of the bulk site, and the temperature is sufficiently low, there will be insufficient thermal energy to allow the excitation to return to the bulk lattice. The emission spectrum is then a characteristic of the trap. Although a trap species may be present in so low a concentration that it cannot be detected in absorption it can often be readily identified in emission.

Figure 5.12 shows the emission spectrum of MnF_2 at 2 K, excited by absorption in the Mn^{2+} ions, in a crystal containing a low concentration of europium ions.[18] The lower portion of the figure shows the broad emission of Mn^{2+} overlaid by sharp, europium, emission lines. The energy, which was absorbed by the manganese ions, has clearly migrated to the europium ions. The upper portion of the figure shows, on an expanded scale, the region where the emission would be expected from the pure electronic transition in MnF_2; this emission is absent. Instead there is emission from traps labelled Mg(II) and Mg(III) with a depth of about 50 cm^{-1} which correspond to Mn^{2+} ions which are respectively the second and third nearest neighbours to trace concentrations of Mg^{2+} impurity ions (which are extremely difficult to remove from the crystal). The bands labelled S(II) and S(III) are the electric-dipole-allowed magnon sidebands associated with these same perturbed sites. They illustrate the same exchange-interaction between Mn^{2+} ions in antiferromagnetic MnF_2 as we described in K_2MnF_4 (Section 5.2.2).

In favourable cases it is possible to measure the rate of energy transfer between sites in a lattice. If the excitation is derived from a short-duration tunable laser pulse it may be possible to selectively excite an ion in one environment and follow the migration to an ion in a different environment by means of the time-evolution of the emission spectrum. Figure 5.13 shows just

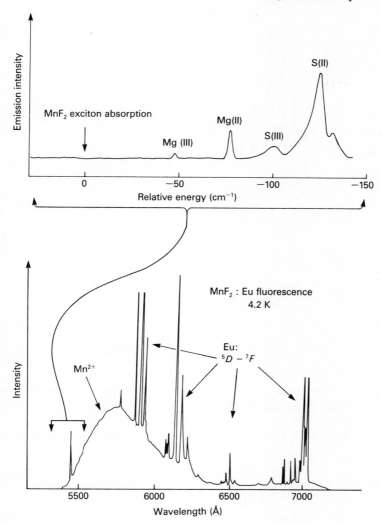

Fig. 5.12. Emission spectrum of MnF_2 doped with a small concentration of europium ions.

such a transfer between uranyl ions in $Cs_2UO_2Cl_4$ which differ in their oxygen isotopic content.[19]

The mobility of the excitation suggests that, if the interactions between neighbouring ions is strong enough, the excitation can be viewed as a collective property of the lattice. The wave-function must then bear the translational symmetry of the lattice and will be characterized by a wavevector k_e. Such an excitation wave is described as an exciton. The energy dispersion with k_e defines an exciton bandwidth which is a measure of the

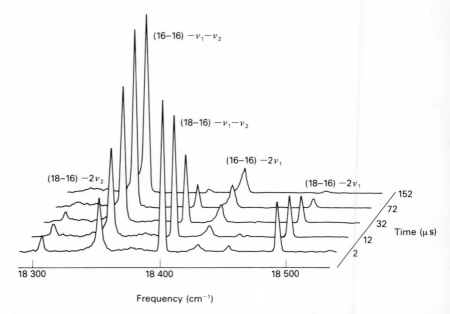

Fig. 5.13. The time-evolution of the 4.2 K emission spectrum of $Cs_2UO_2Cl_4$, excited at the absorption frequency of the $(U^{18}O^{16}O)^{2+}$ ion present about 1 per cent concentration. The labels specify quanta of the O—U—O stretching modes in the electronic ground state.

coulombic and exchange interactions between ions in the lattice.[12] In general, pure electronic transitions will be confined to the point in the exciton band where $k_e = 0$.

In the case where the electronic transitions are intense and electric-dipole-allowed, the coulombic coupling energy is a consequence of the interaction between the transition dipole moments. The interaction energy is proportional to $1/R^3$ where R is the inter-ion separation. As an example consider the linear chains of complex ions, $Pt(CN)_4^{2-}$, occurring in a variety of different salts, all with the same chain-like lattice. The separation between the anions is controlled by the counter-ion. The intense transition in the visible spectrum, whose energy is best characterized by reflection (Section 5.2.4), shifts by as much as 10 000 cm^{-1} to lower energy as the distance between the ions decreases from 3.69 Å to 3.16 Å. In this case the requirement that $k = 0$ restricts the optical transition to the lower extremity of the exciton band, the shift to longer wavelength representing the increase in bandwidth accompanying the reduction in the inter-ion distance. The shift has the correct dependence on R.[20]

By contrast, the excitons formed from the forbidden singlet–triplet transitions of aromatic organic molecules show exciton bandwidths of the order of

$10\,cm^{-1}$ and usually arise from exchange-interactions attributable to the overlap of π-electron systems on adjacent molecules.

5.3 Some applications

The foregoing sections have concentrated on examples of techniques. Here, a few examples are given illustrating interactions between absorbing centres in insulating solids.

The description of the structure of an exciton band is particularly simple in one dimension. How can the exciton bandwidth, and therefore the magnitude of the excitation transfer interaction, be determined? Dibromonapthalene forms single crystals whose structure is such that the interaction is almost entirely one-dimensional. If the six hydrogen atoms are replaced by six deuterium atoms, the pure electronic transition suffers a shift of $65\,cm^{-1}$ to higher energy. Now consider a mixed crystal containing 82 per cent H_6- and 18 per cent D_6-dibromonaphthalene. The chain of bulk hydrogen-containing molecules is randomly interrupted by deuterated molecules whose energy levels are completely out of resonance with those of the bulk. As a consequence, a distribution of chain lengths exists in the bulk material effectively isolated from other chains by the impurity. There will be monomers, dimers, trimers, tetramers, etc., each with their characteristic energy levels; and for each of these only the in-phase combination of the excitations (corresponding to $k_e = 0$ in the bulk) will appear in the optical spectrum. The energy levels are shown schematically in Fig. 5.14. The separation between the monomer absorption and the bulk absorption in this case is just $2J$ where J is the interaction energy between adjacent centres. Figure 5.15 shows the actual spectrum with features for each of the various polymers, from which the exciton bandwidth is found to be $29.6\,cm^{-1}$.[21]

As the concentration of Cr^{3+} ions in ruby is increased, so does the probability that ions will occur adjacent to one another. Because these ions interact, there are distinct spectral features associated with pairs of ions and higher polymers. This situation is illustrated in the spectra shown in Fig. 5.16. The extra features have an intensity whose concentration with respect to monomeric features increases linearly with concentration.[22] Very careful high-resolution work exploiting the linear polarizations in absorption, and in emission, can identify not only the spectrum of the first nearest neighbours but also that of the second, third, and fourth nearest neighbours. The principal exchange energies are found to be 54, 83, 12, and $7\,cm^{-1}$ respectively.[23]

Finally a further illustration is given of the consequences of co-operative interactions in optical spectroscopy. In this case, a ferromagnetically ordered solid is considered, as opposed to the antiferromagnetic example in Section 5.2.2. K_2CrCl_4 is a three-dimensional ferromagnet with an ordering

186 Solid state chemistry: techniques

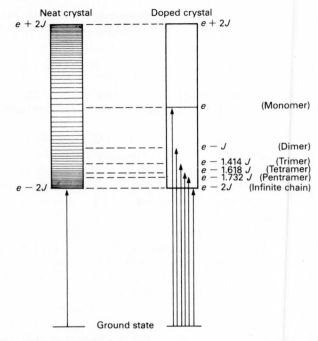

Fig. 5.14. The relation of isolated polymeric cluster energies to the levels in the one-dimensional exciton band.

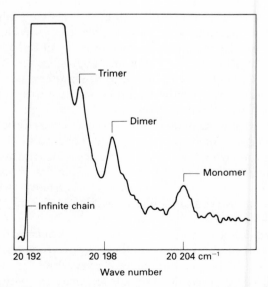

Fig. 5.15. The absorption spectrum at 4.2 K of a crystal containing 82 per cent H_6- and 18 per cent D_6-dibromonaphthalene.

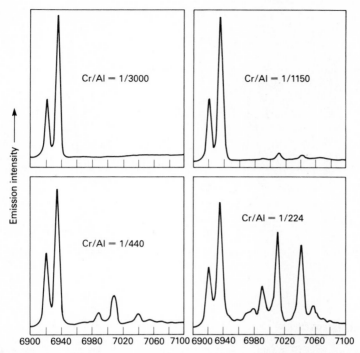

Fig. 5.16. The fluorescence spectrum of ruby at 77 K at various chromium concentrations.

temperature of about 65 K. Being a d^4 system the single-ion ground-state is a quintet. The visible region of the spectrum is expected to show a number of very weak quintet-to-triplet transitions. The actual situation is shown in Fig. 5.17. Two very strong, sharp features in the spectrum decline in intensity very rapidly with temperature so that, at liquid-helium temperature, the overall intensity is perhaps $\frac{1}{100}$ of the intensity at 100 K.[24]

In a ferromagnetic solid the exchange-splitting of all ions within a single domain is of the same sign. The ground state will then be characterized by $M_s = +2$ and the quintet-to-triplet single-ion transition is described as $M_s = +2 \rightarrow +1$. The exchange interaction can only break the forbidden character of the transition if a coupled ion simultaneously undergoes a $M_s = +1 \rightarrow +2$ transition. The $M_s = +1$ levels are not thermally occupied at low temperatures, so this process must be thermally activated. As the temperature rises, the forbidden nature of the single-ion transition is progressively removed. Such an excitation can be described as the annihilation of a thermally activated magnon accompanied by the creation of a pure exciton. The optical consequences of the change in temperature are so dramatic in this case that they are easily detected by the eye as a strong change in colour. Evidently optical properties can be an excellent probe of magnetic order in solids.

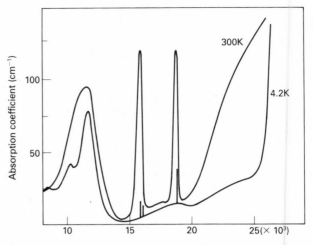

Fig. 5.17. The absorption spectrum of a single crystal of K_2CrCl_4 at 300 K and at 4.2 K.

5.4 References

1. Denning, R. G., Snellgrove, T. R., and Woodwark, D. R. *Mol. Phys.* **32,** 419 (1976).
2. *Idem, ibid.* **30,** 1819 (1975).
3. *Idem, ibid.* **37,** 1109 (1979).
4. Kemp, J. C. *J. opt. Soc. Am.* **90,** 950 (1969).
5. Mason, S. F. *Quart. Rev.* **17,** 20 (1963).
6. Denning, R. G., Foster, D. N. P., Snellgrove, T. R., and Woodwark, D. R. *Mol. Phys.* **37,** 1089 (1979).
7. Weigang, O. E. *J. chem. Phys.* **43,** 3609 (1965).
8. Schatz, P. N., and McCaffery, A. J. *Quart. Rev. chem. Soc.* **23,** 552 (1969); (Err. **24,** 329 (1970).
9. Denning, R. G., Norris, J. O. W., and Brown, D. *Mol. Phys.* **46,** 325 (1982).
10. Henry, C. H., Schnatterly, S. E., and Slichter, C. P. *Phys. Rev.* **A137,** 583 (1965).
11. Schwartz, R. W., Spencer, J. A., Yeakel, W. C., Schatz, P. N., and Maisch, W. G. *J. chem. Phys.* **60,** 2598 (1974).
12. McClure, D. S. in *Optical properties of ions in solids* (ed. B. di Bartolo). Plenum, p. 259 (1975).
13. Cardona, M. *Modulation spectroscopy*, Academic Press (1969).
14. Kaiser, W., Sugano, S., and Wood, D. L. *Phys. Rev. Lett.* **6,** 605 (1961).
15. Muramoto, T., Nakamishi, S., and Hashi, T. *Opt. Commun.* **21,** 139 (1977).
16. Moncuit, C., and Poulet, H. *J. Phys. Rad.* **23,** 353 (1962).
17. Denning, R. G. in *Emission and scattering techniques* (ed. P. Day) D. Reidel Dordrecht, p. 171 (1981).
18. Wilson, B. A., Yen, M. W., Hegarty, J., and Imbusch, G. F. *Phys. Rev.* **B19,** 4238 (1979).

19. Denning, R. G., Ironside, C. N., Thorne, J. R. G., and Woodwark, D. R. *Mol. Phys.* **44,** 209 (1981).
20. Day, P. *J. Am. chem. Soc.* **97,** 1588 (1975).
21. Hochstrasser, R. M., and Whiteman, J. D. *J. chem. Phys.* **56,** 5945 (1972).
22. Schawlow, A. L., Wood, D. L., and Clogston, A. M. *Phys. Rev. Lett.* **3,** 271 (1959).
23. van der Ziel, J. P. *Phys. Rev.* **B9,** 2846 (1974).
24. Day, P., Gregson, A. K., and Leech, D. H. *Phys. Rev. Lett.* **30,** 19 (1973).

6 High-resolution solid-state MAS NMR investigations of inorganic systems

C. A. Fyfe and R. E. Wasylishen

6.1 Introduction

Over the last two decades, many experiments have been developed and implemented to obtain 'high-resolution' NMR spectra of solids. These have been reviewed from both theoretical and practical aspects in several texts[1-3] and a substantial number of review articles.[4-24]

In this chapter we will restrict ourselves to a discussion of the background theory of the technique of 'magic angle spinning' (MAS)[4,5] and its applications. The MAS technique, with and without 'cross-polarization' (CP)[25-7], has become in recent years the most widely used method for obtaining high resolution spectra from solid materials. The discussion will be restricted in the main to applications in inorganic chemistry but applications in other areas of chemistry may be found in the above general references.

6.2 Nuclear interactions in the solid state

High-resolution NMR spectroscopy in solution has become one of the most powerful techniques for the elucidation and investigation of chemical structures via chemical shifts, couplings, and relative intensities of the resonances in their spectra.[28-32] In contrast to the wealth of such information in the solution spectra, the NMR spectra of solid systems show typically very broad, mainly featureless absorptions in which the 'high-resolution' solution characteristics of chemical shifts and of spin–spin couplings are obscured. The main thrust of 'high-resolution' NMR experiments is to manipulate the spin systems to remove or average the characteristic solid state interactions, and to simplify the spectrum so that the 'chemical' information contained in the chemical shifts and other 'solution' interactions may be retrieved and used for structural investigations. The critical feature of the different interactions which can occur for a nuclear spin in the solid state is that they are dependent on the (usually fixed) orientation of the nuclear spin vector to the magnetic field, the random distribution of possible orientations giving rise to severe spectral broadening.

The different interactions that may occur are detailed below.

6.2.1 The Zeeman interaction

This term results from the interaction of the magnetic moment of the nucleus μ_N with the applied static magnetic field H_0, and is described by the Hamiltonian operator $\mathcal{H}_Z$ as in eqn (6.1).

$$\mathcal{H}_Z = -\vec{\mu}_N \vec{H}_0 = -\gamma_N \hbar \vec{H}_0 \cdot \vec{I} \tag{6.1}$$

which for the applied field chosen to lie along the z-direction becomes

$$\mathcal{H}_Z = -\gamma_N \hbar \vec{H}_0 \cdot \vec{I}_z \tag{6.2}$$

where I_z is the component of the nuclear spin in the z-direction, H_0 the applied magnetic field, and γ_N the 'magnetogyric ratio' characteristic of that nucleus. The Zeeman interaction is responsible for the initial splitting of the energy levels of the nucleus, and determines the frequency of observation of a particular nucleus at a given magnetic field strength and, via the Boltzmann distribution between the levels, the fundamental detection sensitivity of that nucleus.

6.2.2 The dipolar interaction

The dipolar interaction between two like spins of type I may be represented as

$$\mathcal{H}_{D(I-I)} = \frac{\gamma^2 \hbar^2}{r^3} \vec{I}_1 \cdot \hat{D} \cdot \vec{I}_2 \tag{6.3}$$

where r is the distance between the two nuclei, and $\hat{D}$ the dipolar coupling tensor. For unlike spins I and S, a similar equation results

$$\mathcal{H}_{D(I-S)} = \frac{\gamma_I \gamma_S \hbar^2}{r^3} \vec{I} \cdot \hat{D} \cdot \vec{S} \tag{6.4}$$

For a single crystal, the interaction between two isolated spins may be evaluated for each of the two cases above. In both cases, considering the observation of a single nucleus I, a spectrum of two lines would be observed with the peak separations given in eqns (6.5) and (6.6) for like and unlike spins respectively,

$$\Delta V = \tfrac{3}{2} \frac{\gamma_I^2 \hbar^2}{r^3} (1 - 3\cos^2\theta) \tag{6.5}$$

$$\Delta V = \frac{\gamma_I \gamma_S \hbar^2}{r^3}(1 - 3\cos^2\theta) \tag{6.6}$$

where θ is the angle between the *internuclear vector* and the magnetic field.

For a particular nucleus, pairwise interactions as described in (6.5) will occur between a single nucleus and all of its neighbours, although the $1/r^3$ dependence will favour near-neighbour interactions. Further, for a polycrystalline material, eqns (6.5) and (6.6) must be averaged over all possible angles of θ corresponding to the random distribution of polycrystallites and therefore of internuclear vector orientations. The net result of these two factors is to produce a very severe broadening of the spectrum, which is field independent. For most systems of spin $\frac{1}{2}$ nuclei, this is the dominant line-broadening interaction and is largest for proton-containing systems as ^{1}H has the largest γ value of any nucleus; and its small size allows for small internuclear distances (r). For proton–proton interactions, dipolar line broadening can range up to about 80 kHz and carbon–proton interactions can give a line broadening of up to 40 kHz in the carbon spectrum.

6.2.3 Chemical shift interaction

This term is caused by the modification of the applied magnetic field experienced by a nucleus by the surrounding electrons and is described as

$$\mathcal{H}_{CS} = \gamma \hbar \vec{I} \cdot \hat{\sigma} \cdot \vec{H}_0 \tag{6.7}$$

where $\hat{\sigma}$ is the chemical shielding tensor which describes the orientation dependence of the interaction (this is discussed in more detail later). The averaging of the interaction over all possible random orientations in a polycrystalline sample produces a line-broadening which in this case is *field-dependent*. Typical values[1-3,21,33] are 0–330 p.p.m. ^{31}P, 0–250 p.p.m. ^{13}C, 0–15 p.p.m. ^{1}H. It should be noted that for a particular isotope, the frequency range covered by this effect is of the same order as, or greater than, the whole isotropic shift range found in solution NMR studies.

6.2.4 Spin–spin coupling

As in solution, an indirect electron-coupled spin–spin interaction can take place between two nuclear spins I and S. in the solid state it is described by

$$\mathcal{H}_{SC} = h\vec{I} \cdot \hat{J}_{IS} \cdot \vec{S} \tag{6.8}$$

and is again orientation-dependent with respect to the two spin vectors $\vec{I}$ and $\vec{S}$. The components of the indirect coupling (electron coupled) $\hat{J}_{IS}$ tensor have only been determined in a few compounds.[34] In most cases, this term will be smaller than the others described above but may be important for the elucidation of chemical structures.

For spin $\frac{1}{2}$ nuclei, the above interactions are the most important. For nuclei with spin $>\frac{1}{2}$, quadrupolar interactions must also be considered as described below.

6.2.5 Quadrupolar interactions

The quadrupolar term arises from the interaction of the nuclear spin with a non-sphericaly symmetric electric field gradient at the nucleus. In the general case, it is described by

$$\mathcal{H}_Q = \vec{I} \cdot \hat{Q} \cdot \vec{I} \tag{6.9}$$

$$\mathcal{H}_Q = \frac{e^2 qQ}{4I(2I-1)}\{3I_z^2 - I^2 + \eta(I_x^2 - I_y^2)\} \tag{6.10}$$

where eQ is the nuclear quadrupolar moment, and the term e^2Qq/h is referred to as the nuclear quadrupole coupling constant. $\hat{Q} = \{eQ/2I(2I-1)h\}\hat{V}$ where $\hat{V}$ is the electric field gradient tensor at the nucleus. Where the Zeeman interaction is larger than the quadrupolar interaction (referred to as the high-field case), the energy shifts due to the quadrupolar interaction may be calculated using first-order perturbation theory.

In the context of inorganic chemistry, quadrupolar nuclei with non-integral spins are particularly important, comprising approximately seventy per cent of all magnetically active nuclei. The further development of the description of the quadrupolar interaction for these systems together with the appropriate line-narrowing experiments will be presented in the next section.

Thus, the total interactions for a nucleus in a field are as described by eqns (6.11) and (6.12) with the approximate magnitudes (in Hz) of the different interactions indicated below the appropriate terms:

$$\mathcal{H} = \mathcal{H}_Z + \mathcal{H}_D + \mathcal{H}_{CS} + \mathcal{H}_{SC} + \mathcal{H}_Q \quad \text{(if } I > \tfrac{1}{2})$$
$$(10^6 - 10^9) \quad (0 - 10^5) \quad (0 - 10^5) \quad (0 - 10^4) \quad (0 - 10^9) \tag{6.11}$$

$$\mathcal{H} = \mathcal{H}_Z + \frac{\gamma^2 \hbar^2}{r^3}\vec{I}_1 \cdot \hat{D} \cdot \vec{I}_2 + \gamma \hbar \vec{I} \cdot \hat{\sigma} \cdot \vec{H}_0 + h\vec{I} \cdot \hat{J} \cdot \vec{S}. \tag{6.12}$$

The most critical characteristic feature of the interactions described in eqn (6.12) is their orientation dependence which arises from the fixed molecular and spin orientations in the solid state, which means that the interactions are between *vector* quantities $\vec{I}, \vec{H}_0$, etc. In each case, the interaction between the vectors must be described by a 3×3 matrix or 'tensor', $\hat{D}, \hat{\sigma}, \hat{J}, \hat{Q}$. By choice of a suitable coordinate system, each of these tensors may be reduced to diagonal form, that is having non-zero elements only on the diagonal a_{11}, a_{22}, a_{33}. These are called the 'principal elements' of the tensor.

The fundamental difference between solution and solid-state NMR spectra is that, in the former case, the fast rotational and translational motion of the molecules averages the interactions. In each case, the average value ($a_{iso.}$) is equal to one-third of the sum of the principal elements (or 'trace') of the diagonalized matrix.

$$a_{iso.} = \tfrac{1}{3}(a_{11} + a_{22} + a_{33}). \tag{6.13}$$

In the case of $\hat{\mathcal{H}}_D$ and $\hat{\mathcal{H}}_Q$ the tensors are 'traceless', and their isotropic average values are *exactly* zero—these are not directly observed *at all* in solution NMR spectra although they may contribute indirectly via relaxation mechanisms. In the case of $\hat{\sigma}$ and $\hat{J}_{IS}$, non-zero averages result, and these are the isotropic chemical shifts and spin–spin couplings observed in high-resolution solution spectra and used for structure determinations and investigations.

The basic problem in obtaining high-resolution solid-state spectra is thus to implement experimental procedures which (for non-quadrupolar nuclei), remove the dipolar interaction completely and produce the isotropic average values for the other interactions (as occurs in solution). These procedures are described in the next section.

6.3 Techniques for 'line narrowing' in solids: CP–MAS experiments

6.3.1 Dilute spin systems

In the description of the development of MAS and CP–MAS experiments for line narrowing in solid-state NMR spectra, it is convenient to divide nuclear spin systems in solids into two types: 'abundant' and 'dilute' spin systems. In abundant spin systems, of which by far the most common example is that of protons, there is a high density of nuclei of high isotropic abundance. In this case, the interactions are represented as

$$\mathcal{H} = \mathcal{H}_Z + \mathcal{H}_{D(H-H)} + \ldots \tag{6.14}$$

In this case, the dipolar term is dominant, other terms being negligible, and the proton NMR spectra of solid systems in general show only a single broad featureless absorption. Although various techniques have been developed for the removal of the homonuclear H–H dipolar coupling,[1-5,7] their finite efficiencies, combined with the very small range of proton chemical shifts (10 p.p.m.) have precluded their general application as routine analytical techniques. (See, however, the recent experiments of Gerstein and co-workers.[35a,b])

A much simpler situation exists for 'dilute' spin systems. In this case, there is only a *low concentration* of the nucleus being observed. This may occur from there being a low isotopic abundance of the particular nucleus under study, or from a low absolute concentration of the particular nucleus in the sample, whatever the abundance of the NMR active isotope.

The total Hamiltonian in this case may be represented, for a nucleus X, as

$$\mathcal{H} = \mathcal{H}_Z + \mathcal{H}_{D(H-X)} + \mathcal{H}_{D(X-X)} + \mathcal{H}_{CS}. \tag{6.15}$$

There are simplifying features in the 'dilute' spin case described by eqn (6.15) compared to the abundant spin system discussed above. Thus, the heteronuclear H–X dipolar interactions are large, but they are between different nuclei, and may now be removed by decoupling at the proton resonance frequency while observing at the X-nucleus frequency. As discussed above, the proton spectrum is generally broad, and very powerful decoupling fields will be necessary, but the technical aspects of this problem have been solved. The critical simplifying feature which comes from the 'dilute' nature of the spin system is that the *homonuclear* dipolar interactions $\mathcal{H}_{D(X-X)}$ either do not exist or are very small (remembering the $1/r^3$ dependence of the dipolar interaction on the internuclear distance, the dilution does not have to be extreme to cause a large reduction in this term). As indicated above, the dilution may occur from a low isotopic abundance of an abundant atom as in the case of carbon (the first system studied in detail by these techniques) where the abundance of the ^{13}C isotope is 1.1 per cent; but *any* nucleus can satisfy the requirements of 'dilution', even 100 per cent abundant isotopes such as ^{31}P or ^{19}F, as long as there is not a high density or concentration of these nuclei in the sample, especially directly bonded or in close proximity to each other. Successful experiments have been performed on a whole variety of nuclei including ^{13}C, ^{29}Si, ^{31}P, ^{113}Cd, ^{15}N, ^{27}Al, and many others.[36] Even the ^{1}H nucleus may satisfy the requirements if it is present as only a small percentage 'impurity' in a perdeuterated molecule.[37]

6.3.2 Chemical shift anisotropy: magic angle spinning (MAS) techniques

After removal of hetero- and homonuclear dipolar interactions, the major remaining interaction for spin $\frac{1}{2}$ nuclei is the chemical shift anisotropy. The orientation dependence of the chemical shift described in eqn (6.7), where $\hat{\sigma}$ is a tensor quantity, means that for a powder sample, the characteristic absorptions for all possible random orientations of the nuclei with respect to the field will be observed simultaneously giving a characteristic broad absorption or 'anisotropy pattern', from which the principal elements of the shielding tensor may be obtained directly. Two common situations are encountered as shown in Fig. 6.1.

If the shielding of the nucleus is completely asymmetric (i.e. different in each of the three orthogonal directions of the principal axis system) a general pattern such as that shown in Fig. 6.1(a) results with the values of the principal elements and the isotropic shift as indicated in the figure. In some

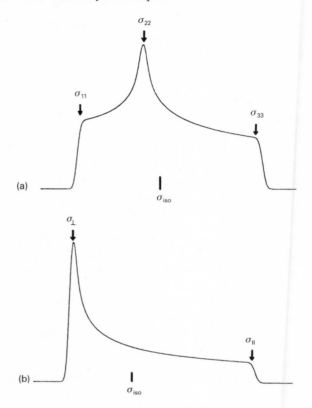

Fig. 6.1. Schematic representation of shift anisotropy patterns where σ_{11}, σ_{22}, σ_{33}, $\sigma_\perp$, $\sigma_\parallel$, and σ_{iso} have the values indicated for (a) the general case, (b) axially symmetric case where $\sigma_{11} = \sigma_{22} = \sigma_\perp$.

cases, due to symmetry of bonding, two of the elements may be identical giving an 'axially symmetric' shielding pattern as shown in Fig. 6.1(b) where again the principal elements are as indicated. It should be noted that the shielding anisotropies contain considerable information of direct chemical interest, as they represent the three-dimensional chemical shielding of the nucleus, and, if they can be related to the molecular coordinate system, they will provide incisive information regarding the detailed nature of the bonding of the nucleus. (Such studies require the use of single crystals,[1,33,38–40] and little work of this type has been done to date on inorganic systems. See for example references 41–3.)

The chemical shielding anisotropy powder pattern may be averaged to the isotropic value by the technique of 'magic angle spinning' (MAS). In this experiment, illustrated schematically in Fig. 6.2, the sample is spun mechanically *around* an axis inclined at an angle θ to the magnetic field axis at a

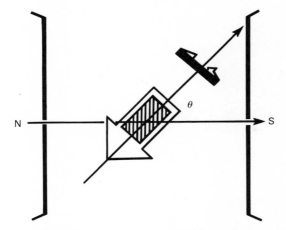

Fig. 6.2. Schematic representation of the experimental arrangement for sample spinning. In MAS, $\theta = 54°44'$.

frequency comparable to the frequency spread of the shift anisotropy. Spinning rates must be of the order of several kHz, and special types of spinning apparatus have been developed,[3,4,21] based largely on the original work of Andrew,[44] and Lowe.[45]

The effect of the spinning experiment illustrated in Fig. 6.2 is to modify the shift anisotropy by a factor of the form $(3\cos^2\theta - 1)$. Thus, if θ is chosen to be 54°44', the so-called 'magic angle', this term becomes zero, and the remaining part of the expression is the isotropic chemical shift (as in solution). MAS may also remove any small residual dipolar interactions. The effect of MAS is illustrated in Fig. 6.3 for the case of the ^{31}P spectrum of triphenyl phosphine oxide [I].

Spinning at a rate approximately equal to the frequency spread of the anisotropy yields a single isotropic average peak (bottom spectrum) while spinning at slower rates yields a spectrum where the isotropic peak is flanked by a series of 'spinning side bands' separated by the spinning frequency, whose intensities reflect approximately the profile of the shift anisotropy pattern (top spectrum). In fact, spectra of this type may be used to obtain the principal elements of the shielding tensor while still retaining the high resolution aspects of the experiment—important where there is an unresolvable overlap of anisotropy patterns in the stationary spectrum, but where the isotropic peaks (and their associated sidebands) are clearly resolvable[46,47] (also see Section 6.6).

6.3.3 Cross polarization (CP) techniques

At this stage of our line-narrowing experiment, the use of high-power proton

198 *Solid state chemistry: techniques*

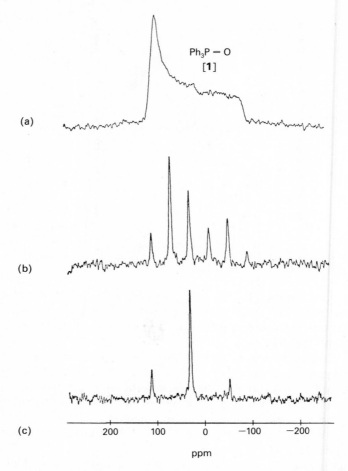

Fig. 6.3. ^{31}P CP–MAS spectra of triphenylphosphine oxide, $Ph_3P=O$ at 34 MHz with (a) $v_{rot.}=0$, (b) $v_{rot.}=1.5\,kHz$, (c) $v_{rot.}=3.0\,kHz$. Reproduced, by permission, from ref. 21.

decoupling to remove the (H–X) dipolar interactions, magnetic dilution to preclude (X–X) homonuclear dipolar interactions, and MAS to average the shift anisotropies to their isotropic values, yields all of the spectral resolution that will be obtained. The technique of cross polarization (CP)—first introduced by Pines, Gibby, and Waugh—may now be used to *increase the signal-to-noise ratio (S/N) of the spectrum* of the *dilute* nucleus X, but will *not affect the resolution* of the spectrum.

The basic idea of the pulse sequence (illustrated in Fig. 6.4) is to enhance the magnetization of the dilute nucleus X from the magnetization reservoir *of the abundant proton spin system*. In the first step of the sequence the proton

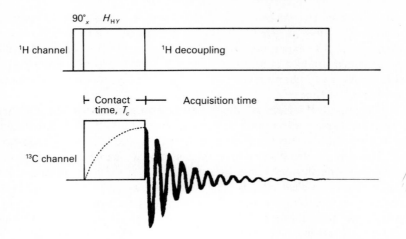

Fig. 6.4. Schematic representation of the pulse timing and magnetization changes for the cross-polarization experiment (for details see text).

spin magnetization is rotated by 90° to align with the Y' axis in the rotating frame. The phase of this on resonance pulse is now shifted by 90° so that it (H_H) and the proton magnetization are both aligned along the Y' axis. The proton spins will now precess around H_H, and are said to be 'spin-locked', the fast decay due to T_2 processes which would normally occur being prevented by the spin-locking field H_H (a slower decay occurs according to a time constant $T_1\rho$, the relaxation in the rotating frame). Simultaneous with the phase-shift of the proton pulse, an on-resonance RF pulse H_X is applied to the dilute nucleus X, the magnitudes of H_H and H_X being chosen so that they satisfy the 'Hartmann–Hahn' condition (eqn (6.16)):[1,3,21,48]

$$\gamma_H H_H = \gamma_X H_X. \tag{6.16}$$

The X-nucleus magnetization now grows by transfer of magnetization from the proton reservoir during the 'contact time' τ_1 where the Hartmann–Hahn condition is satisfied as shown in Fig. 6.4. At the end of this time, the X-pulse is turned off, and the X-nucleus decay recorded in the time period τ_2, the proton field H_H acting as a decoupling field during this period. After a delay time τ_3, the sequence is repeated. In principle, several 'contacts' of the X-spins could be made during a single proton spin-locking pulse, but in practice, single contacts have been used in the vast majority of published data.

The use of the CP sequence increases the S/N of the X-nucleus spectrum by two mechanisms: (1) there is an enhancement of the X-magnetization, up to a theoretical maximum factor of γ_H/γ_X (4 for ^{13}C, 5 for ^{29}Si, and 4.5 for ^{113}Cd); (2) the relaxation time of the proton magnetization is generally shorter than those of the X-nuclei. As can be seen from Fig. 6.4, the X-nucleus magnetization depends only on growth from the proton magnetization during the 'contact time'. The experiment may thus be repeated after the much shorter time τ_3 which enables substantial recovery of the proton magnetization, giving an indirect enhancement of the S/N of the X-nucleus spectrum in a given period of time.

The cumulative effects of the different components of the CP–MAS experiment—first implemented by Schaefer and Stejskal[49]—are illustrated in Fig. 6.5 for the ^{13}C spectrum of hexamethylbenzene chromium tricarbonyl [2].[50] The top spectrum (6.5(a)) is of a stationary sample recorded under normal 'high-resolution' conditions employing low power proton decoupling; it shows no clear signals. The second spectrum (6.5(b)) is again of a stationary sample, but now recorded using high-power proton decoupling; it shows some resolution, but is substantially broadened by the chemical shielding anisotropies of the various non-equivalent carbon nuclei. The methyl resonance has the smallest anisotropy and is clearly observed, but the aromatic and carbonyl absorptions are broadened to the extent that they are almost unobservable. The use of CP increases the S/N of the spectrum as shown in the third spectrum (6.5(c)) which is identical to the second except for the signal intensities. The fourth spectrum (6.5(d)) is recorded using both high-power decoupling and magic angle spinning, and shows the carbonyl, aromatic, and methyl resonances clearly resolved. The bottom spectrum (6.5(e)) is recorded using high-power decoupling, magic angle spinning, and also cross-polarization. The S/N in the last spectrum is greatly increased, and the relative intensities of the carbonyl and aromatic carbons which have the longest spin-lattice relaxation times increase substantially with respect to the methyl resonance. The *resolution* of the spectrum is, however, *not affected*.

The use of the CP–MAS experiment thus yields spectra of good-to-moderate resolution for a whole variety of dilute nuclei in solid samples. In the particular case of inorganic systems, however, there are two further experimental approaches which are of particular importance. These are simple MAS experiments at very high magnetic field strengths,[36] and techniques for the observation of quadrupolar nuclei with non-integral spins.[51]

6.3.4 MAS spectra at high magnetic fields

In many inorganic systems, including some such as glasses, alloys, silicas, zeolites, and clays which are of considerable commercial importance, there are no hydrogen atoms covalently bonded to the inorganic matrix. In these cases, it is not possible to utilize the CP technique, and high-power proton decoupling is not necessary. Thus, the experiment reduces to one of a simple X-nucleus pulse experiment with MAS to remove the shift anisotropy term and any small dipolar interactions. *This experiment may be easily performed using a conventional high-resolution spectrometer*—equipment routinely avail-

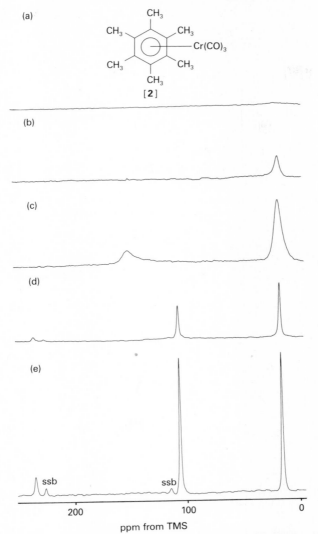

Fig. 6.5. ^{13}C solid-state NMR spectra of hexamethylbenzene chromiumtricarbonyl, $(C_{12}H_{18})Cr(CO)_3$. (a) Static sample, scalar ^{1}H decoupling, 6000 FID, 1 s delay, 20 Hz line-broadening. (b) Static sample, dipolar decoupling, 6000 FID, 1 s delay, 20 Hz line-broadening. (c) Static sample, dipolar decoupling, cross polarization, 10 ms contact, 6000 FID, 1 s delay, 20 Hz line-broadening. (d) MAS, dipolar decoupling, 6000 FID, 1 s delay, 20 Hz line-broadening. (e) MAS, dipolar decoupling, cross-polarization, 6000 FID, 1 s delay, 20 Hz line-broadening.

able to most chemists. Figure 6.6 shows examples of spectra of a variety of nuclei in different inorganic systems obtained in this way.[36] In addition, as illustrated in Fig. 6.7 for the ^{29}Si spectra of two typical zeolite samples, there are often advantages in carrying out these experiments at the highest possible

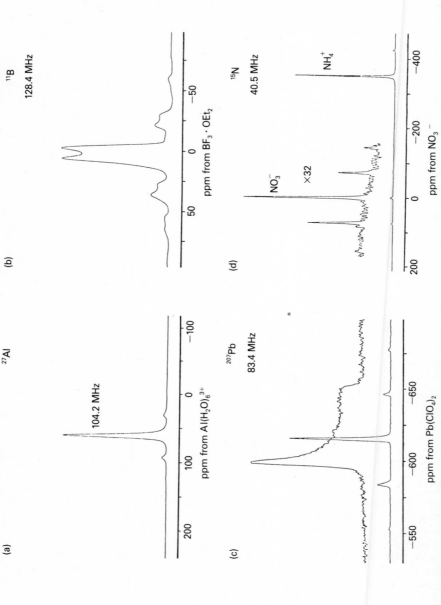

Fig. 6.6. Solid-state magic-angle spinning NMR spectra. (a) ^{27}Al spectra of zeolite-Y at 104.2 MHz (2656 scans, no line-broadening, 0.1 s relaxation delay). (b) ^{11}B spectrum of Corning 7070 glass at 128.4 MHz (13 775 scans, 25 Hz line-broadening, no relaxation delay). (c) ^{207}Pb spectra of Pb(NO$_3$)$_2$ at 83.4 MHz. The powder pattern (420 scans, 20 Hz line-broadening, 2 s relaxation delay) is shown above the spinning spectrum (230 scans, 20 Hz line-broadening, 2 s relaxation delay). (d) ^{15}N spectrum of ^{15}NH$_4$^{15}NO$_3$ at 40.5 MHz (2078 scans, 40 Hz line-

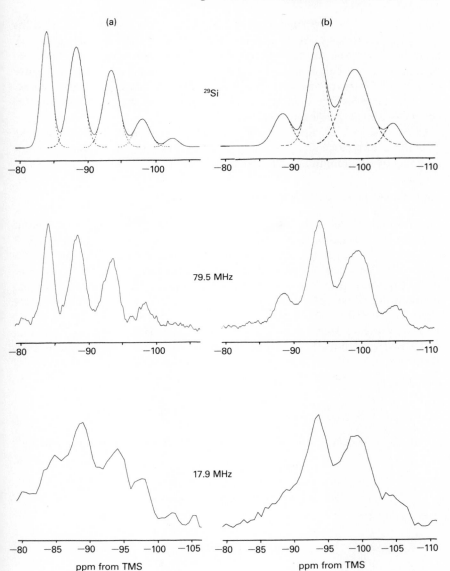

Fig. 6.7. ^{29}Si MAS NMR spectra of (a) analcite and (b) zeolite-Y. Top: deconvoluted spectra. Middle: 79.5 MHz, 8 K (Analcite, 1586 scans; zeolite-Y 5858 scans). Bottom: 17.9 MHz, 2 K spectra (analcite, 7406 scans; zeolite-Y 3762 scans). Typical relaxation delays were 5 s (79.5 MHz) and 1 s (17.9 MHz), and minimal line-broadening (analcite, 20 Hz; zeolite-Y, 10 Hz) was applied. (Reproduced, by permission, from ref. 36.)

field strength. The experiment is no more difficult at higher fields, and for spin $\frac{1}{2}$ nuclei there can be a substantial gain in resolution as seen (Fig. 6.7). Some of this is due to the superior stability and homogeneity of superconducting solenoid-based magnets compared to iron-core systems, but minor contributing interactions not removed by MAS (for example, dipolar interactions to quadrupolar nuclei) will be minimized at high fields, while the chemical shift dispersion is maximized.

In the case of quadrupolar nuclei with non-integral spins such as ^{11}B, ^{27}Al, ^{17}O, there are particular advantages in working at the highest possible magnetic field strength as will be discussed below.

6.3.5 Quadrupolar nuclei with non-integral spins

Quadrupolar nuclei of the above type are particularly important in inorganic systems, in fact making up the majority of all nuclei observable by NMR.[29] The effect of the quadrupolar interaction on the Zeeman splitting in such systems is illustrated in Fig. 6.8 for a spin $\frac{3}{2}$ nucleus such as ^{11}B.[52] As can be seen (Fig. 6.8), the $+\frac{1}{2}\leftrightarrow -\frac{1}{2}$ transition is unaffected by the quadrupolar interaction to first order (both levels are shifted by the same amount in the same direction), and it is this transition which is observed; the other two allowed transitions are usually too broad and shifted too far from resonance to be observed directly. The line shape due to the $+\frac{1}{2}\leftrightarrow -\frac{1}{2}$ transition is distorted and shifted due to the second-order quadrupole interaction.[51,53-5] The shift in the centre of gravity of this signal is given by[53]

$$\omega_{CG}-\omega_0=\frac{1}{30}\frac{\omega_Q^2}{\omega_0}\{I(I+1)-\tfrac{3}{4}\}(1+\tfrac{1}{3}\eta^2) \qquad (6.17)$$

where $\omega_Q=3e^2qQ/2I(2I-1)h$, eq is the z-component of the electric field

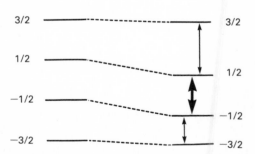

Fig. 6.8. Schematic representation of the energy levels of a nucleus with spin $I=\frac{3}{2}$ due to the Zeeman and first-order quadrupolar interactions. Note that the $I=\frac{1}{2}\leftrightarrow I=-\frac{1}{2}$ transition indicated in the figure is independent of the quadrupolar interaction to first order.

tensor, eQ is the quadrupole moment of the nucleus, η is the asymmetry parameter, and ω_0 is the frequency (rad s^{-1}) associated with the Zeeman field H_0 (eqn 6.2).

The distortion of the line shape also depends on ω_Q^2/ω_0 and η, and in addition depends on the angle between the axis of sample rotation and the magnetic field. The important feature of the second-order quadrupolar interaction is that, as indicated in eqn 6.17, it is inversely field-dependent, since ω_0 is directly proportional to the applied field. Therefore undesired shifts and line shape distortions are minimized by working at high fields. This is illustrated in Fig. 6.9 which shows the ^{27}Al MAS NMR spectra of a zeolite sample at 23.5 and 104.2 MHz. The high-field spectrum is much less distorted, and the apparent chemical shift is much closer to the correct isotropic shift value that would be measured in solution. In the case of quadrupolar nuclei, the nuclear spin is not quantized along the Zeeman field direction, and MAS greatly reduces, but does not completely eliminate the interaction.[51,53-5] In the case where the second-order quadrupolar interaction completely dominates the spectrum—as for example the case of ^{51}V in VO$_3^-$ and V$_2$O$_5$ investigated by Oldfield and co-workers[51,56,57]—spinning at an angle other than 54°44' may be employed to reduce quadrupolar interaction;[51] but other interactions such as the shift anisotropy may increase, and an experimental compromise must often be sought. In any event, use of the highest possible field will be advantageous, and these experiments for the most part may be carried out again using a conventional high-resolution spectrometer.

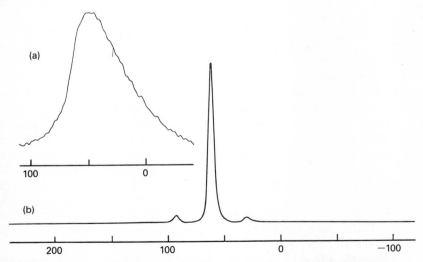

Fig. 6.9. ^{27}Al MAS NMR spectra of zeolite Y. (a) At 23.5 MHz (7771 FID, 0.1 s delay, 25 Hz line-broadening). (b) At 104.2 MHz (2656 FID, 0.1 s delay, 0 Hz line-broadening).

6.4 General features of high-resolution solid-state NMR spectra

From the work of the last two decades, particularly from that in the last five years in which there has been a tremendous growth in the application of MAS and CP–MAS techniques, a reasonably clear picture has evolved of utility of these measurements, especially in relation to X-ray diffraction studies of solids. In general, the NMR and diffraction techniques are complimentary.

In the case of crystalline compounds whose crystal and molecular structures can be determined by diffraction techniques, NMR studies may yield additional information and, since the isotropic NMR chemical shifts may be obtained *both* in solution *and* in the solid state, they may be used as a 'bridge' between the solid-state structures and those which exist in solution. For example, they will provide 'benchmark' values for the chemical shifts of complexes which are well defined in the solid state, but which dissociate and/or are involved in fast multiple exchange equilibria in solution. They may also be used to place structure and reactivity correlations developed for inorganic systems on a sounder basis, by relating them directly to the X-ray structures in the solid by using the solid-state spectra. In other systems, as will subsequently be seen, the X-ray structural data may be limited in nature, even for crystalline compounds, and the NMR data can provide a more complete picture of the solid state structure.

In the case of amorphous systems such as glasses and surfaces, the solid-state spectra are particularly useful as the lack of long-range order precludes the use of diffraction techniques, and the NMR data provide the best available information regarding their structures, which are often largely unknown. The present chapter, however, will focus on applications to crystalline materials.

In a consideration of the application of high-resolution solid-state NMR techniques, some general differences between the solution and solid-state spectra must be recognized. First, although very considerable line-narrowing has been a achieved, the resolution in solid-state spectra is usually not as good as in solution. For extremely crystalline compounds, linewidths of 0.5 p.p.m. may be achieved, but in heterogeneous and/or amorphous systems they may range up to about 5–10 p.p.m. Second, although CP–MAS techniques do, indeed, yield isotropic chemical shift values, it must always be kept in mind that these are the isotropic shifts *for the solid state*, and as such may reflect the effects of the rigid lattice in various ways; they may be more complex than solution NMR spectra where rapid overall and internal rotations average out many of the non-equivalences observed in the solid state.

For example, Fig. 6.10 shows the solution and solid-state ^{31}P spectra of the transition-metal complex *cis*-[PtCl$_2$(PPh$_2$CH$_3$)$_2$] [*3*].[58] The solution spectrum shows a single central resonance and two satellites due to coupling of the

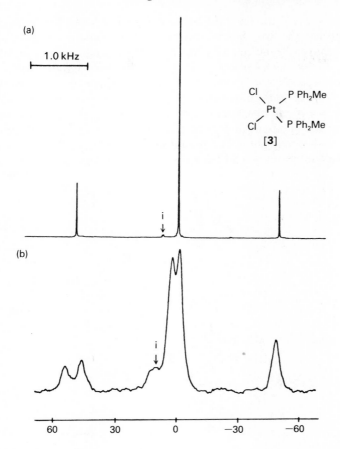

Fig. 6.10. ^{31}P NMR spectra of *cis*-[PtCl$_2$(PPh$_2$Me)$_2$]. (a) Solution spectrum, ^{1}H scalar decoupling. (b) CP–MAS solid-state spectrum with ^{1}H dipolar decoupling. (Reproduced, by permission, from ref. 58.)

phosphorus atoms to the 33.7 per cent abundant isotope ^{195}Pt (^{195}Pt–^{31}P = 3783 Hz), reflecting the *cis*-geometry of the complex as indicated. In the solid-state spectrum, however, there are *two* central resonances of equal intensity, each with its own set of satellites as indicated in Fig. 6.10(b) (the magnitudes of the couplings, 3877 and 3623 Hz indicate that the *cis*-geometry has been preserved). There are two possible explanations for the difference in the spectra, both involving the effect of the crystal lattice. First, there might be two crystallographically *inequivalent* sites in the unit cell, and the environments of the molecules at these two sites would be inequivalent and would give rise to slightly different chemical shifts for all of their component nuclei. Second, crystallization and the packing of the molecules

208 *Solid state chemistry: techniques*

in the lattice could force a fixed conformation on the molecule, particularly with respect to the two phosphine ligands which could make them inequivalent and remove the molecular plane of symmetry indicated in formula (*3*). In solution, this symmetry plane exists because there is rotation of the attached phosphine ligands about the Pt–P bonds making them equivalent 'on average'. In this particular case, the latter explanation is correct, as confirmed by the X-ray determined structure (Fig. 6.11)[59] which shows clearly the loss of the apparent molecular plane of symmetry observed in solution because of the imposition by the molecular packing in the crystal of fixed phosphine conformations which are not mirror images of each other.

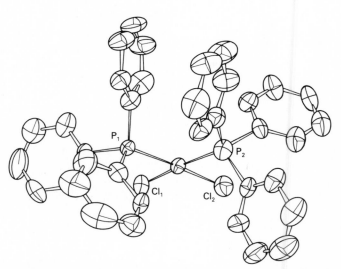

Fig. 6.11. Solid-state structure of *cis*-[PtCl$_2$(PPh$_2$Me)$_2$] as determined by X-ray diffraction techniques, showing the lack of a plane of symmetry through the central Pt atom due to the fixed conformations of the phosphine ligand whose central phosphorus atoms are indicated for clarity. (Reproduced, by permission, from ref. 59.)

A cautionary example, which shows that in some extreme cases solution and solid-state structures may be only tenuously related, comes from the early work of Andrew and co-workers[60,61] on the ^{31}P solid state MAS spectrum of PCl$_5$. The spectrum, shown in Fig. 6.12, has two absorptions which, from their chemical shift values, may be identified as arising from PCl$_4^+$ and PCl$_6^-$. Thus, in this instance, crystallization has completely changed the structure and indeed the basic nature of the compound from its solution structure of PCl$_5$.

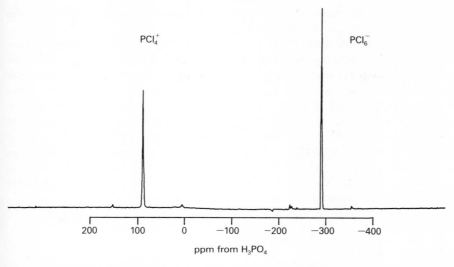

Fig. 6.12. ^{31}P MAS spectrum of 'solid PCl$_5$', showing resonances due to the species PCl$_4^+$ and PCl$_6^-$ as indicated. (Reproduced, by permission, from ref. 61.)

The above example is an extreme one, and, in general, there is a reasonably close correspondence between solution and solid-state structures and chemical shifts, but it emphasizes that care must always be exercised when relating the solid-state and solution data, especially when no diffraction data is available. Different conformations may exist *within* a single crystal, and different polymorphs may exist for a single molecular solid (and may occur simultaneously). Even when diffraction data is available, it must be kept in mind that this will have been obtained for a *single* crystal which may not reflect the bulk (about 0.5 g) of the sample used in the NMR experiment, or even the most common solid-state form of the compound. Not all of these problems have been encountered to date in inorganic systems, but they have been seen in other areas, and must be anticipated, especially since the large shift ranges of many inorganic nuclei indicate an extreme sensitivity to environmental effects.

6.5 Investigations of crystalline inorganic systems

The power of the MAS technique in providing high-resolution solid-state spectra was clearly demonstrated by the early work of Andrew[4,5,60-6] who investigated the ^{31}P and ^{19}F spectra of a series of small molecular solids including PCl$_5$, P$_4$S$_3$, K$^+$PF$_6^-$, K$^+$AsF$_6^-$, Mg$_3$P$_2$, and Zn$_3$P$_2$, and the spectra of metals such as cadmium and aluminium. The case of PCl$_5$ has been discussed in the previous section. Typical results from the other studies are presented in Fig. 6.13 and 6.14.

In the ^{19}F spectrum[61,62] of $K^+AsF_6^-$, MAS gives very substantial line-narrowing, producing the isotropic average chemical shift but also giving the indirect spin-spin coupling of 905 Hz from the central As atom which yields four lines of equal intensity from coupling to its four possible spin states (^{75}As is a spin $\frac{3}{2}$ nucleus). Although MAS is very successful in these cases, they are somewhat unusual systems in that they all exhibit molecular re-orientations which greatly reduce the dipolar and shift interactions, making their subsequent averaging by MAS much easier. Further, the nuclei being observed (^{31}P, ^{19}F) have large chemical shift dispersions making 'resolution' easier to obtain. However, these early examples are very important as they clearly indicated the potential utility of these measurements.

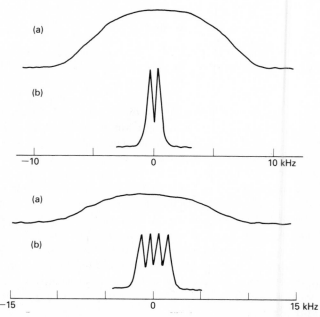

Fig. 6.13. (a) Static and (b) MAS ^{19}F NMR spectra of (top) $K^+PF_6^-$ and (bottom) $Na^+AsF_6^-$. The MAS spectra show the effect of scalar coupling to the central nucleus. (Reproduced, by permission, from ref. 62.)

In the case of metals, MAS averages the shift anisotropy as shown in Fig. 6.14 for cadmium metal,[61,65] making it possible to accurately measure the Knight shift of the metal. Again, a somewhat unusual situation exists as the line-broadening of the resonance comes from a single and a relatively small interaction. In general, for example ^{31}P in most inorganic systems, the experiment will be more complex.

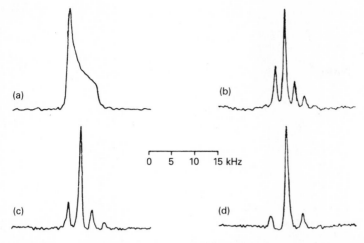

Fig. 6.14. Static and MAS spectra of ^{113}Cd in cadmium metal. (a), static; (b), (c), (d) respectively rotated at 2.1, 2.6 and 3.6 KHz. (Reproduced, by permission, from ref. 65.)

6.5.1 Coordination compounds

In much more recent work, a number of groups have used high-resolution solid-state NMR to investigate the structures of transition-metal-based complexes via the ^{31}P spectra of attached phosphine ligands. From solution NMR studies the ^{31}P spectra of attached ligands are known to reflect the structure, geometry, and bonding of the complex via their chemical shift and coupling constant values; a large number of structural correlations have been developed with solution NMR parameters.[67,68] In general, it is found that the types of correlations found in solution carry over to the solid state. For example, Fig. 6.15 shows the ^{31}P CP–MAS spectra[69] of *cis*- and *trans*-isomers of two square-planar platinum complexes of general formula PtCl$_2$(PR$_3$)$_2$ [4]. It is known from solution studies that $^1J(^{195}$Pt–^{31}P) is about 2500 Hz for the *trans* P—Pt—P arrangement.[67,68] As can be seen from the spectra, these correlations carry over into the solid, and the ^{31}P spectra of attached phosphine ligands may be used as probes of the molecular geometry. A general correspondence of ^{31}P shifts and couplings has been found for a large series of phosphine ligands and their complexes with various Pt and Pd species.[69]

A similar study by Maciel and co-workers[70] reports solid-state and solution data on the ^{31}P spectra of five different Rh(I) disphosphine catalysts and their corresponding ligands [5–9]. Again the chemical shifts in solution and in the solid state are generally comparable, but diphosphine complexes which give rise to chemically equivalent ^{31}P resonances in solution often yield two chemically shifted peaks in the solid. In fact, one of the ligands itself, 'chiraphos' [9] shows a single resonance at −8.7 p.p.m. (with respect to

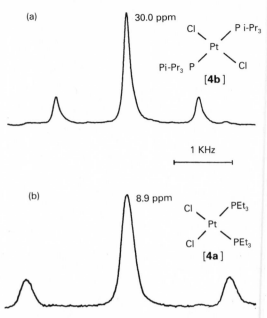

Fig. 6.15. ^{31}P CP–MAS spectra of (a) *trans*-[PtCl$_2$(Pi-Pr$_3$)$_2$] and (b) *cis*-[PtCl$_2$(PEt$_3$)$_2$]. (Reproduced, by permission, from ref. 69.)

H$_3$PO$_4$) in solution, but two peaks in the solid state. Similar considerations involving lattice effects as were discussed in Section 6.4 above are relevant here, but no diffraction data are available in these cases from which to ascribe a specific mechanism for the peak multiplicities.

^{31}P CP–MAS spectra have been obtained for 'Wilkinson's catalyst' RhCl(PPh$_3$)$_3$[10], by Veeman and co-workers[71] (see Fig. 6.16). In solution, the

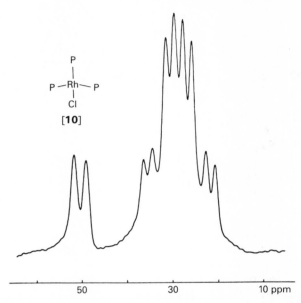

Fig. 6.16. ^{31}P–CP MAS spectrum of solid Wilkinson's catalyst, RhCl(PPh$_3$)$_3$ at 72.8 MHz. The spectrum shows indirect couplings between the phosphorous nuclei approximating to an ABX pattern, further split by coupling to the metal nucleus. (Reproduced, by permission, from ref. 71.)

two P atoms *trans* to each other are equivalent, and an A$_2$X pattern is observed in the ^{31}P spectrum. In the solid state, the two *trans* P-atoms are no longer equivalent, consistent with the X-ray structure, and an ABX pattern is observed as indicated in Fig. 6.16, further coupled to the rhodium atom $J(^{103}\text{Rh–P}_X) = 185$ Hz, $^1J(^{103}\text{Rh–P}_B) \approx {}^1J(^{103}\text{Rh–P}_A) \approx 139$ Hz, and $\Delta\delta_{AB} = 7.4$ p.p.m., $^2J(\text{P}_A\text{–P}_B) = 365$ Hz. These derived parameters do not exactly reproduce spectra obtained at lower fields, and there is some ambiguity in that several polymorphs of this material have been identified.[72]

A ^{31}P spectrum of the gold cluster [Au(PPh$_3$)$_8$$^{3+}$][NO$_3$$^-$]$_3$ [11] also shows two resonances for the phosphorus atoms in the ligands but there is no X-ray structural data available to determine the reason.

Veeman and co-workers[71] have also studied the copper based complexes (Ph$_3$P)$_3$CuNO$_3$ [12] and (Ph$_3$P)$_3$CuCl [13]. Both complexes show large couplings to the two isotopes of copper (^{63}Cu, spin $\frac{3}{2}$, 69.1 per cent; ^{65}Cu, spin $\frac{3}{2}$, 30.9 per cent). The copper nucleus is quadrupolar, and the phosphorus–copper couplings are not simply the indirect spin-spin couplings $^1J(\text{Cu–P})$ as the quadrupolar constants of the copper nuclei are comparable in magnitude to the Zeeman interaction; MAS does not completely average the dipolar and indirect coupling interactions. A subsequent complete and general analysis of the frequency dependence of the spectra by the same group has yielded an unambiguous interpretation of the spectra.[73]

Thus, results have been obtained already on crystalline transition-metal-

214 *Solid state chemistry: techniques*

Note: This structure is by inference from the X-Ray structure of [Au$_9$ (P(p–CH$_3$C$_6$H$_4$)$_3$)$_8$] [PF$_6$]$_3$

[11]

[12]

[13]

[Me$_3$Pt (acac)]$_2$

[14]

based complexes, and they promise to yield significant information on catalyst systems in the future. Some results on amorphous systems are presented in the next section. An obvious experiment in this area is the direct observation of spectra from the central metal nucleus, many of which have been investigated in solution. In fact, because of the enhancement possible from CP techniques, solid-state spectra may well be more sensitive for metal nuclei with small magnetogyric ratios and/or long spin-lattice relaxation times. To date, attempts to observe these nuclei have been largely unsuccessful, although the spectrum of a relatively symmetrical platinum compound [14], has been recently[74] reported. The problem is probably due to the large chemical shift anisotropies of these nuclei, and their very marked sensitivity to environmental effects as reflected in their large shift dispersions found in solution. For a large anisotropy, a very large number of spinning sidebands (see Fig. 6.3) would be produced but, for sensitive nuclei such as ^{195}Pt, the reduction in S/N should not by itself preclude their observation.

A more difficult problem is that of the potentially huge shift ranges which will make it difficult to find the signals, given the fixed observation 'window' of about 100 kHz of most spectrometers. The solution spectra may not be a reliable guide, as the metal atoms are the most likely to be affected by solution. This problem may be solved by the use of ultra-fast digitizers which permit observation of frequency ranges of up to 4 MHz. It is thus anticipated that, in the future, many studies of transition metal complexes may well centre on investigations of the central metal atoms themselves.

^{113}Cd studies have been made of a number of crystalline cadmium salts.[75,76] Typical of the data is the spectrum of 3CdSO$_4$.8H$_2$O shown in Fig. 6.17. In general, the results are in agreement with the expected structures but, in this case, the NMR data has prompted a re-investigation of the diffraction data, and the establishment of a new crystal and molecular structure for 3CdSO$_4$.8H$_2$O.[76] There are very significant gains in using the CP technique for ^{113}Cd. Apart from the direct gain from the cross-polarization from ^{1}H (γ^1H/γ^{113}Cd = −4.5), the ^{113}Cd relaxation times can be very long indeed (up to

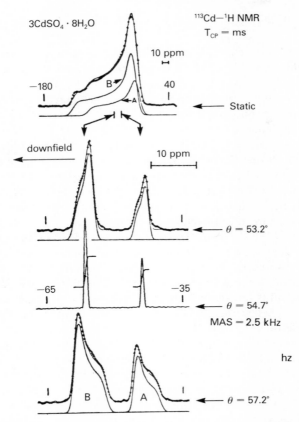

Fig. 6.17. ^{113}Cd CP–MAS spectra of 3CdSO$_4$.8H$_2$O. (Reproduced, by permission, from ref. 76.)

300 s) compared to the ^{1}H relaxation times (about 0.5 s), and the use of recycle times based on the ^{1}H relaxation yields a tremendous increase in efficiency.

Other studies have built up a reliable data base for the interpretation of ^{113}Cd chemical shifts by recording the shifts of cadmium complexes of well-defined geometry and ligand coordination (tetrahedral, octahedral, and trigonal bipyramidal) directly in the solid states.[77–81] These may be used as 'benchmark' values to interpret ^{113}Cd chemical shifts in solution which are very difficult to interpret due to the existence of multiple fast-exchange equilibria involving both ligands and solvent molecules which can occur.[77,78]

In the first and most complete study of this type, Maciel and co-workers[77] used the solid-state ^{113}Cd chemical shifts of the tetrahedral CdX$_4^{2-}$ species (X = Cl, Br, I), as reference values to aid in the interpretation of the concentration dependence of the solution shifts which are determined by the equilibria shown in eqn (6.18):

$$CdX_4^{2-} \rightleftharpoons CdX_3^- + X^- \rightleftharpoons CdX_2 + 2X^- \rightleftharpoons CdX^+ + 3X^- \rightleftharpoons Cd^{2+} + 4X^-.$$
(6.18)

Reliable values of the chemical shifts of the different Cd species present, and of the associated equilibrium constants, were obtained using the solid-state shifts of CdX_4^{2-} species as a starting point in the iteration.

One of the reasons for the effort to develop a detailed understanding of the dependence of Cd shifts on geometry, and on the nature of the coordinated ligands, is so that they can be used to investigate cadmium-containing bioinorganic systems.[79-82] Cd(II) occurs naturally in some metalloproteins but may be substituted for other metal atoms in a whole range of materials of biological importance. Thus, Gerstein and co-workers[79] have studied the decanuclear cation $Cd_{10}(SCH_2CH_2OH)_{16}^{4+}$ [15] (see Fig. 6.25) as a model for the cadmium environments in proteins, for example in metallothionen. The solid-state structure of the reference compound, and its solution NMR spectrum, are both known. The solid-state and solution spectra were identical, and the authors concluded that the Cd atom was in a 'CdS_4' site.

Ellis and co-workers[82] have probed metal porphorin structures in a study of Cd porphorin in the solid state, both as the free porphorin [16] in which the central metal is four-coordinate, and also as the pyridine adduct where the metal is five-coordinate [17]. These data serve as a basis for the interpretation of the ^{113}Cd shifts of these complexes in solution in the presence of donor ligands. It was concluded from the solid-state data that the observed solution shifts probably arise from large but opposing shifts in the tensor elements from the electronic effect of the donor atom binding to the metal, and from the geometric change induced by this coordination where the metal atom is moved out of the porphorin molecular plane as indicated schematically in [17].

[16] [17]

6.5.2 Aluminosilicates

In a different area, ^{29}Si MAS NMR has been used to investigate silicates,[83] and ^{29}Si and ^{27}Al MAS NMR for the structures of zeolites and other aluminosilicates.[21,22,84-6] In an investigation of the ^{29}Si solid-state NMR spectra of silicates, Lippmaa and co-workers[83] established that the chemical shifts fell into relatively clear ranges, just as in solution, depending on the number of further silicon atoms attached at the vertices of a given SiO_4 tetrahedron, and that the solid-state spectra could be used to determine structure. The current nomenclature for the five possible silicon environments is Q^0, Q^1, Q^2, Q^3, Q^4 where the superscript indicates the number of attached silicons. In related work, studies have recently been carried out on silicon-based polymers.[87]

In the case of zeolites, the ^{29}Si spectra are influenced by whether the four groups at the vertices of an SiO_4 tetrahedron are Si or Al. Zeolites are very loose framework aluminosilicates of the very general formula given in eqn. (6.19) where AlO_4 tetrahedra have been substituted for SiO_4 tetrahedra. For every Al-centred unit present, a positive charge must be present to preserve electrical neutrality, and the cavities and channels of these materials contain sufficient cations to compensate the number of Al atoms present as indicated in eqn (6.19) together with associated water molecules.

$$M_y^+(SiO_2)_x(AlO_2^-)_y \cdot nH_2O. \qquad (6.19)$$

Figure 6.18 shows structures of three typical, common zeolites. Zeolite-A is that commonly used as a 'molecular sieve', and faujasite (also known as

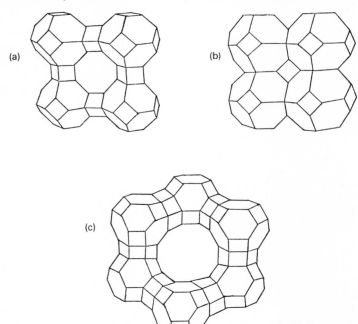

Fig. 6.18. Structures of three common zeolites based on the 'sodalite cage' building block identified in each of the structures. Each corner is the location of either a silicon or aluminium atom, joined by oxygen bridges (not shown). (a) Zeolite-A. (b) Sodalite. (c) Faujasite. (Reproduced, by permission, from ref. 86.)

218   *Solid state chemistry: techniques*

zeolites X and Y, depending on composition) has been widely used as a cracking catalyst in the petroleum industry. The usefulness of these materials comes from their loose framework structure, which have both channels and cavities to confer size and shape selectivity in reactions, and a large concentration of highly acidic sites for acid catalysis (in the second zeolite in Fig. 6.18, sodalite, the 'cavity' in the centre of the structure is exactly the shape of the building unit and thus is filled, giving a relatively dense structure with no channels or cavities; it is not of any catalytic importance).

Zeolites represent a situation in which diffraction measurements give only limited structural information and the NMR data are of particular importance.[88] Although the samples are usually of extremely high crystallinity, they are usually microcrystalline, and it is difficult to obtain samples of optimal size for X-ray diffraction measurements. The problem is compounded by the fact that Si and Al are adjacent elements in the periodic table and have almost identical scattering factors. Thus, X-ray diffraction measurements of these compounds define the gross framework structures very well, as shown in Fig. 6.18, but do not define the positions of the individual silicon and aluminium atoms. These are conventionally distributed according to the

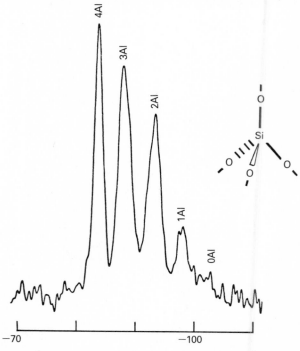

Fig. 6.19. ^{29}Si MAS spectrum at 79.6 MHz of the zeolite, analcite, showing five absorptions characteristic of the five possible permutations of Si and Al atoms attached at the corners of the SiO$_4$ tetrahedron as indicated. (Reproduced, by permission, from ref. 84.)

constraints of Lowenstein's Rule which postulates that Al—O—Al linkages do not occur, and which, in some cases, can yield an unambiguous Si/Al distribution.

The power of ^{29}Si NMR as applied to these systems is that the resonances are sensitive to the local environment of the ^{29}Si nucleus, and thus give a detailed description of the microstructure of these materials in terms of the distribution of silicon environments, and indirectly the distribution of silicon and aluminium atoms in the lattice.[84,89] The ^{29}Si spectrum of a typical zeolite is shown in Fig. 6.19, and it shows signals due to the five possible local environments of silicon generated by the permutation of Si and Al atoms at the vertices of the SiO$_4$ tetrahedron, i.e. Si[4Al], Si[3Al, Si], Si[2Al, 2Si], Si[Al, 3Si] and Si[4Si]—as indicated in Fig. 6.19.

Lippmaa, Engelhardt, and co-workers[83,90] first showed, from an examination of a series of aluminosilicates of relatively well-defined structures, that (at least to a first approximation) the chemical shift ranges of ^{29}Si for the different environments were independent of the cation or specific structure, but reflected the local environment, and thus could be used for structure elucidation. (It must be borne in mind that NMR spectra always represent the total averaging of environments over the whole lattice.) Figure 6.20

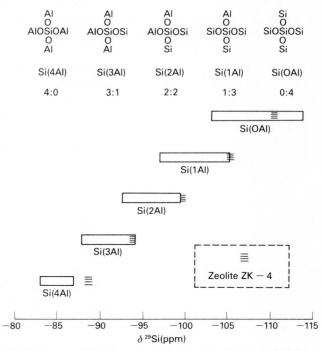

Fig. 6.20. Characteristic ^{29}Si chemical shift ranges found in zeolite structures[83,90] together with the shift values for zeolite ZK-4. (Adapted from ref. 83 and reproduced, by permission, from ref. 85.)

indicates the frequency ranges found from the original work of Lippmaa and Engelhardt as well as the shifts of a specific zeolite, ZK-4. ^{27}Al NMR spectra of these systems show only single resonances because every Al has the environment Al[4Si], although the Al chemical shifts of individual zeolites have characteristic values. The main chemical distinction afforded by the ^{27}Al spectra is between (1) octahedral coordination which gives a peak at about 0 p.p.m. (with reference to $Al(H_2O)_6^{3+}$.aq.) and (2) tetrahedral (lattice) coordination which occurs in the range of 50–65 p.p.m. Since ^{27}Al is a quadrupolar nucleus, care must be taken to work at high fields in order to minimize quadrupolar effects, and give reasonably reliable chemical shift information. It is not necessary to ^{1}H decouple in these experiments as the protons present are generally mobile and do not contribute greatly to the final ^{27}Al linewidths. Quantitatively reliable ^{29}Si and ^{27}Al spectra can be obtained straightforwardly at 400 MHz proton-frequency using a conventional high-resolution spectrometer.[36]

A number of groups[92-4] have investigated the ^{29}Si spectra of faujasite zeolites of various Si/Al ratios. As the Si/Al ratio increases, there is a gradual increase in the relative intensities of the higher field peaks as expected from the peak assignments given above (Fig. 6.20). Various groups[92-5] have attempted detailed interpretations of the lattice distributions of Si and Al atoms from this data. An important feature of the spectra is that they allow the Si/Al ratio to be calculated directly from the ^{29}Si spectrum alone, via the effect of the aluminium atoms which are present on the silicon nuclei, i.e. they measure only lattice aluminium. This is in contrast to the information from bulk chemical analyses which measure the total aluminium content of the sample whether it is present in the lattice or trapped in the pores and channels. Complementary data in this regard may be obtained by direct measurement of the ^{27}Al spectra. The quantitative relationship is given by eqn (6.20), where I_{nAl} indicates the intensity of the ^{29}Si peak corresponding to a Si[nAl,(4 − n)Si] environment.

$$\frac{\text{Si}}{\text{Al}} = \frac{\sum_{n=0}^{4} I_{nAl}}{\sum_{n=0}^{4} 0.25 n I_{nAl}} \tag{6.20}$$

Chemical conversions of zeolite frameworks may also be followed by these spectra. Figure 6.21 shows the ^{29}Si spectra of a faujasite sample of Si/Al $\approx$ 2.6, before and after treatment with $SiCl_4$ which is postulated[96,97] to cause dealumination with replacement of Al by Si according to

$$SiCl_4 + Na_y(SiO_2)_x(AlO_2)_y \rightarrow Na_{(y-1)}(SiO_2)_{(x+1)}(AlO_2)_{y-1}$$
$$+ NaCl + AlCl_3 + Na^+AlCl_4^-. \tag{6.21}$$

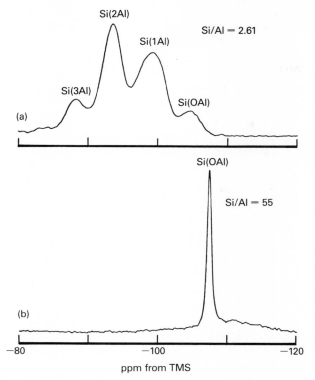

Fig. 6.21. ^{29}Si MAS NMR spectra at 79.6 MHz of (a) faujasite of Si/Al = 2.61 (zeolite-Y), and (b) the same sample after successive dealuminations with SiCl$_4$ and washings. (Reproduced, by permission, from ref. 7.)

The sharp ^{29}Si resonance confirms that the dealumination has occurred, and diffraction measurements show that the faujasite lattice has been retained. The ^{27}Al spectra (Fig. 6.22) confirm the basic mechanism of eqn. (6.21). The top spectrum (Fig. 6.22(b)), taken after reaction with SiCl$_4$ but before washing, shows that the tetrahedral (lattice) aluminium has greatly decreased, but there is appearance of a very intense absorption at about +100 p.p.m., which can be identified as due to AlCl$_4^-$ from the high-temperature reaction of NaCl and AlCl$_3$ as indicated in eqn. (6.21). This peak is removed by subsequent washing, but much of the aluminium remains trapped in the cavities as octahedral aluminium species. Careful further washing and if necessary, ion exchange is needed to remove more completely non-lattice aluminium (bottom spectrum, Fig. 6.22(d)). Several studies[98,99] have reported similar spectral investigations of the ultrastabilization/dealumination of faujasites (which is of extreme importance for imparting thermal stability in use as catalysts), and of the dealumination of a number of other zeolite systems.[98b-g]

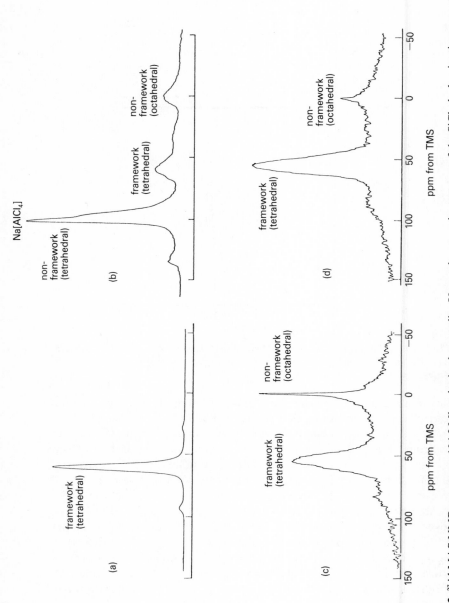

Fig. 6.22. ^{27}Al MAS NMR spectra at 104.2 Mhz obtained on zeolite-Y samples at various stages of the SiCl$_4$ dealumination procedure. (a) Starting Faujasite sample. (b) Intact sample after reaction with SiCl$_4$ before washing. (c) Sample (b) after washing with distilled water. (d) After several washings. (Reproduced, by permission, from ref. 85.)

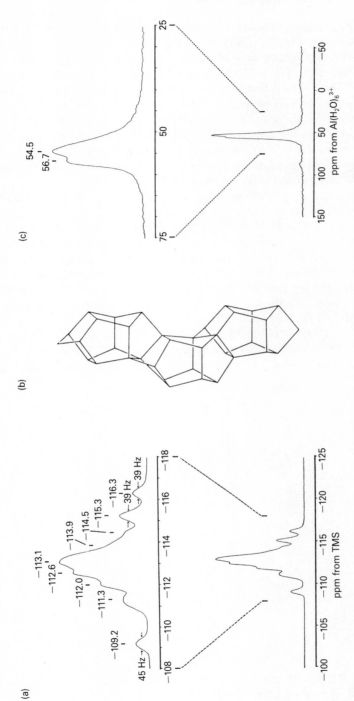

Fig. 6.23. MAS NMR spectra of silicalite. (a) ^{29}Si MAS NMR spectra at 79.6 MHz. (b) The pentasil building unit on which this system is based. (c) The ^{27}Al spectrum at 104.2 MHz. (Reproduced, by permission, from ref. 100.)

In suitable cases, considerable resolution can be obtained in the spectra of these materials. Figure 6.23(a) shows the ^{29}Si spectrum of a very pure sample of silicalite, thought to be a completely silicon-containing sample based on the pentasil unit (Fig. 6.23(b)) arranged in a similar structure to that of the zeolite ZSM-5. The different peaks observed are all due to Si[4Si] units, and reflect atoms at non-equivalent lattice sites (100). The relative intensities imply twenty-four such sites in the unit cell. The ^{27}Al spectrum (Fig. 6.23(c)) clearly indicates the presence of tetrahedral aluminium, implying that it is present in the lattice. ^{27}Al is a very sensitive nucleus, and quantitative analysis by comparison with known aluminium-containing materials indicates that the Si/Al ratio of this ultrapure sample is about 1000/1.

A cautionary and timely note, in the use of NMR in the deduction of solid-state structures, is given by the investigation of the structure of zeolite-A. This zeolite has a Si/Al ratio of 1, and application of the constraint of Loewenstein's Rule, to the framework structure determined by X-ray diffraction, implies that there is exact alternation of silicon and aluminium atoms in the lattice. There is thus a single silicon environment, Si[4Al]. As observed by Lippmaa and Engelhardt,[83] and subsequently confirmed by other workers,[101] zeolite-A does indeed show a single sharp resonance, but its chemical shift value of about 89 p.p.m. (with reference to TMS) suggests that the coordination is Si[3Al,Si] according to the chemical shift ranges of Fig. 6.20. Structures with this coordination, in which Loewenstein's Rule was systematically broken, have been proposed by a number of research groups.

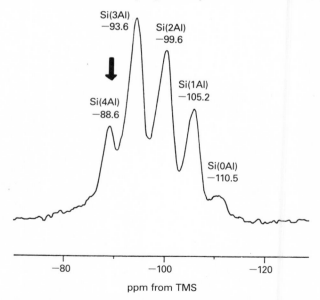

Fig. 6.24. ^{29}Si MAS NMR spectrum at 79.6 MHz of zeolite ZK-4. Both zeolite ZK-4 and zeolite-A have the same framework structure, and the multiplet of peaks in ZK-4 gives an unambiguous assignment of the coordination of the single resonance of zeolite-A. (Reproduced by permission from ref. 104.)

The basic problem is that, since only a single resonance is observed, the assignment of the coordination rests entirely on its chemical shift value and the validity of the characteristic shift ranges which have been established. This problem has recently been resolved by Fyfe, Thomas, and co-workers[102,103] and by Melchior and co-workers,[104] who independently investigated the zeolite ZK-4, which has the same structure as zeolite-A (but which is synthesized using an organic template), and has Si/Al ratio greater than unity. The spectrum of this material must now show five peaks as with other zeolite systems, and their assignment is therefore unambiguous. The spectrum, which is shown in Fig. 6.24 clearly indicates that the peak at $\delta = -89$ p.p.m. must be due to Si[4Al] and there is now general agreement on the structure of zeolite-A. The chemical shifts of ZK-4 are shown in Fig. 6.20 where it can be seen that the Si[4Al] peak falls outside the established range, possibly because of the distortion of the T–O–T (tetrahedral) angles in the structure.

6.6 Extensions of the CP–MAS experiment

Many extensions and modification of the 'simple' CP–MAS experiment have been developed and implemented, mainly to ^{13}C NMR studies of organic compounds. These have been reviewed in detail and only the most applicable are indicated here.

A useful technique for discriminating between protonated and non-protonated carbon atoms has been introduced by Opella et al.[105] and will be of use in organometallic systems. More recently, Saika and co-workers,[106] and subsequently Grant and Zilm,[107] have shown that it is possible to retain heteronuclear proton–X couplings without sacrificing the high-resolution characteristics of the solid-state spectra—again a technique most applicable to ^{13}C in organometallic systems.

As discussed in Section 6.2, the chemical shift anistropy reflects the three-dimensional shielding of the nucleus, and thus is a potentially rich source of information regarding the chemical bonding. The problem is that for systems with several non-equivalent nuclei, these patterns will overlap and be unresolvable (the shift anisotropy in many cases will be larger than the whole isotropic chemical shift ranges). This may be overcome by spinning slower than the frequency range of the anisotropy pattern (as illustrated in Fig. 6.3 where spectral resolution is maintained), and the shielding tensor elements may be evaluated from the sideband intensities.[46,82] An alternative approach is to spin the sample slightly off the 'magic angle'—in which case a scaled-down (and exactly calculable) anisotropy pattern is observed. An application of this approach to ^{113}Cd spectra by Gerstein and co-workers[108] is presented in Fig. 6.25. It should be emphasized, however, that the complete description of the shielding tensor must include the orientation of the principal shielding elements with respect to the molecular coordinate system, and a single-crystal study will usually be necessary to obtain this information.

There are two very important experiments which are not really extensions

226 *Solid state chemistry: techniques*

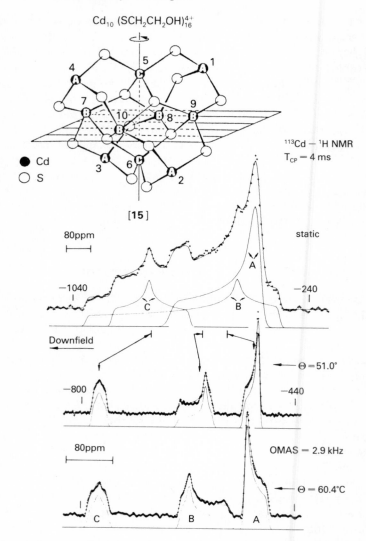

Fig. 6.25. Off-angle spinning ^{113}Cd spectra of solid $Cd_{10}(SCH_2CH_2OH)_{16}^{4+}$, together with the calculated shift anisotropy patterns indicated. (Reproduced, by permission, from ref. 135.)

of the CP–MAS technique, but which will be of considerable importance in the application of high-resolution solid-state NMR to inorganic systems.

First, there is the realization that high-field MAS experiments using conventional high-resolution spectrometers are quite adequate and, in some cases, superior to more complex experiments for many inorganic materials.[36] Several examples have been given in the preceding sections, and it brings the experiment easily within the capabilities and available resources of most inorganic chemists.

Observations have not been reported for many inorganic nuclei which, in principle, should be easily detected in the solid state. This may be due to the large shift ranges, and (this is the second area of experiment) the use of fast digitizers to expand the available spectral 'window' should greatly facilitate the further investigation of inorganic systems.

6.7 References

1. Mehring, M. High resolution NMR spectroscopy in solids. In *NMR basic principles and progress* (eds. P. Diehl, E. Fluck, and R. Kosfeld) Vol. 11. Springer Verlag, New York (1976).
2. Haeberlen, U. *Adv. magn. Reson.* Suppl. 1 (1976).
3. Fyfe, C. A. *Solid state NMR for chemists*. C. F. C. Press (1984).
4. Andrew, E. R. *Prog. NMR Spectroscopy* **8**, 1 (1971).
5. Andrew, E. R. In *Int. Rev. Sci. phys. Chem.* Series 2, Vol. 4 (Magnetic Resonance) (ed. C. A. McDowell), p. 173. Butterworths, London (1975).
6. Griffen, R. G. *Anal. Chem.* **49**, 951A (1977).
7. Vaughan, R. W. *Ann. Rev. phys. Chem.* **29**, 397 (1978).
8. Schaefer, J. and Stejskal, E. D. High resolution ^{13}C NMR of solid polymers. In *Topics in carbon-13 NMR spectroscopy* (ed. G. C. Levy), Vol. 3, p. 283. Wiley-Interscience, New York (1979).
9. Torchia, D. A. and VanderHart, D. L. High-power double-resonance studies of fibrous proteins, proteoglycans, and model membranes. In *Topics in carbon-13 NMR spectroscopy* (ed. G. C. Levy), Vol. 3, p. 325. Wiley-Interscience, New York (1979).
10. Miknis, F. P., Bartuska, V. J., and Maciel, G. E. *Amer. Lab.* Nov., 19 (1979).
11. Lyerla, J. R. High resolution carbon-13 NMR studies of bulk polymers. In *Contemporary Topics in Polymer Science* (ed. M. Shen), Vol. 3, p. 143. Plenum Publishing Corp., New York (1979).
12. Schaefer, J., Stejskal, E. O., Sefcik, M. D., and McKay, R. A. *Phil. Trans. R. Soc.* **A299**, 593 (1981).
13. Garroway, A. N., VanderHart, D. L., and Earl, W. L. *Phil. Trans. R. Soc.* **A299**, 609 (1981).
14. Balimann, G. E., Groombridge, C. J., Harris, R. K., Packer, K. J., Say, B. J., and Tanner, S. F. *Phil. Trans. R. Soc.* **A299**, 643 (1981).
15. Opella, S. J., Hexem, J. G., Frey, M. H., and Cross, T. A. *Phil. Trans. R. Soc.* **A299**, 665 (1981).
16. Gerstein, B. C. High resolution nuclear magnetic resonance of solids. In *Magnetic resonance in colloid and interface science* (eds. J. P. Fraissard and H. A. Resing), p. 175, Reidel Publishing Company (1980).
17. Griffin, R. G. *Methods in Enzymology* **72**, 108 (1981).
18. Duncan, T. M. and Dybowski, C. *Surf. Sci. Rep.* **1**, 157 (1981).
19. Yannoni, C. S. *Acc. chem. Res.* **15**, 201 (1982).
20. Lyerla, J. R., Yannoni, C. S., and Fyfe, C. A. *Acc. chem. Res.* **15**, 208 (1982).
21. Wasylishen, R. E. and Fyfe, C. A. High-resolution NMR of solids. In *Annu. Rep. NMR Spectrosc.* (ed. G. A. Webb) **12**, 1. Academic Press, New York (1982).
22. (a) Fyfe, C. A., Bemi, L., Clark, H. C., Davies, J. A., Gobbi, G. C., Hartman, T.

S., Hayes, P. J., and Wasylishen, R. E. In *Inorganic chemistry: toward the 21st century* (ed. M. H. Chisholm) ACS Symposium Series **211**, 405 (1983).
(b) Appleman, B. R. and Dailey, B. P. *Adv. magn. Reson.* **7**, 231 (1974).
23. Gerstein, B. *Phil. Trans. R. Soc.* **A299**, 521 (1981).
24. (a) Schaefer, J., Stejskal, E. O., and McKay, R. In *NMR spectroscopy: new methods and applications* (ed. G. C. Levy) ACS Symposium Series **191**, 187 (1982).
(b) Dawson, W. H., Kaiser, S. W., Inners, R. R., Doty, F. D., and Ellis, P. D. ibid. p. 219.
(c) Gierasch, L. M., Frey, M. H., Hexem, J. G., and Opella, S. J. ibid. p. 233.
(d) Maciel, G. E. and Sullivan, M. J. ibid. p. 319.
(e) Jelinski, L. W., Dumars, J. J., Schilling, F. C., and Bovey, F. A. ibid. p. 3.
25. Pines, A., Gibby, M. C., and Waugh, J. S. *J. chem. Phys.* **56**, 1776 (1972).
26. Pines, A., Gibby, M. C., and Waugh, J. S. *Phys. Lett.* **15**, 373 (1972).
27. Pines, A., Gibby, M. C., and Waugh, J. S. *J. chem. Phys.* **59**, 569 (1973).
28. Becker, E. D. *High resolution NMR—theory and chemical applications* (2nd edn). Academic Press, New York (1980).
29. Harris, R. K. and Mann, B. E. (eds) *NMR and the periodic table*. Academic Press, London (1978).
30. Rinaldi, P. L., Levy, G. C., and Choppin, G. R. *Rev. inorg. Chem.* **II**, 53 (1980).
31. Jolly, P. W. and Mynott, R. *Adv. organometal. Chem.* **19**, 257 (1981).
32. Dechter, J. J. *Prog. inorg. Chem.* **29**, 285 (1982).
33. Appleman, B. R. and Dailey, B. P. *Adv. magn. Reson.* **7**, 231 (1974).
34. Tutunjian, P. N. and Waugh, J. S. *J. chem. Phys.* **76**, 1228 (1982).
35. (a) Gerstein, B. C. *Phil. Trans. R. Soc.* **A299**, 521 (1981).
(b) Ryan, L. M., Tayler, R. E., Paff, A. J., and Gerstein, B. C. *J. chem. Phys.* **72**, 508 (1980).
36. Fyfe, C. A., Gobbi, G. C., Hartman, J. S., Lenkinski, R. E., O'Brien, J., Beange, E. R., and Smith, M. A. R. *J. magn. Reson.* **47**, 168 (1982).
37. Eckman, R. *J. chem. Phys.*, **76**, 2767 (1982).
38. Lauterbur, P. C. *Phys. Rev. Lett.* **1**, 343 (1958).
39. Ackerman, J. L., Tegenfeldt, J., and Waugh, J. S. *J. Amer. chem. Soc.* **96**, 6843 (1974).
40. Veeman, W. S. *Phil. Trans. R. Soc.* **A299**, 475 (1981).
41. Lutz, O. and Nolle, A. *Z. Physik* **36B**, 323 (1980).
42. Krieger, A. I., Lundin, A. G., Moskivich, Yu. N., and Sukhovskii, A. A. *Phys. Stat. Sol.* **58**, k81 (1980).
43. Stoll, M. E., Vaughan, R. W., Saillant, R. B., and Cole, T. *J. chem. Phys.* **61**, 2896 (1974).
44. Andrew, E. R., Bradbury, A., and Eades, R. G. *Nature* **182**, 1659 (1958).
45. Lowe, I. J. *Phys. Rev. Lett.* **2**, 285 (1959).
46. Herzfeld, J. and Berger, A. E. *J. chem. Phys.* **73**, 6021 (1980).
47. Stejskal, E. O., Schaefer, J., and McKay, R. A. *J. magn. Reson.* **25**, 569 (1977).
48. (a) Hartmann, S. R. and Hahn, E. L. *Bull. Amer. phys. Soc.* **5**, 498 (1960).
(b) Hartmann, S. R. and Hahn, E. L. *Phys. Rev.* **128**, 2042 (1962).
49. Schaefer, J. and Stejskal, E. O. *J. Am. chem. Soc.* **98**, 1031 (1976).
50. Veregin, R. E. Unpublished results (1984).
51. Ganapathy, S., Schramm, S., and Oldfield, E. *J. chem. Phys.* **77**, 4360 (1982).
52. Abragam, A. *The principles of nuclear magnetism*. Clarendon Press, Oxford (1961).
53. Zumbulyadis, N., Henrichs, P. M., and Young, R. H. *J. chem. Phys.* **75**, 1603 (1981).
54. Behrens, H. -J. and Schnable, B. *Physica* **114B**, 185 (1982).

55. Kundla, E., Samoson, A., and Lippmaa, E. *Chem. Phys. Lett.* **83,** 229 (1981).
56. Oldfield, E., Kensey, R. A., Montez, B., Ray, T., and Smith, K. A. *J. chem. Soc., chem. Comm.* 1982, p. 254 (1982).
57. Meadows, M. D., Smith, K. A., Kensey, R. A., Rothgeb, T. M., Skarjune, R. P., and Oldfield, E. *Proc. Nat. Acad. Sci., USA* **79,** 1351 (1982).
58. Bemi, L., Clark, H. C., Davies, J. A., Drexler, D., Fyfe, C. A., and Wasylishen, R. *J. organometal. Chem.* **224,** C9 (1982).
59. Anderson, G. K., Clark, H. C., Davies, J. A., Ferguson, G., and Parvez, M. *J. Crystallog. Spectrog. Res.* **12,** 449 (1982).
60. Andrew, R. E., Bradbury, A., Eades, R. G., and Jenks, G. J. *Nature* **188,** 1096 (1960).
61. Andrew, E. R. *Phil. Trans. R. Soc.* **A299,** 505 (1981).
62. Andrew, E. R., Farnell, L. F., and Gledhill, T. D. *Phys. Rev. Lett.* **19,** 6 (1967).
63. Andrew, E. R., Firth, M., Jasinski, A., Randall, P. J. *Phys. Lett.* **31A,** 446 (1970).
64. Andrew, E. R., Hinshaw, W. S., Tiffen, R. S. *Phys. Lett.* **46A,** 57 (1973).
65. Andrew, E. R., Hinshaw, W. S., Tiffen, R. S. *J. magn. Reson.* **15,** 191 (1974).
66. Andrew, E. R., Hinshw, W. S., Jasinski, A. *Chem. Phys. Lett.* **24,** 399 (1974).
67. Pregosin, P. S. and Kunz, R. W. *^{31}P and ^{13}C NMR of transition metal phosphine complexes.* Springer Verlag, Berlin (1979).
68. Davies, J. A. Multinuclear magnetic resonance methods in the study of organometallic compounds. In *The chemistry of the metal–carbon bond* (eds S. Patai and F. R. Hartley). Wiley-Interscience, New York, in press.
69. Bemi, L., Clark, H. C., Davies, J. A., Fyfe, C. A., and Wasylishen, R. E. *J. Amer. chem. Soc.* **104,** 438 (1982).
70. Maciel, G. E., O'Donnell, D. J., and Greaves, R. *Adv. Chem. Series* **196** (eds E. D. Alyea and D. W. Meek). American Chemical Society (1981).
71. Diesveld, J. W., Menger, E. M., Edzes, H. T., and Veeman, W. S. *J. Amer. chem. Soc.* **102,** 7935 (1980).
72. Bennett, M. J. and Donaldson, P. B. *Inorg. Chem.* **16,** 659 (1977).
73. Menger, E. M. and Veeman, W. S. *J. magn. Reson.* **46,** 257 (1982).
74. Doddrell, D. M., Barron, P. F., Clegg, D. E. and Bowie, C. *J. chem. Soc.,* 1982, p. 575 (1982).
75. Cheung, T. T. P., Worthington, L. E., DuBois Murphy, P., and Gerstein, B. C. *J. magn. Reson.* **41,** 158 (1980).
76. DuBois Murphy, P. and Gerstein, B. C. *J. Amer. chem. Soc.* **103,** 3282 (1981).
77. Ackerman, J. J. H., Orr, T. V., Bartuska, V. J., and Maciel, G. E. *J. Amer. chem. Soc.* **101,** 341 (1979).
78. Mennitt, P. G., Shatlock, M. P., Bartuska, V. J., and Maciel, G. E. *J. phys. Chem.* **85,** 2087 (1981).
79. DuBois Murphy, P., Stevens, W. C., Cheung, T. T. P., Lacelle, S., Gerstein, B. C., and Kurtz, D. M. *J. Amer. chem. Soc.* **103,** 4400 (1981).
80. Turner, R. W., Rodesiler, P. F., and Amma, E. L. *Inorg. Chim. Acta* **66,** L13 (1982).
81. Rodesiler, P. F. and Amma, E. L. *J. chem. Soc., chem. Comm.,* 1982, p. 182 (1982).
82. Jakobsen, H. J., Ellis, P. D., Inners, R. R., and Jensen, C. F. *J. Amer. chem. Soc.* **104,** 7442 (1982).
83. Lippmaa, E., Magi, M., Samoson, A., Engelhardt, G., and Grimmer, A. -R. *J. Amer. chem. Soc.* **102,** 4889 (1980).
84. Thomas, J. M., Ramdas, S., Millward, G. R., Klinowski, J., Audier, M., Gonzalez-Calbet, J., and Fyfe, C. A. *J. solid-state Chem.* **45,** 368 (1982).
85. Fyfe, C. A. and Thomas, J. M. (To be published.)

86. Whan, D. A. *Chem. in Britain* **17,** 532 (1981).
87. Engelhardt, G., Jancke, H., Lippmaa, E., and Samoson, A. *J. organometall. Chem.* **210,** 295 (1981).
88. Thomas, J. M. In *Inorganic chemistry: toward the 21st century* (ed. M. H. Chisholm) ACS Symposium Series **211,** 445 (1983).
89. Klinowski, J., Ramdas, S., Thomas, J. M., Fyfe, C. A., and Hartman, J. S. *J. chem. Soc., Faraday Trans.* (2) **78,** 1025 (1982).
90. Lippmaa, E., Magi, M., Samoson, A., Tarmak, M., and Engelhardt, G. *J. Amer. chem. Soc.* **103,** 4992 (1981).
91. Muller, D., Gessner, W., Behrens, H. -J. and Scheler, G. *Chem. Phys. Lett.* **79,** 59 (1981).
92. Klinowski, J., Thomas, J. M., Fyfe, C. A., Gobbi, G. C. *Nature* **296,** 533 (1982).
93. Melchior, M. T., Vaughan, D. E. W. and Jacobson, A. J. *J. Amer. chem. Soc.* **104,** 4859 (1982).
94. Engelhardt, G., Lohse, U., Lippmaa, E., Tarmak, M., and Magi, M. *Z. anorg. allg. Chem.* **482,** 49 (1981).
95. Peters, A. W. *J. phys. Chem.* **86,** 3489 (1982).
96. Klinowski, J., Thomas, J. M., Audier, M., Vasudevan, S., Fyfe, C. A., and Hartman, J. S. *J. chem. Soc., chem. Comm.,* 1981, p. 570 (1981).
97. Klinowski, J., Thomas, J. M., Fyfe, C. A., Gobbi, G. C., and Hartman, J. S. *Inorg. Chem.* **22,** 63 (1983).
98. (a) Klinowski, J., Thomas, J. M., Anderson, M. W., Fyfe, C. A., and Gobbi, G. C. *Zeolites* **3,** 5 (1983).
    (b) Fyfe, C. A., Gobbi, G. C., Murphy, W. J., Oyubko, R. S., and Slack, D. A. *J. Amer. Chem. Soc.,* **106,** 4435 (1985).
    (c) Fyfe, C. A., Gobbi, G. C., Murphy, W. J., Oyubko, R. S., and Slack, D. A. *Chem Lett.,* p. 1547 (1983).
    (d) Fyfe, C. A., Gobbi, G. C., Kennedy, G. J., De Schutter, C. T., Murphy, W. J., Oyubko, R. S., and Slack, D. A. *Chem. Lett.,* p. 163 (1984).
    (e) Fyfe, C. A., Gobbi, G. C., and Kennedy, G. J. *Chem. Lett.,* p. 1551 (1983).
    (f) Thomas, J. M., Klinowski, J., and Anderson, M. *Chem. Lett.* p. 1555 (1983).
    (g) Fyfe, C. A., Gobbi, G. C., and Kennedy, G. J. *J. phys. Chem.,* **88,** 3248 (1948).
99. Engelhardt, G., Lohse, U., Samoson, A., Magi, M., Tarmak, M., and Lippmaa, E. *Zeolites* **2,** 59 (1982).
100. Fyfe, C. A., Gobbi, G. C., Klinowski, J., Thomas, J. M., and Ramdas, S. *Nature* **296,** 530 (1982).
101. Thomas, J. M., Klinowski, J., Fyfe, C. A., Hartman, J. S., and Bursill, L. A. *J. chem. Soc., chem. Comm.* 1981, p. 678 (1981).
102. Cheetham, A. K., Fyfe, C. A., Smith, J. V., and Thomas, J. M. *J. chem. Soc.,* 1982, p. 823 (1982).
103. Thomas, J. M., Fyfe, C. A., Ramdas, S., Klinowski, J., and Gobbi, G. C. *J. phys. Chem.* **86,** 3061 (1982).
104. Melchior, M. T., Vaughan, D. E. W., Jarman, R. H. and Jacobson, A. J. *Nature* **298,** 455 (1982).
105. Opella, S. J. and Frey, M. H. *J. Amer. chem. Soc.* **101,** 5854 (1979).
106. (a) Terao, T., Miura, H., and Saika, A. *J. chem. Phys.* **75,** 1573 (1981).
     (b) Terao, T., Miura, H., and Saika, A. *J. magn. Reson.* **49,** 365 (1982).
107. Zilm, K. W. and Grant, D. M. *J. magn. Reson.* **48,** 524 (1982).
108. DuBois Murphy, P., Stevens, W. C., Cheung, T. T. P., Lacelle, S., Gerstein, B. G., and Kurtz, D. M. Jr., *J. Amer. chem. Soc.* **103,** 4400 (1981).

7 Computational techniques and simulation of crystal structures

C. R. A. Catlow

7.1 Introduction

Solid-state chemistry has been increasingly concerned with complex and disordered solids, for example, minerals and non-stoichiometric compounds. Other chapters in this book have demonstrated the difficulties in experimental investigations of such systems—difficulties that arise first from the large number of parameters to be determined in structural descriptions of such systems, secondly from the problems in extracting local structural information, and thirdly from the large variety of mechanisms contributing to physical properties, e.g. spectroscopic, magnetic, and transport properties. For this reason, the possibility of modelling solids, with the aid of high-speed computers, is becoming an increasingly important adjunct to experimental studies in solid-state chemistry. Computational methods are well established in studies of liquids where similar problems are often encountered. In solid-state studies, computer modelling techniques have now achieved a high degree of precision in calculating energetic and dynamic properties of a large number of oxide and halide crystals; and the methods are now being broadened to encompass more covalent systems including minerals. This chapter aims to outline the methodology of the calculations, the factors that limit their reliability, their achievements to date, and, in conclusion, to hint at the scope for future developments.

7.2 Methodology

7.2.1 Scope and aims of the simulation studies

The basis of any computer simulation technique is the specification of a potential model, i.e. a description in mathematical terms—either numerical or analytical—of the energy of the system as a function of particle coordinates. Given adequate mathematical and computational procedures, the reliability of the results of the simulations depends on the extent to which such models represent the system under study. The development of reliable potential models has indeed become the key factor in the field of computer modelling. Their derivation is discussed in Section 7.2.3. Once models are available, the techniques may be used to investigate the following types of property.

(a) Structural properties

This class of calculation includes the prediction of crystal structures, i.e. the unit cell parameters (cell edges and angles) and atomic coordinates of a structure. The calculations are of two types: first, one may wish to predict *distortions* from some idealized model based on, for example, a regular polyhedral structure or on close packing; alternatively the aim may be to discriminate between different structural models. Examples of both type of prediction will be discussed in Section 7.3.

A second important structural problem concerns the structural rearrangement (or *relaxations*) about elements of disorder which may be *point defects*—vacancies, interstitials, or substitutionals—or *extended defects* such as the shear-planes discussed in Section 7.3; the techniques applied to extended defects may be used to investigate *surfaces* and *interfaces*. Relaxation effects often play a vital rôle in stabilizing disordered structures, and although information on relaxation can be obtained from modern diffraction and microscopy techniques, experimental investigations are difficult, and guidance from simulation studies is of great value. Several examples will be discussed in Section 7.3.

A third structural property that has been studied very recently is the amplitude of thermal motions, which determine the Debye–Waller factors used by crystallographers to describe the effects of thermal motions on the effective scattering factors of atoms in crystals. As the magnitudes of Debye–Waller factors are commonly a source of uncertainty in structural studies, the ability of simulations to make reliable calculations of these parameters promises to be a useful aid to structure determination.

(b) Calculations of physical properties

The principal properties with which we are concerned here are the dielectric and elastic constants, and the lattice dynamical properties, i.e. the phonon dispersion curves. The commonest use of such calculations is as a test of the interatomic potential model: calculated properties are compared with experimentally determined values to gauge the adequacy of a proposed model. Such calculations may also be of more direct value when the experimental data are uncertain and inaccessible. Examples are provided by high-pressure phases of minerals, knowledge of whose elastic properties is of considerable relevance to a number of geophysical problems, but where experimental studies are very difficult.

(c) Defect properties

This area of study concerns the calculation of the energies of critical defect parameters: principally formation, migration, and interaction energies; the calculations necessarily involve simulations of the structural relaxations around the defects referred to above. In the present context, the term defect should be understood in the broadest possible sense: it includes point defects, line defects (e.g. dislocations), and planar defects (e.g. shear planes and surfaces). Simulation studies have added considerably to our knowledge of detailed properties of defects, and are of value in the analysis of transport, thermodynamic, and structural properties of disordered phases. Indeed it is this class of simulation study which first established the reliability of solid-state simulations. We note that most calculations reported to date refer to energy terms. The demanding problems posed by the calculation of entropy terms are being actively investigated at present, and some success has been achieved in calculating the vibrational entropy changes accompanying the formation of point defects in solids.[1] It will, however, be several years before these calculations attain the precision of those for defect energetics.

Another general feature of most simulation studies of inorganic compounds is that they have been based on *static* models, that is descriptions which do not include any explicit description of thermal vibrations. Recently, however, use has been made of the more computationally demanding dynamical simulations; these techniques may indeed be used in the calculation of Debye–Waller factors referred to above, and in simulating ion transport in high temperature solids.

7.2.2 Simulation techniques

(a) Perfect lattice simulations

For applications in inorganic chemistry, the most important perfect-lattice property amenable to the technique discussed in this chapter is the lattice energy, E_L. (For ionic crystals, the *lattice energy* is defined as the energy of the crystal with respect to component ions at infinity. For non-ionic systems the term *cohesive energy* is often used; the reference point here is component *atoms* at infinity.) The most general expression for this quantity for a static lattice is:

$$E_L = \sum_{ij} \frac{q_i q_j}{r_{ij}} + \sum_{ij} \Phi_{ij}(r_{ij}) + \sum_{ijk} \Phi_{ijk}(r_{ijk}) + \dots \qquad (7.1)$$

where the summations refer to all pairs of ions i and j and all trios of ions ijk in the crystal. In principle, terms involving larger numbers of ions could be included; but in practice it is rare to take the summations beyond the 'three-body' terms—indeed, in most studies, these latter terms are omitted.

The first term on the right-hand side of eqn (7.1) is, of course, the sum of the Coulomb interactions between pairs of ions i and j separated by a distance r_{ij}. The summation of this term presents problems due to slow convergence. However, convergence is improved in most modern work by an ingenious technique developed by Ewald,[2] the essence of which is a transformation of the summation into reciprocal space.

The remaining terms on the right-hand side of the equation refer to short-range interactions. These include both the repulsive forces due to the overlap of ion charge clouds, and the attractive terms due to dispersive interactions† and to covalence in semi-ionic systems. The terms of the type $\Phi_{ij}(r_{ij})$ are the *two-body, central-force* contributions to the short-range energy; they vary only with the distances between pairs of ions, and have no angular dependent components. They are unquestionably the dominant component of the short-range energy in all ionic and semi-ionic systems. 'Three-body' terms of the type $\Phi_{ijk}(r_{ijk})$, which are functions of the coordinates of three atoms, are known to have significant effects on the vibrational properties of ionic materials; but their contribution to the lattice energy is small. We should note, however, that these terms become of greater importance with increasing deviation from ionic bonding. The derivation of reliable descriptions of the short-range interaction plays a central rôle in computer simulation studies of inorganic systems; a detailed discussion is presented in Section 7.2.3. Computationally, however, evaluation of these terms presents few difficulties.

In addition to the short-range and Coulomb contribution discussed above, there are small contributions to measured lattice energies from lattice vibrational terms. These are of course omitted from static descriptions of the lattice with which we are concerned here. (For a detailed discussion of these factors, refer to the reviews of Tosi[3] and Waddington,[4] who also describe much of the earlier work on lattice energy calculations.)

The other perfect crystal properties referred to in the preview included elastic and dielectric constants, and phonon dispersion curves. Calculations of elastic constants and dispersion curves require first and second derivatives of the lattice energy with respect to the coordinates of the atoms in the unit cell, and with respect to the cell dimensions themselves. Knowledge of these derivatives also enables us to calculate the forces acting on atoms within a unit cell, and on the cell as a whole (which are known as basis and bulk strains, respectively). Given a fully adequate interatomic potential (which, if the ionic model is appropriate, means a fully accurate representation of the short-range interactions) these latter forces will be zero—a condition which turns out to be a useful way of investigating interatomic potentials, as will be discussed later in this section.

Dielectric constants depend on ionic polarization as well as on short-range forces. Polarization that arises from the distortion of electron charge clouds at the ions by applied fields has been described by a number of models. The simplest is the point polarizable ion (PPI) model, in which a polarizability a is assigned to each ion, with the dipole moment μ in a field of magnitude E, being given by:

† Dispersive or 'London' terms are induced-dipole–induced-dipole interactions which show an attractive r^{-6} dependence on internuclear separation.

$$\mu = \alpha E. \tag{7.2}$$

This model, however, performs poorly when used in calculating lattice dynamical and dielectric properties of solids.[5] Its failure can be attributed to the neglect of a crucial factor which arises when ions are polarized within solids—that is, the *coupling* between polarization and short-range repulsion. Since polarization is due to the distortion of the valence shell electron distribution by an electric field, short-range repulsion, which arises from overlap of valence shell orbitals on different atoms, will be affected by polarization. In general, this coupling acts so as to 'dampen' polarization. The simple PPI model, which omits this effect, generally overestimates polarization within a crystal. Thus when dielectric constants are evaluated using short-range potentials that yield accurate elastic constants, the values of the static dielectric constants ε_0 are invariably overestimated.

The simplest and most successful of the models that include this vital coupling between short-range repulsive forces and ionic polarization is the *shell model*, originally developed by Dick and Overhauser.[6] The model consists of a simple mechanical representation of the ionic dipole, as illustrated in Fig. 7.1: the polarizable valence shell electrons are represented by a mass-less shell which is connected to the core by an harmonic spring. The development of a dipole moment is effected by the displacement of the shell relative to the core, as shown in Fig. 7.1.

Despite its simplicity, the shell model has enjoyed remarkable success in modelling dynamical and defect properties of ionic halides and oxides;[5] and most of the work discussed subsequently in this chapter is based on shell model treatments of polarizability. The spring constants K and shell charges Y are generally treated as variable parameters in the model, and are

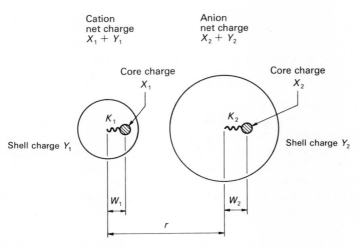

Fig. 7.1. Schematic representation of the shell model.

commonly evaluated from 'fitting' to measured dielectric constants—a procedure discussed in greater detail in Section 7.2.3.

Within the static lattice approximation, techniques are therefore available for the calculation of all the crystal properties which depend on interatomic forces. Moreover, efficient and automated computer codes (of which the PLUTO program[7] is the notable) are available for these calculations. However, as discussed in Section 7.1, perfect lattice simulation becomes very much more powerful when lattice energy calculations are combined with minimization procedures. The concept is simple: structural parameters, i.e. atomic coordinates and cell dimensions, are adjusted until a minimum energy configuration is attained. The minimization procedures generally, however, involve considerable mathematical sophistication, and care must be devoted to minimization techniques if large complex structures are to be handled with reasonable amounts of computer time.

The simplest type of minimization method involves a '*search*' procedure, i.e. the parameter space is scanned until the minimum is located. The method is in most cases inefficient and is rarely used in modern work. The second class of methods are known as *gradient* techniques; they involve use of calculated first derivatives ($g_i = \partial E_i/\partial x_i$) of the energy function with respect to the parameters x_i that are being varied. The parameters are then adjusted iteratively, according to a variety of prescriptions, of which the simplest is:

$$x^{(p+1)} = x^p - g^p \delta \qquad (7.3)$$

where $x^{(p+1)}$ is a vector, the components of which are the variables for the $(p+1)$th iteration; x^p and g^p refer to variables and their gradients on the pth iteration; δ is a scalar parameter, whose choice may be crucial in determining the speed of convergence (i.e. the number of iterations required to reach the minimum energy configuration).

Convergence may be greatly accelerated by making use of second, in addition to first, derivatives. Of these *Newton* type methods, the most widely used in energy minimization studies is the *Newton–Raphson* procedure in which variables are updated according to the formula:

$$x^{(p+1)} = x^p - (W^p)^{-1} g_p \qquad (7.4)$$

where W is a matrix of the second derivatives of the energy functions, i.e. $W_{ij} = \partial^2 E/\partial x_i \partial x_j$. These procedures are generally rapidly converging, and can reduce the required number of iterations by about three to ten. This is achieved at a cost, however, of requiring calculation and inversion of the second-derivative matrix W. The extra computation required here may easily outweigh the advantage obtained in the improved convergence. Fortunately, however, it is possible to *update* W without too large a loss of accuracy; a commonly used formula is:

$$(\mathbf{W}^{(p+1)})^{-1} = (\mathbf{W}^p)^{-1} + \frac{(x^{(p+1)} - x^p)(x_T^{(p+1)} - x_T^p)}{(g_T^{(p+1)} - g_T^p)(x^{(p+1)} - x^p)}$$
$$- \frac{(\mathbf{W}^p)^{-1}(g^{(p+1)} - g^p)\ (g_T^{(p+1)} - g_T^p)(\mathbf{W}^p)^{-1}}{(g_T^{(p+1)} - g_T^p)(\mathbf{W}^p)^{-1}(g^{(p+1)} - g^p)}. \tag{7.5}$$

The use of such matrix-updating procedures in the Newton–Raphson minimizations forms the basis of most energy-minimization studies in solid-state chemistry and physics.

The methods have, however, one major disadvantage, that is the necessity of storing the matrix $\mathbf{W}$, the dimension of which is $3N \times 3N$ where N is the number of atoms whose coordinates are being adjusted; normally, this is the number of atoms in the unit cell. (If a description of ionic polarization is included using, for example the shell model, then a $6N \times 6N$ matrix must be used). For structures containing more than 200 atoms per unit cell, the resulting matrix may be greater than can be handled by even the largest modern computers. For this reason, gradient methods are used in calculations on very big structures, for example some of the minerals discussed in Section 7.3.4.

Greater details of minimization techniques are given in references (5) and (8); and the former reference contains a detailed discussion of all aspects of perfect lattice simulation. Many of the principles and procedures described above are, however, applicable to the class of calculations discussed in the next section.

(b) Defect simulations

Calculations of defect structures and energies introduce one vital feature additional to those for the perfect-lattice methods, discussed in the previous section. This is the occurrence of *relaxation* of lattice atoms around the defect species. The effect is large because the defect generally provides an extensive perturbation of the surrounding lattice, and, in the case of ionic crystals, the relaxation field is long-range as the perturbation provided by the defect is mainly Coulombic in origin.

Realistic defect simulations must therefore include an accurate treatment of lattice relaxation. The method generally adopted is based on the '*two-region strategy*'. The idea is simple. The crystal surrounding the defect is divided into two regions, as illustrated in Fig. 7.2 The inner region (region I) immediately surrounding the defect contains typically 150 ions, all of which are simulated *atomistically*, i.e. using specified interatomic potentials. All coordinates are adjusted (and if the shell model is used, this means core and shell coordinates for each ion) until the minimum energy configuration is attained. (Newton–Raphson procedures, of the type described in the previous section, have again proved to be most successful in defect calculations.)

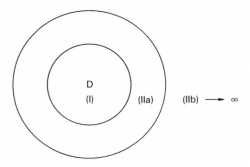

Fig. 7.2. Two-region strategy for defect calculations. D represents the defect at the centre of region I. Note that region II extends to infinity.

An atomistic simulation of this type is essential for region I since the forces exerted by the defect on the surrounding lattice are strong; and only an atomistic model can correctly describe the response of the surrounding crystal. In contrast, in the more distant regions of the crystal, i.e. region II, which extends to infinity, the forces are relatively weak and may be treated by methods based on continuum theories. One of the simplest of these, which has been widely used in ionic crystal studies, is the Mott–Littleton approximation.[9] Here, we assume that the response of region II is essentially a dielectric response to the effective charge of the defect. We then use a simple result of continuum dielectric theory, which gives the polarization **P** of the crystal, per unit cell, at a distance **r** from the defect as:

$$\mathbf{P} = \frac{V}{4\pi} \frac{q\mathbf{r}}{r^3} \left(1 - \frac{1}{\varepsilon_0}\right) \tag{7.6}$$

where V is the unit cell volume, q is the effective charge of the defect. **P** may then be divided up into atomistic components according to the nature of the interatomic potential model that is used; thus the division will be into core and shell displacements for shell model potentials. Equation (7.6) becomes more complex in materials in which the dielectric properties are anisotropic (see Catlow and Mackrodt[5]); but the same basic approach is still employed.

Following the division of the crystal into the two regions, the following equation for the total defect formation energy may be written:

$$E = E_1(x) + E_2(x, y) + E_3(y) \tag{7.7}$$

where E_1 is a function of the coordinates x (and dipole moments) of the ions solely within the region, E_3 depends solely on the displacements y of the ions within region II, and $E_2(x, y)$ arises from interaction between the two regions.

It is generally assumed that E_3 is an harmonic function of the displacement y, i.e.

$$E_3(y) = \tfrac{1}{2} y \, \mathbf{A} \, y \tag{7.8}$$

where $\mathbf{A}$ is a force constant matrix. We note that this assumption is compatible with the Mott–Littleton and related treatments for calculating the displacements in region II. If we then impose the equilibrium condition with respect to the displacement y, i.e. $(\partial E/\partial y)_x = 0$, we are led to the relationship

$$(\partial E_2/\partial y)_x = -\mathbf{A} \, y. \tag{7.9}$$

By using eqn (7.9) we are then able to remove any explicit dependence of the defect energy on E_3, which is a considerable advantage.

In practice therefore the calculations proceed as follows:[10]

(1) The energy minimization in region I is performed yielding the equilibrium values of x. (In practice the minimization is terminated when the forces on the ions in the inner region vanish. This force balance procedure is not identical to a full energy-minimization of the total defect energy, although the differences are unimportant; for greater discussion see references (5) and (8)).
(2) The Mott–Littleton calculation is performed to evaluate directly the displacements within an inner region of region II, known as region IIA surrounding region I (see Fig. 7.2). The displacements are calculated as the sum of those due to all the charged defects in region I.
(3) $E_1(x)$ is calculated by explicit summation. Care must be exercised in handling the Coulomb term.
(4) E_2 and its derivatives are calculated explicitly for all ions within region IIA.
(5) For the remainder of region II, Mott–Littleton displacements are calculated assuming that the crystal responds only to the net charges on the defect. These displacements, when used to calculate E_2 and its derivative, yield analytical expressions which may be summed out to infinity.

The major computational effort turns out to be in the energy-minimization performed on region I. The development of efficient minimization procedures of which the Newton–Raphson methods appear to be most successful, played a central rôle in the development of this field. Several studies evaluating the accuracy of the techniques have now been reported. Particular attention has been paid to the question of the sensitivity of the calculated energy to the size of region I. Typical results are illustrated in Fig. 7.3. They show that once the inner region has attained a size of about 100 ions, the calculations have 'converged', i.e. they are not sensitive to further expansion of the region. Indeed it is now clear that provided a sufficiently large region I

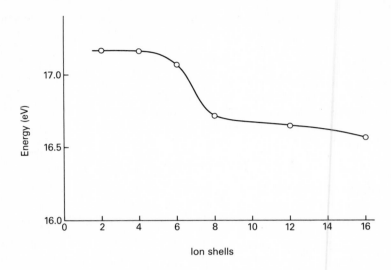

Fig. 7.3. Plot of calculated anion vacancy formation energy in UO_2 vs. number of shells of surrounding ions. (After Catlow.[72])

is employed, the calculations are essentially exact, within the approximations arising from the interatomic potential model, and from the static description of the lattice. As noted, for most purposes, static models are adequate. Dynamics simulations are, however, finding an increasingly important role in solid-state studies, and a brief account of their methodology is presented below.

(c) Dynamical simulation techniques

In these calculations, an ensemble, generally containing about 100–500 atoms within some fixed cell is taken, and periodic boundary conditions are applied (that is the ensemble is repeated infinitely in order to generate an unbounded system with no surfaces). For solids the cell may correspond to the unit-cell, and more commonly a super-cell. At the beginning of the simulation the ions are supplied with kinetic energy according to the temperature desired in the simulations. In solids the starting positions will be close to the crystallographically determined ones (assuming these are known). The system is then allowed to evolve in time by solving, in an iterative fashion, the classical equations of motion using the specified interatomic potentials. The simulation thus generates a series of *configurations* each of which is characterized by a set of particle coordinates and velocities, and which are separated in time by a period τ, usually known as the 'time-step'. Thus for each configuration, the forces acting on all particles are calculated. Provided τ is sufficiently small, from knowledge of the forces the velocities appropriate to the next configuration may be accurately calculated; and from knowledge of velocities, the positions may be updated.

Details of the up-dating algorithms will be found in reference (11), which gives a general account of dynamical simulation methods as applied to ionic systems. Aspects of the methods which relate to interatomic potentials and summation techniques resemble the same aspects discussed earlier for the static simulations; although it should be noted that the inclusion of ionic polarization in dynamical simulations greatly increases the computer time required for the calculations, and, as a result, most studies have omitted polarization effects.

The preliminary stages of dynamical simulations involve 'equilibration' of the system, which generally takes of the order of 1000 time-steps. During this period the system attains a distribution of velocities, and a partition of energy between potential and kinetic forms, which corresponds to thermal equilibrium. Once this has been attained, the time evolution of the equilibrated system is studied. Analysis of the results of the simulation yields dynamical information—e.g. diffusion coefficients, amplitudes of thermal vibration (yielding Debye–Waller factors)—and, of course, structural information; in particular, radial distribution functions may be readily calculated. Examples of such applications are given in references (53)–(55).

Limitations on the accuracy of dynamical simulations follow from the size of the periodically repeated ensemble, although this is not generally a serious problem provided one is not simulating long-wavelength properties. The more serious limitations, as is generally the case in simulation studies, stem from uncertainties in the interatomic potentials. Potentials used in both static and dynamical studies are discussed in the following section.

7.2.3 Interatomic potentials

Most simulation studies of inorganic solids have been concerned with strongly polar materials. For this reason, ionic model potentials have received the greatest attention. Most potentials developed in recent years have the following features in common:

(1) Use of integral ionic charges. This is more of a convenience than a necessity; partial charges can be used, although there may then be problems in correctly modelling the cohesive properties of the crystal.

(2) Use of *two-body, central force* models in describing the short-range potentials; that is, the right-hand side of eqn (7.1) is truncated after the second term. The potentials may be in numerical form (i.e. tabulations of Φ_{ij} as a function of r_{ij}). More commonly, however, simple analytical functions are used. The most popular is the Buckingham potential, in which an exponential repulsion term is supplemented by an attractive $1/r^6$ term, i.e.

$$\Phi_{ij}(r_{ij}) = A\exp(-r_{ij}/\rho_{ij}) - C/r_{ij}^6. \tag{7.10}$$

Other analytical expressions, e.g. the Morse potential, have also been used.

(3) Representation of ionic polarization by the shell model discussed in Section 7.2.2. This is particularly important for defect calculations. Indeed, the omission of polarization or its treatment by, for example, the PPI model is known to lead to unreliable results. Shell model potentials can, however, yield accurate values of elastic and dielectric properties of an ionic solid; and since the response of the surrounding crystal is largely an elastic and more particularly dielectric response to the perturbation provided by the defect, shell model potentials lead to reliable calculated defect energies.

Potential models of the type discussed above include variable parameters, e.g. A, ρ, and C in eqn (7.10) characterizing the short-range interactions, and the shell parameters Y and K for each ion. Two procedures are available for determining these variables; the first, known as *empirical fitting*, consists of adjusting the potential parameters until the best agreement between calculated and experimental crystal properties is obtained. This is normally achieved by using the perfect lattice simulation programs in conjunction with a least-squares-fitting routine. The data included in the fitting procedure generally include elastic and dielectric constants (where available), and structural information. The latter are incorporated via the bulk and basis strains discussed in Section 7.2.2, that is minimization of the bulk and basis strains corresponds to achieving, as far as possible, the condition that the crystal is in equilibrium at the observed structure with the specified potential. In complex, low-symmetry crystals, the information on potentials available in the crystal structure may be very considerable—a feature that has been exploited in studies of mineral systems, as discussed in Section 7.3.3.

For a large number of oxide and halide crystals, empirically derived potentials can yield very good agreement between calculated and experimental crystal data; results for a high-symmetry, simple halide, CaF_2, and for a lower-symmetry, more complex oxide, TiO_2, are given in Table 7.1. Despite this success there is an inherent weakness in empirically derived potentials in that they necessarily only extract information on the potential at interatomic spacings close to those in the perfect lattice; and their use for separations widely different from perfect-lattice values relies on the validity of the analytical forms of the potential over a considerable range of internuclear distances—an assumption which it is not possible to verify except indirectly, by the success of the empirically derived potentials in studying, for example, defect properties which depend on the potentials at non-perfect lattice spacings.

These limitations provide a strong incentive for the derivation of reliable theoretical methods for calculating interatomic potential parameters. Notable progress has been made in recent years by Mackrodt and co-workers[12] in calculating short-range potentials for ionic solids. In some cases, they have been able to use *ab initio*, Hartree–Fock methods to obtain short-range interactions as a function of internuclear separation. Results of an earlier

Table 7.1 *Calculated and experimental crystal properties for ionic crystals*
(a) CaF_2. (After Catlow and Norgett.[73])

	Calculated value	Experimental value
r_0	(2.722)	2.722
C_{11}	(16.9)	17.124
C_{12}	(4.80)	4.675
C_{44}	3.23	3.624
ε_0	(6.42)	6.47
ε_∞	(2.01)	2.05
ω_{TO}	(259.2)	270.0
ω_R	310.7	330.5
a_+	(0.984)	0.979
a_-	(0.765)	0.759
C_{111}	−107.8	−124.6
C_{112}	−33.8	−40.0
C_{123}	−17.5	−25.4
C_{144}	−9.3	−12.4
C_{166}	−23.2	−21.4
C_{456}	−7.8	−7.5
H_L	−28.06	−26.76

Key: r_0: lattice constant (Å)
C_{11}, C_{12}, C_{44}: second-order elastic constants (10^{11} dyne cm^{-2})
ε_0, ε_∞: dielectric constants at zero and high frequency
ω_{TO}, ω_R: transverse optic and Raman frequencies (cm^{-1})
a_+, a_-: cation and anion polarizabilities (Å^3)
C_{111}, C_{112}, C_{123}, C_{144}, C_{166}, C_{456}: third-order elastic constants (10^{11} dyne cm^{-2})
H_L—lattice formation energy (eV).
(Bracketed values used in fitting.)

(b) TiO_2. (After Catlow and James.[47])

Crystal property	Calculated value	Experimental value
static dielectric constants†		
$\varepsilon_\parallel$	157.32	170
$\varepsilon_\perp$	94.76	86
high-frequency dielectric constants		
$\varepsilon_\parallel$	7.99	8.43
$\varepsilon_\perp$	6.28	6.83
elastic constants/(10^{11} dyne cm^{-2})‡		
$C_{11} = C_{22}$	25.33	27.01
C_{33}	77.92	48.19
C_{12}	17.80	17.66
C_{13}	20.90	14.80
C_{44}	9.22	12.39
C_{66}	22.12	19.30

†The subscript ∥ indicates a quantity measured parallel to the *c*-axis of the rutile structure: ⊥ indicates a quality perpendicular to this axis.
‡dyne = 10^{-5} N.

study[77] are shown in Fig. 7.4 for the case of the F^--F^- short-range potential. It has been found that, in such calaculations, it is essential to include a proper representation of the effect of the Madelung potential of the surrounding lattice ions on the wave functions of the interacting species.

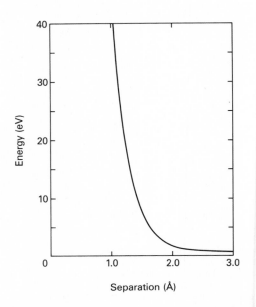

Fig. 7.4. Non-coulombic component of the interaction energy for two interacting F^- ions. (After reference 77.)

At present, the applications of *ab initio* techniques are limited in view of the very considerable computer time required when heavier ions are considered. For this reason, extensive use has been made in the calculation of short-range potentials of more approximate theoretical methods. The most widely used methods employ 'electron gas' approximations which are based on the following procedure:

(i) The wave functions, and hence the electron density distribution, of the isolated interacting ions are obtained, usually by a Hartree–Fock procedure, as good Hartree–Fock wave functions are available now, even for heavy atoms and ions.

(ii) The interaction energy $\Phi_{ij}(r_{ij})$ of pairs of ions is then written as a function of the electron densities of the interacting ions using the expression

$$\Phi_{ij}(r_{ij}) = E_N + E_C + E_{EX} + E_{KE} \tag{7.11}$$

where E_N, E_C, E_{EX}, and E_{KE} are respectively the nuclear repulsion, the

Coulomb, exchange, and correlation contributions to Φ_{ij}. The values of E_N and E_C may be written straightforwardly as:

$$E_N = \frac{Q_i Q_j}{r_{ij}} \quad (7.12)$$

$$E_C = \iint \frac{\rho_i(x_i)\rho_j(x_j)}{d_{ij}} dV_i dV_j + \int \frac{\rho_i(x_i) Q_j dV_i}{r'_{ij}} + \int \frac{\rho_j(x_j) Q_i dV_j}{r''_{ij}} \quad (7.13)$$

where ρ_i is the electron density due to atom i at a point x_i, ρ_j is the density due to atom j at point x_j, d_{ij} is the distance between points i and j; and the double integral is over all the points x_i and x_j; r'_{ij} and r''_{ij} are the distances between x_i and nucleus j and x_j and nucleus i, respectively. The expressions for the kinetic energy and exchange terms are obtained from the electron gas formalism,[5] and are given by:

$$E_{EX} = -\tfrac{3}{4}\left(\frac{3}{\pi}\right)^{\frac{1}{3}} \int \rho^{\frac{4}{3}} dV \quad (7.14)$$

$$E_{KE} = (\tfrac{3}{10})(3\pi^2)^{\frac{2}{3}} \int \rho^{\frac{5}{3}} dV. \quad (7.15)$$

Some effects of electron correlation may be crudely incorporated into the potential; but it is not possible to include a representation of 'dispersion' effects, i.e. the longer-range terms which arise from correlated instantaneous fluctuations of electron density. The method is clearly crude; in addition to the approximation inherent in the electron density formalism, no allowance is made for the distortion of the electron charge clouds which is caused by overlap. Nevertheless electron-gas methods have been surprisingly successful in generating short-range potentials which have led to good results when applied to the calculation of perfect and defective crystal properties. Indeed, electron-gas and empirically derived potentials are often very similar, as illustrated in Fig. 7.5 for the case of the Al^{3+}–O^{2-} potential. Significant differences do occur in some cases. An important example concerns the short-range interaction between O^{2-} ions, which the electron-gas methods calculate to be far more repulsive—a result with considerable consequence for the calculated formation energies of O^{2-} ions in interstitial sites.

No theoretical methods are available yet for the calculation of shell parameters. A problem arises with the empirical parameterization in some cases where unrealistic values are obtained; indeed for UO_2, positive cation shell charges are obtained! When this occurs, the physical basis of the model has clearly broken down, and results of calculations employing these potentials should be treated with some caution.

In general, however, for strongly ionic halides and oxides, it is possible to derive reliable potential models. The uncertainties increase with the degree of covalence where the approximations inherent in two-body potentials of the type discussed above become more serious. Nevertheless even for systems where covalence is appreciable, useful results may still be obtained, as will become apparent in the next section.

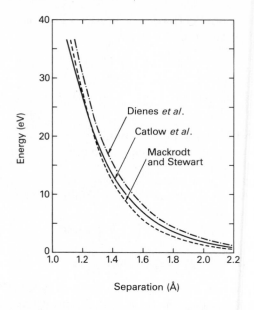

Fig. 7.5. A comparison of the non-coulombic potentials for $Al^{3+} - O^{2-}$ in $a\text{-}Al_2O_3$, taken from Chapter 10 of reference 5, which gives references to original work.

7.3 Applications

The aim of computer simulation methods is to guide, interpret, and stimulate experimental work. But before these aims can be achieved, it is necessary to establish the reliability of the techniques. Our discussion of applications therefore starts with an account of the area which has most clearly established the quantitative reliability of the techniques: the calculation of *point defect parameters*. We continue with an account of more qualitative applications in the field of *heavily disordered solids*—either non-stoichiometric or doped solids. A special class of disordered materials, i.e. *superionic conductors*, are next considered; these materials, whose defining characteristic is the occurrence of exceptionally high ionic mobilities, have attracted considerable attention in recent years owing to a number of important potential technological applications, principally in the development of secondary batteries and fuel cells. The modelling of *complex crystal structures*, including transition-metal oxides and minerals, is an important and growing field in simulation studies. After describing this application we continue with a brief discussion of *surface* and *interfacial properties*, a field which is certain to grow rapidly in the future owing to its importance in, for example, catalysis and several aspects of materials science.

7.3.1 Point defect energies

There are three principal types of point† defect: *vacancies*, i.e. atoms or ions that are missing from normal lattice sites, *interstitials*, i.e. atoms present at sites that are not normally occupied in the perfect structure, and *substitutionals*, i.e. foreign atoms present at lattice sites. These species control many of the most important physical properties of solids, including transport properties (electrical conductivity and diffusion) and thermodynamic properties involving equilibrium between the solid and the gas phase. In addition, point defects associated with electronic states—electrons or holes—may strongly affect the spectroscopic properties of the solid. Finally, solid-state reactions are commonly effected via the agency of point defects.

There are three ways in which point defects are introduced into crystals. The first we shall refer to as 'thermally'. Here we are concerned with the equilibrium concentration of defects which basic statistical mechanical considerations show must be present in *all* crystals above absolute zero. This defect population, which, as discussed below, may be treated by standard chemical thermodynamics, increases rapidly with temperature, and indeed may become a major structural feature of the solid at high temperatures. The second origin of point defects is as a response to impurities. In particular, *aliovalent substitutional* ions—that is, substitutionals whose charge differs from that of the host lattice cation—invariably induce a charge-compensating defect population; this point will be amplified below. Thirdly, non-equilibrium concentrations of point defects may be created by mechanical or irradiation damage. This latter area is outside the scope of our present article; the practical applications are, however, considerable and are discussed in reference (13).

Thermal defect populations may be treated by a theory borrowed from chemical thermodynamics which is based on a mass action formalism. The essential concept is that of defect reactions of which there are two types: the first is known as the *Frenkel* disorder reaction which involves displacing atoms from lattice to interstitial sites; the second—*Schottky* disorder—involves the creation of charge-balancing populations of vacancies by displacing lattice ions to the surface of the crystal. The simplest example is NaCl, where the Schottky pair comprises a sodium and a chloride ion vacancy.

The mass action formalism is based on the representation of these basic modes of disorder in terms of pseudo-chemical reactions; thus the Frenkel disorder reaction may be written as:

$$\underset{\text{(Unity)}}{\text{LAT}} \rightleftharpoons \underset{(x_V)}{\text{VAC}} + \underset{(x_I)}{\text{INT}} \qquad (7.16)$$

† A *point* defect may be defined as a deviation from the regular periodicity of the lattice which is localized at a particular site or group of sites, in contrast to extended *defects* which, as the term implies, extend indefinitely in one or two dimensions.

where by LAT we mean a perfect-lattice atom which is displaced to an interstitial site (INT) thus leaving a vacancy (VAC), the concentrations of the two latter species being x_V and x_I respectively. Standard chemical thermodynamics then allows us to write the following expressions governing the defect concentration:

$$x_V\, x_I = \exp(-G_F^p/kT) \qquad (7.17)$$

where G_F^p is the Gibbs free energy of Frenkel pair formation. (The superscript 'p' implies that the process takes place at constant pressure.) Expressing G_F^p in terms of its component enthalpies (h_F^p) and entropy (s_F^p) term, we have:

$$x_V\, x_I = \exp(s_F^p/k)\exp(-h_F^p/kT). \qquad (7.18)$$

Similar expressions govern the concentrations of vacancies introduced by the Schottky disorder reaction; and the thermodynamic treatment of the type we have just discussed will lead to the definition of enthalpies and entropies of Schottky pair formation (h_S^p and s_S^p).

The levels of thermally generated (or 'intrinsic') disorder are clearly therefore controlled by the thermodynamic quantities s^p and h^p. The enthalpy terms are, however, amenable to calculation by the techniques described in Section 7.2.2. (Calculations in fact yield values of the internal energy u^v rather than the enthalpy. The latter can, however, be calculated using the expression $h^p = u^v + (\beta T/K)V^p$ where β is the expansivity, and K the isothermal compressibility of the solid; V^p is the volume of defect formation defined by $V^p = (\partial G/\partial p)_T$.) Values for a number of strongly ionic materials are presented in Table 7.2. They are compared with experimental data deduced from transport measurements, i.e. the study of ionic conductivity or diffusion. The latter are generally effected by the migration of point defects whose concentration show an 'Arrhenius' like variation with temperature (i.e. a linear variation of the logarithm of the concentration with T^{-1}) allowing the defect formation energies to be deduced from plots of the logarithm of the diffusion coefficient, or of the conductivity, against T^{-1}; the Arrhenius energy deduced from such plots will of course include an activation energy term for the defect migration whose importance is discussed below. Typical data are illustrated in Fig. 7.6. Temperature-dependent transport data may, however, be analysed to yield formation and migration energies by techniques discussed by Jacobs,[14] and by Corish and Jacobs.[15,16] The data reported in Table 7.2 were obtained by such analyses.

The comparison between calculated and experimental defect parameters is continued in Table 7.3 which reports activation energies. Vineyard[17] showed that 'absolute rate theory' could be applied to defect transport processes. The calculated values are based on the difference between the energies of proposed saddle point configurations and those of the normal configuration of the defect; a typical saddle point configuration is shown in Fig. 7.7.

A third, important set of defect parameters that may also be deduced from

Computational techniques and simulation of crystal structures 249

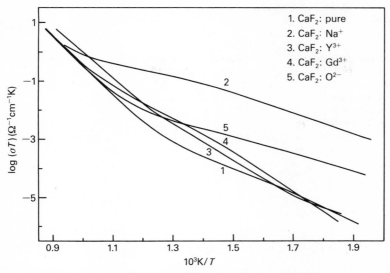

Fig. 7.6. Arrhenius plot for the conductivity of CaF_2 (see Jacobs[14]).

Table 7.2 *Calculated and experimental defect formation energies in ionic materials (taken from review of Mackrodt[25])*

Materials	Theory (eV)	Experiment (eV)
LiF (Schottky)	2.37	2.34–2.68
NaCl (Schottky)	2.22	2.20–2.75
	2.32	—
KBr (Schottky)	2.27	2.37–2.53
RbI (Schottky)	2.16	2.1
MgF_2 (Anion Frenkel)	3.12	—
CaF_2 (Anion Frenkel)	2.75	2.7
SrF_2 (Anion Frenkel)	2.38	2.5
BaF_2 (Anion Frenkel)	1.98	1.91
$CaCl_2$ (Anion Frenkel)	4.7	—
AgCl (Cation Frenkel)	1.4	1.45–1.47
γ-CuCl (Cation Frenkel)	1.05	1.11
Li_2O (Cation Frenkel)	2.28	—
MgO (Schottky)	7.5	~5–7
BaO (Schottky)	3.4	—
a-Al_2O_3 (Schottky)	5.14†	3.7
a-Fe_2O_3 (Schottky)	4.46†	—
MnO (Schottky)	4.6	—
FeO (Schottky)	6.5	—
$MgAl_2O_4$ (Schottky)	4.15	—
ZnO (Anion Frenkel)	2.51	—
UO_2 (Anion Frenkel)	5.47	5.1

† Energy per defect, i.e. energy to form Schottky quintet divided by five.

250   *Solid state chemistry: techniques*

Table 7.3 *Calculated and experimental defect activation energies*

Migration mechanism	Material	Activation energy (eV)	
		Calculated	Experimental
Cation vacancy	NaCl	0.67	0.65
Cation vacancy	MgO	2.18	2.2–2.6
Anion vacancy	BaF_2	0.46	0.55
Anion interstitial	BaF_2	0.72	0.76
Cation interstitial	AgCl	0.09	0.05

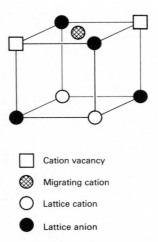

☐ Cation vacancy

⊗ Migrating cation

○ Lattice cation

● Lattice anion

Fig. 7.7. Saddle point configuration for vacancy migration in NaCl-structured solids.

analyses of transport data are the binding energies for defect clusters. At lower temperatures the defect population in a crystal is invariably dominated by the impurities that are accidentally or deliberately introduced into the crystal. These charge-compensating defects will generally interact with the dopants via Coulombic and elastic forces. Such interactions often lead to the formation of well-defined clusters which, in a number of systems, have been well characterized by spin resonance and, in certain cases, by diffraction techniques. Defect clustering is indeed a dominant theme in the structural chemistry of heavily disordered solids, as will be discussed in the following section. Here, we are more concerned with the simpler clusters present in crystals with only low levels of disorder; an example is shown in Fig. 7.8.

Computational techniques and simulation of crystal structures

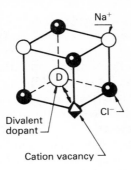

Fig. 7.8. Simple defect cluster comprising divalent cation substitutionals and charge-compensating vacancies in the NaCl structure.

Cluster formation influences ion transport because the formation of clusters immobilizes the defect unless the cluster itself is mobile. Cluster formation may be readily incorporated into the mass-action treatment of defect equilibria;[18, 19] and binding energies may therefore be extracted from the temperature dependence of the transport coefficients. The calculated binding energies are obtained simply from the difference between the energies of the clusters and those of the isolated component defects. Calculated and experimental data are reported in Table 7.4.

The results collected in Tables 7.2, 7.3, and 7.4 show an impressive measure of agreement between calculated and experimental defect parameters. They are typical of those which may be obtained for strongly ionic halides and oxides. Their importance is twofold: first, as argued, they show the quantitative reliability of the simulation techniques; second, owing to this reliability, the methods have become a valuable tool in defect physics and chemistry.

To illustrate the latter point, we take two examples. The first concerns oxygen migration in CeO_2. This fluorite-structured oxide has been widely studied, and, when doped with trivalent ions (e.g. Sc^{3+}, Y^{3+} and the rare-earth

Table 7.4 *Calculated and experimental binding energies for simple defect clusters*

Cluster	Material	Binding energy (eV)	
		Calculated	Experimental
Sr^{2+} substitutional—vacancy pair	KCl	0.66	0.65
Sc^{3+} substitutional vacancy pair	CeO_2	0.62	0.67
La^{3+} substitutional—anion interstitial	BaF_2	0.53	0.33

cations), it shows exceptionally high oxygen ion mobilities, and it can be included in the class of superionic conductors. It is well established that this behaviour is common to the fluorite-structured oxides doped with low valent ions, and follows from the low activation energies for the oxygen vacancies that are introduced as charge compensators for the low-valence dopant. The dopant ions and the vacancies interact, and it is clearly established that well-defined clusters are formed.[20] At low dopant concentrations there is now good evidence,[20, 21] that these clusters have the simple structure illustrated in Fig. 7.9 in which the vacancy is in the nearest neighbour site with respect to the dopant ion. Cluster formation in these materials has been extensively studied as it is a major factor limiting the oxygen mobility. Recent work suggested that the magnitude of the dopant-defect binding energy depended strongly on the radius of the dopant ion—a result which was, at first, surprising as it has generally been considered that the Coulomb term is the dominant component of cluster binding energies in ionic crystals. However, calculations[22] accurately reproduced the variation with dopant in the binding energies that had been proposed on the basis of conductivity studies. Comparison between theory and experiment is illustrated in Fig. 7.10 for the four dopants investigated experimentally: Sc^{3+}, Y^{3+}, Gd^{3+}, and La^{3+}. The major effect of ion size is clear, as is the fact that this is reproduced by the calculations.

The calculated binding energies for the doped CeO_2 system have thus confirmed an interpretation of experimental data that might otherwise have been controversial. Our second example concerns a system where the calculations have been able largely to settle a long-standing controversy. Diffusion in MgO has been very extensively studied; the material is one of the simplest of the ceramic oxides and one, moreover, where good single crystals are available. Interpretations of the experimental data have, however, raised considerable problems; and a particular difficulty has concerned the importance of intrinsic Schottky disorder. Calculations found a high value for the

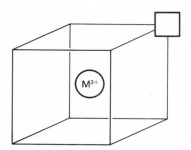

Fig. 7.9. Cluster comprising one trivalent cation substitutional and one anion vacancy in doped fluorite oxides.

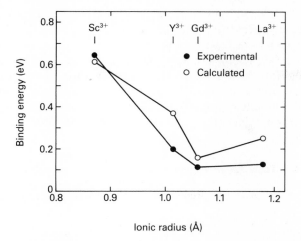

Fig. 7.10. Calculated and experimental binding energies for the cluster shown in Fig. 7.9, for a variety of dopants.

energy of formation of Schottky pairs; see Table 7.2. The magnitude of the formation energy is indeed such that thermally generated disorder would not be present in significant concentrations except possibly at temperatures close to the melting point. In general, however, the defect population will be dominated by the charge compensation introduced by impurities. Such interpretations are now generally accepted in analyses of transport data on MgO.[23, 24]

Several other examples are available where simple defect calculations of the type discussed above have assisted the interpretation and analyses of experimental data; the interested reader is referred to the reviews of Mackrodt.[25, 26]

7.3.2 Defect aggregation and elimination in heavily disordered solids

The commonest cause of high levels of disorder in crystalline solids is the deviation of the composition of the compound from stoichiometry. Non-stoichiometry in solids arises from the possibility of variable cation valence in transition-metal, actinide, and certain rare-earth compounds. Variation in cation valence has similar consequences to the introduction of aliovalent dopant ions; that is, a charge-compensating defect population is created. Thus in, for example, FeO, non-stoichiometry arises from the oxidation of Fe^{2+} ions to the trivalent state, and charge compensation is effected by the creation of cation vacancies; the metal-deficient phase is therefore usually written as $Fe_{1-x}O$. The commoner non-stoichiometric oxides are listed in

Table 7.5, which also gives the nature of the variation in cation valence and of the charge-compensating defects.

Extensive deviations from composition occur in many non-stoichiometric phases; thus in $Fe_{1-x}O$, for example, x may attain a value of about 0.15. High defect concentrations are therefore present in these systems, which as a consequence adopt a number of intriguing structural modes for stabilizing the disorder. These may usefully be divided into two broad categories: the first are based on aggregation, and the second elimination.[27]

Table 7.5 *Important non-stoichiometric compounds*

Non-stoichiometric phase	Redox process†	Charge-compensating defects
$Fe_{1-x}O$ (cf. $Mn_{1-x}O$, $Co_{1-x}O$, $Ni_{1-x}O$)	$Fe^{2+} \rightarrow Fe^{3+}$	Cation vacancies
CeO_{2-x} (cf. PrO_{2-x}, TbO_{2-x}, PuO_{2-x})	$Ce^{4+} \rightarrow Ce^{3+}$	Anion vacancies
SnO_{2-x}	$Sn^{4+} \rightarrow Sn^{2+}$	Anion vacancies
$UO_{2\pm x}$	$U^{4+} \rightarrow U^{3+}$ $U^{4+} \rightarrow U^{5+}$	Anion vacancies Anion interstitials
TiO_{2-x} (cf. VO_{2-x})	$Ti^{4+} \rightarrow Ti^{3+}$	Shear planes‡
WO_{3-x} (cf. MoO_{3-x})	$W^{6+} \rightarrow W^{5+}$	Shear planes‡

† Formal ionic charges are assumed.
‡ In the *near*-stoichiometric region of these phases there is evidence for point-defect compensation, whose nature (anion vacancies or cation interstitials) is uncertain.

(a) Defect clustering

As examples of non-stoichiometric phases in this class we consider fluorite- and rock-salt-structured compounds for which detailed experimental data are available, and whose strongly ionic bonding should allow reliable computer modelling studies.

(i) Anion-excess fluorites

Lattice cation oxidation in the fluorite-structured compound will result in charge compensation by anion interstitials. The fluorite structure, illustrated in Fig. 7.11, contains large sites (marked in the figure) for the accommodation of interstitials. UO_{2+x}† is the classic example of an anion-excess fluorite

† UO_2 is amongst the most versatile non-stoichiometric compounds, showing both anion-deficient and anion-excess composition regions.

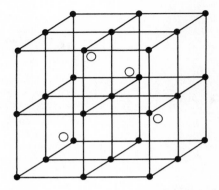

Fig. 7.11. The fluorite structure. Cations indicated by open circles; anions by filled circles. Interstitial sites at centres of unoccupied anion cubes.

oxide; but analogous phases may be prepared by doping the fluorite-structured halides (e.g. CaF_2) with trivalent ions: the high-valence substitutional cation will be compensated by interstitials in just the same way as the oxidized lattice cation. Indeed the doped halides may be described as simulating the non-stoichiometric oxide—an argument that is discussed in greater detail in reference (27).

Diffraction data[28,29] has indeed shown that the structural analogy between UO_{2+x} and the doped alkaline-earth fluorides is very close; and much of our discussion will concentrate on the latter systems whose study poses fewer experimental difficulties than that of UO_{2+x}, and where the possibility of varying the trivalent dopant adds an intriguing additional variable.

Like UO_{2+x} (where x may vary from 0 to 0.25) high levels of interstitial disorder may be introduced in the doped alkaline-earth fluorides; solid solutions with trivalent ion concentrations of about 30–40 per cent may be prepared. At low dopant concentrations the defect cluster structure is simple: spin resonance studies show that the simple pair clusters are formed in which the anion interstitials occupy nearest neighbour (nn) or next nearest neighbour (nnn) sites with respect to the dopant ion; the interstitials appear to occupy sites close to the body centre position of the vacant anion cubes in the fluorite structure. In contrast, diffraction data on more concentrated defect solutions ($\geqslant 5$ mole per cent trivalent dopant) reveal strongly distorted interstitial structures in which the excess anions are displaced along [110] and [111] directions from the centres of the cubic interstitial sites; in addition, vacancies were detected at the lattice anion sites. Following Willis[28,29] who obtained very similar results for $UO_{2.13}$, Cheetham et al.[30] suggested the formation of interstitial clusters, known as 2:2:2 and 4:3:2 clusters, to explain their data on 5 per cent and 10 per cent Y^{3+}-doped CaF_2. The clusters, illustrated in Fig. 7.12 comprise both [110] and [111] interstitials as well as vacancies.

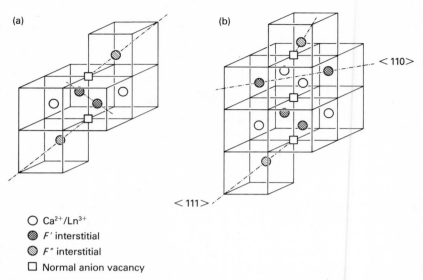

○ Ca²⁺/Ln³⁺
◉ F' interstitial
◐ F" interstitial
☐ Normal anion vacancy

Fig. 7.12. Dopant–interstitial aggregates in doped CaF_2. (a) 2:2:2 cluster, (b) 4:3:2 cluster.

One of the earlier successes of the computer modelling studies was their use in furnishing a plausible explanation of the formation of these interstitial clusters.[31] The 2:2:2 cluster was found essentially to be a complex of two dopant ions and two interstitials; see Fig. 7.13. This is, however, stabilized by a coupled lattice—interstitial relaxation mechanism. The interstitials move inwards, along the [110] direction towards the dopant cations; this inward interstitial relaxation is accommodated by [111] relaxations of neighbouring lattice anions towards vacant interstitial sites. This latter process is responsible for the observation of both the anion vacancies and the relaxed lattice ions.

Similar models may be advanced to explain the formation of the 4:3:2 interstitials which may be described as an aggregate of four dopant ions with

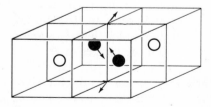

Fig. 7.13. Planar cluster comprising two dopant ions and two F⁻ interstitials. The cluster is converted into the 2:2:2 cluster by lattice and interstitial relaxation indicated by the arrows in the diagram.

charge-compensating interstitials. In both cases the calculations provide strong support for the models proposed. Thus the dopant dimer shown in Fig. 7.13 is not bound with respect to two simple dopant–interstitial pairs; but the calculations showed that when the coupled lattice–interstitial relaxation mode was allowed, the cluster was strongly bound (by about 1 eV).

Are even more complex dopant–interstitial aggregates possible in doped fluorite halides? X-ray diffraction studies of Greiss and co-workers[32, 33] suggested that in certain systems large super-clusters involving six dopant ions might form;† a possible model for such a cluster is shown in Fig. 7.14. Calculations[34] found that the stability of such clusters was competitive with that of the smaller aggregates shown in Fig. 7.12, and that the relative stability of the different types of cluster depended on the ratio of the dopant ion radius to the lattice parameter. A few illustrative results are collected in Table 7.6. They show that, for large dopant ions (e.g. La^{3+}) dissolved in a host with a relatively small lattice parameter (e.g. CaF_2), the 2:2:2 and 4:3:2 type of cluster are favoured. However, decreasing the radius of the dopant ion leads to an increase in the relative stability of the cubo-octahedral cluster.

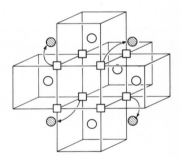

Fig. 7.14. Cubo-octahedral cluster comprising six dopant ions around central interstitial site. Anion sites at corners of cube are vacant; interstitials are situated above each cube-edge.

These predictions have been tested experimentally by neutron diffraction techniques. As discussed in Chapter 2, the information provided by Bragg diffraction techniques is limited to *average* unit cell contents. However, even with this restriction it should be possible to distinguish, from diffraction experiments, between clusters of the types shown in Figs 7.12 and 7.14, since the former include [111] interstitials, which are absent from the latter. Thus single-crystal diffraction experiments on 10 per cent Pr-doped $SrCl_2$[35] and 5

† It should be noted that in this work, the samples used had been annealed for long periods at high temperatures which is almost certainly important in allowing large clusters to form.

Table 7.6 Binding energies of dopant-interstitial clusters in doped CaF_2

Dopant	Binding energy per dopant ion (eV)†		
	2:2:2 cluster (Fig. 7.12(a))	4:3:2 cluster (Fig. 7.12(b))	Cubo-octahedral cluster (Fig. 7.14)
La^{3+}	1.01	1.19	0.86
Gd^{3+}	0.88	1.12	1.55
Er^{3+}	0.91	1.17	1.50

† All energies given with respect to isolated dopant ions.

per cent Er-doped CaF_2 revealed only the presence of [110] interstitials which is compatible with the formation of the octahedral super-clusters predicted to be most stable for these systems; in contrast, for 5 per cent La–CaF_2,[36] [111] in addition to [110] interstitials are detected in accordance with the calculations, which suggest that the smaller clusters will be favoured for this system. Detailed examination of the diffraction experiments found, however, that the agreement between theory and experiments could be improved by an alternative cluster shown in Fig. 7.15 which had been predicted to have especially high stability. This aggregate is formed by the capture by a 2:2:2 of an additional interstitial. Its presence in La-doped CaF_2 is strongly supported by other physical techniques[37] which indicate that a dipole species is present; the cluster shown in Fig. 7.15 will have a dipolar moment unlike the 2:2:2 cluster.

(ii) Rock salt-structured transition-metal oxides

A large number of studies have been devoted to the non-stoichiometric oxides: $Mn_{1-x}O$, $Fe_{1-x}O$, $Co_{1-x}O$, and $Ni_{1-x}O$. $Fe_{1-x}O$ shows the greatest deviation from stoichiometry† with, as noted, a maximum value of x at

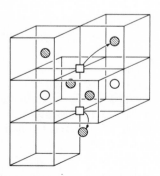

Fig. 7.15. '2:2:2' cluster plus additional F^- interstitial.

† Stoichiometric FeO is, in fact, unstable; and the non stoichiometric phase is only stable at high temperatures.

about 0.15. In contrast, for $Co_{1-x}O$ and $Ni_{1-x}O$, x may only attain values of about 10^{-2}, while $Mn_{1-x}O$ shows intermediate behaviour with x values at about 0.13. Diffraction[38,39] and transport studies have clearly established that these non-stoichiometric phases are cation deficient. The diffraction results (available only for $Fe_{1-x}O$), however, posed an intriguing problem by showing that metal interstitials in addition to vacancies were present. Thus the detailed neutron diffraction study of Cheetham et al.,[39] which was performed at the high temperatures (1400°C) at which the $Fe_{1-x}O$ phase is thermodynamically stable, found a ratio of vacancies-to-interstitials of about 4 for lower values of x; this decreases to about 3 as x increases.

An explanation of the presence of metal interstitials in these phases was provided by the calculations[74] which found very high stability for the cluster shown in Fig. 7.16(a), in which four cation vacancies are situated at the lattice sites surrounding an Fe^{3+} interstitial. (The cluster shown in Fig. 7.16(a) needs to be surrounded by Fe^{3+} ions at lattice sites for stability.) The stability of this unusual cluster arises from Coulombic interactions between the vacancies and the interstitial. The calculations suggested, moreover, that more complex modes of aggregation were possible in which the tetrahedral vacancy clusters share common edges and corners; the proposed models are illustrated in Figs 7.16(b)–(d), and calculated binding energies are reported in Table 7.7.

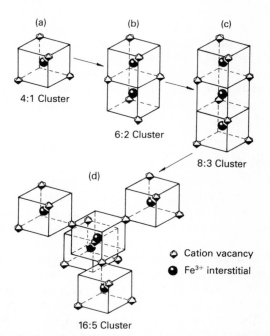

Fig. 7.16. 4:1 cluster in $Fe_{1-x}O$. (a) Single cluster; (b)–(d), more complex clusters.

Table 7.7 Binding energies of vacancy aggregates in $Fe_{1-x}O$[74]

Cluster	Binding energy per vacancy (eV)
4:1 (Fig. 7.16(a))	1.98
6:2 (Fig. 7.16(b))	2.42
8:3 (Fig. 7.16(c))	2.52
16:5 (Fig. 7.16(d))	2.38

The cluster models proposed in Figs 7.16(a)–(d) provide a convincing explanation of the neutron diffraction results of Cheetham et al.,[39] since the larger, more complex aggregates—which would be expected to form at higher deviations from stoichiometry—have a lower vacancy/interstitial ratio. As noted, this ratio is observed to decrease from the value of 4 with increasing x; the latter is of course clearly compatible with the formation of 4:1 clusters.

Support for the formation in the more heavily defective region of the $Fe_{1-x}O$ phase of the larger aggregates of the 6:2 and 8:3 types (Figs 7.16(b)–(d)) has been provided by a neutron diffraction study of the magnetic structure of the oxide.[40] We should note, however, that the status of the large corner-shared aggregate shown in Fig. 7.16(d) is less certain. Although the calculations found this to be strongly bound (and its formation would indeed be plausible as it is an element of the inverse spinel structure adopted by Fe_3O_4) recent microscopy studies[41] on quenched samples have suggested that alternative edge-shaped models might be more stable. Detailed features of the cluster structure of this fascinating phase may therefore remain uncertain; and there is also uncertainty as to the extent to which the defect structure of the other rock-salt-structured non-stoichiometric phases resembles that of $Fe_{1-x}O$.[42,43]

(b) Systems based on defect elimination

The formation of large, strongly bound clusters clearly provides one way of stabilizing the extensive disorder introduced by large deviations from stoichiometry. An alternative mechanism is provided by defect 'elimination', the nature of which is illustrated in Fig. 7.17. The diagram shows a section through the ReO_3 structure (a distorted version of which is adopted by the non-stoichiometric oxide WO_{3-x}). Imagine that the oxide is reduced, the loss of oxygen being accommodated by the formation of oxygen vacancies. These are then aligned as shown in Fig. 7.17(a); extension of the diagram in to three dimensions would, of course, generate a vacancy disk. However, if the bottom half of the structure is then sheared with respect to the top half in the direction shown in the diagram, lattice oxygen ions will be superimposed on

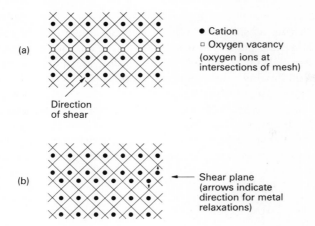

Fig. 7.17. Schematic representation of shear-plane formation in ReO_3-structured oxides. (a) Alignment of oxide vacancies; (b) elimination of vacancies to form shear-plane.

the vacancies. The latter defects will therefore have been eliminated; but, as shown in Fig. 7.17(b), the resulting configuration is now characterized by a fault (which will be planar on extending the diagram into three dimensions) on the metal sublattice. In this planar fault—known (owing to its origin) as a *shear plane*—the nature of the MO_6 octahedra has been changed. The corner sharing of the octahedra, characteristic of the ReO_3 structure, has been replaced by edge sharing.

Shear plane formation has been observed in the non-stoichiometric oxides WO_{3-x}, MoO_{3-x} which are based on the ReO_3 structure,† and the rutile-structured phases TiO_{2-x} and VO_{2-x}; a general discussion of the latter experimental data is given in reference 44. The clearest evidence is provided by electron microscopy studies. In a limited number of cases detailed information on shear plane structures is also available from diffraction studies, as will be discussed below.

The first question posed by the observation of these extended defects concerns their *stability*, and their *relationship to point defects*. What factors determine whether shear-plane formation by defect elimination occurs? And in those non-stoichiometric oxides where shear planes have been observed, are point defects present, and if so in what concentrations?

A second problem relates to shear plane *orientations*. The shear process illustrated in Fig. 7.17 will generate a shear plane situated along an [001] plane of the parent ReO_3 structure. Other orientations are, however,

†Stoichiometric MoO_3 does not adopt the ReO_3 structure. Nevertheless, the *non-stoichiometric* phases contain shear planes within an ReO_3-structured matrix.

possible. They correspond to different arrangements of the groups of edge-sharing octahedra, as shown in Fig. 7.18 which illustrates the [104], [103], [102], and [001] shear plane in the ReO$_3$ structure. All four orientations have been observed in WO$_{3-x}$; indeed the orientation is found to change with composition following the sequence [102]→[103]→[104]→[001] as x increases. Similar behaviour is found in TiO$_{2-x}$ and VO$_{2-x}$ where different shear plane orientations, [121] and [132], are observed, with orientations varying with the composition. These observations obviously raise the question of the factors controlling the differences in energy between the different shear plane orientations, and the reasons for the change in orientation with composition.

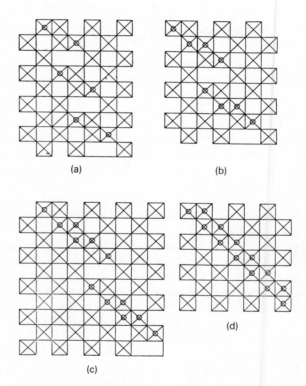

Fig. 7.18. Shear-plane orientations in WO$_3$. (a) [102] plane. (b) [103] plane. (c) [104] plane. (d) [001] plane. Cations are at centres of octahedra; the marked octahedra, according to calculations, contain reduced cations.

Possibly the most fascinating problem in the study of these systems concerns the remarkable *long-range ordering* of shear planes that is commonly observed. Electron microscopy studies[45] on both the rutile- and ReO$_3$-structured oxides have revealed, in several cases, regular-spaced shear-planes often with large spacings: about 100–150 Å. Moreover, homologous series of well-defined, ordered compounds are known, based on regularly spaced shear planes, with each member of the series having a characteristic shear-plane spacing; examples are the Ti_nO_{2n-1} and W_nO_{3n-2} series. The origin of this phenomenon must be either kinetic, i.e. arising from the mechanism of reduction of the oxides, or it must be due to unusually long-range interactions between the extended defects.

Computer modelling studies have contributed to our understanding of all of the problems discussed above. Let us take first the question of shear-plane stability. Calculations of shear plane energies have been carried out both on isolated shear planes, and on shear-plane super-cells;[46,47,48] the latter will, of course, include any shear plane interaction terms—a valuable additional feature, as will be discussed below. The results, however, clearly show that relaxations of the cations neighbouring the shear plane are a key factor in stabilizing the extended defects. The reason for this may be understood in a simple qualitative manner. Shear plane formation causes an increase in cation–cation repulsions between the ions within the plane. This term is large and, in most cases, would be sufficient to outweigh the favourable energy term arising from the elimination of the point defects. However, if extensive cation relaxations can occur, the unfavourable repulsion energy may be sufficiently reduced to stabilize the extended defects. The rôle of the simulations has been to provide quantitative support for the qualitative argument. The calculations have shown that if relaxations of the cations neighbouring the shear planes are allowed, then the extended defects are indeed more stable than point defects, as shown in Table 7.8. However, on performing calculations in which such relaxations were not included, the extended defects are found to be considerably higher in energy than the point-defect species; results are given in reference (48).

Calculations on ReO_3-structured oxides also confirmed the importance of relaxations in the stabilization of shear planes. And although other factors may be significant, there now seems to be a strong body of evidence

Table 7.8 *Calculated energy of shear plane formation on reduction relative to that for point defect formation (see references 47, 48)*

Material	Shear plane orientation	Relative energy per eliminated oxygen ion (eV)
TiO_{2-x}	[132]	−2.0
	[121]	−1.9
WO_{3-x}	[102]	−2.13
	[103]	−1.33
	[104]	−1.22
	[001]	−1.26

favouring the relaxation model outlined above. The results summarized in Table 7.8 also yield information on the second problem raised earlier in this section—that of the relationship between extended- and point-defect structures. In oxides such as TiO_{2-x} and WO_{3-x}, shear planes are more stable than point defects; but the difference in energy (generally about 2 eV) is sufficiently small that, at elevated temperatures, point defects will exist in appreciable concentrations in equilibrium with the extended-defect structures. The calculated values of vacancy concentrations are reported in Table 7.9. The vacancy concentrations are seen to be appreciable; they could have a significant effect on transport and thermodynamic properties of the materials.

Table 7.9 *Calculated vacancy concentration in equilibrium with shear planes at 1000 K*

Material	Shear plane orientation	Equilibrium vacancy concentration
TiO_{2-x}	[132]	1.1×10^{-3}
	[121]	1.2×10^{-3}
WO_{3-x}	[102]	0.2×10^{-3}
	[103]	3.7×10^{-3}
	[104]	5.6×10^{-3}
	[001]	4.8×10^{-3}

Vacancies are assumed to be the predominant point defect, as is predicted by the calculations.

The third question raised earlier in our discussion concerned the question of the shear-plane orientations. Calculations show that ion relaxations are also important in this context. The calculated shear-plane energies are considerably different for unrelaxed configurations, but ion relaxations remove most of the differences between the energies of different configurations. Indeed the remaining differences after relaxation are of the same order of magnitude as the interaction energies discussed below. Shear-plane interactions, which would be expected to become more important at higher deviations from stoichiometry, could therefore lead to changes in orientation with composition, as is observed.

Finally, the calculations have provided valuable insight into the problems of shear-plane ordering. By calculating the lattice energy of super-cells containing ordered arrays of shear planes, it is possible to extract a shear-plane interaction energy. Results shown in Fig. 7.19 were obtained from calculations[48, 75] in which the energies of all the super-cells examined were minimized with respect to both ion coordinates and cell dimensions. This procedure was found to be essential if reliable results for the interaction function were to be obtained. The results confirm that the shear planes

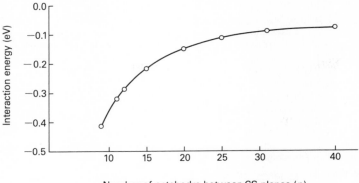

Fig. 7.19. Shear-plane interaction function for [102] shear planes in WO_{3-x}. (After Cormack et al.[48, 75])

interact attractively over large distances—a result which suggests that the observation of shear-plane ordering is thermodynamic in origin. Moreover, detailed analyses of the results of the calculations find that the interaction, which is essentially elastic in origin, arises from a constructive 'interference' between the relaxations around neighbouring shear planes.

7.3.3 Superionic conductors

Superionic or fast-ion conductors are materials with an ionic conductivity of the same order of magnitude as molten salts. They have considerable actual and potential applications in high-energy-density batteries, and in fuel cells; as a consequence they have been the subject of intensive research in the last decade. Table 7.10 lists some of the commoner superionic materials. Many of them are relatively simple inorganic oxides and halides, e.g. PbF_2 and ZrO_2. In all cases, the superionic behaviour is associated with disorder, and a distinction is made in Table 7.10 between those systems in which the disorder is generated thermally (usually in a high-temperature phase), and those compounds where the disorder is induced by high levels of impurity ions.

The greatest contribution made by simulation methods to the study of superionic conduction is in the elucidation of migration mechanisms and related factors governing ion migration rates, such as the strength of dopant–defect interactions. However, simulation techniques have also provided useful information on structural properties of superionics which often raise intriguing problems. Simulation studies have now been reported for most of the classes of superionics listed in Table 7.10, and an example will be given from each class.

Table 7.10 *Some crystalline superionic conducting materials.*
(a) Materials showing superionic properties above a phase transition temperature

Material	Crystal structure	Transition temperature (K)
$SrCl_2$	Fluorite	1000
PbF_2	Fluorite	700
BaF_2	Fluorite	1230
CuI	Body-centred cubic*	450
AgI	Body-centred cubic*	420
Ag_2S		430

*Above transition temperature.

(b) Materials without phase transitions

Material	Structure
Li_3N	Hexagonal layer structure with bridging Li^+ ions
β-Al_2O_3($Na_2O \cdot 11Al_2O_3$)	Layer structure with conducting Na^+ ions concentrated in conduction planes
ZrO_2/CaO CeO_2/M_2O_3 (M = trivalent cation)	Fluorite structure with divalent and trivalent cation substitutionals

(a) High-temperature fluorite-structured crystals

At temperatures within a few hundred degrees of the melting point, all fluorite-structured systems that have been studied show a diffuse phase transition manifested by a specific heat anomaly; an example is shown in Fig. 7.20 for the case of $SrCl_2$.[49] Above the phase-transition temperature the materials are superionic conductors; conductivity data are shown in Fig. 7.21 for PbF_2. The phase transition is unquestionably associated with the generation of disorder; and since, as discussed in Section 7.3.1, the characteristic disorder of the fluorite is of the anion Frenkel type, we would expect disorder to be generated on the anion sub-lattice. Thus the high-temperature superionic properties of fluorites are due to anion mobility.

Despite their structural simplicity, the physics of the superionic fluorites appears to be complex. Earlier studies proposed that the disordering of the anion sub-lattice, which occurs during the phase transition, could be described as 'sub-lattice' melting. Although the exact meaning of this term was not defined, it clearly implies a major disordering of the sub-lattice. Recent studies suggest that such models are not acceptable. Single crystal studies[51] of high-temperature $SrCl_2$ show that over 90 per cent of anions are still located near the regular anion sites. Calculations support this conclusion. Thus the enthalpy of the diffuse phase transition, obtained by integrat-

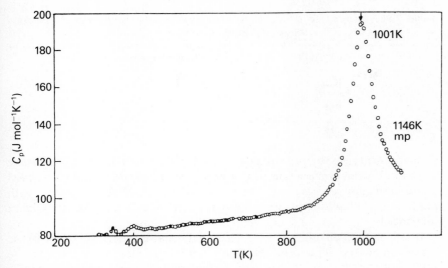

Fig. 7.20. Specific heat vs. temperature for $SrCl_2$.

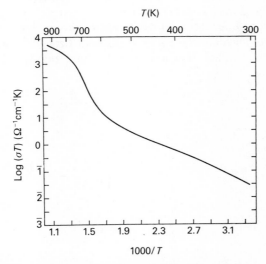

Fig. 7.21. Arrhenius plot for conductivity of PbF_2. (After Carr et al.[50])

ing under the specific heat curve, is generally about 0.1 eV (see Table 7.11). Calculated Frenkel energies are, in contrast, 2–3 eV. Frenkel energies refer, of course, to isolated vacancies and interstitials; interactions between these species might be expected to lower the energy of disordering of the anion sublattice. Calculations showed that such effects could indeed reduce interstitial

Table 7.11 *Energies of diffuse phase transitions to superionic state in Fluorite-structured crystals.* (See Catlow.[76])

Crystal	Transition energy (eV)
K_2S	0.17
$SrCl_2$	0.15
PbF_2	0.10

formation energies, by (up to) a factor of 2. This would, however, still leave interstitial formation energies far in excess of the experimentally observed phase transition energy; and the results of the calculations accord with models for the high-temperature fluorites based on only a few per cent anion interstitials.

If only limited disorder is being generated, what is the nature of the phase transition to the superionic state? A plausible suggestion is that, at the transition temperature, the vacancy and interstitial concentrations are sufficiently large to allow the formation of large vacancy–interstitial aggregates of the type shown in Fig. 7.22. These resemble the type of cluster models proposed for the doped systems described in Section 7.3.2, except that the interstitial aggregate is stabilized by vacancies rather than by dopant ions. The formation of such clusters rationalizes both Bragg[51, 78] and quasi-elastic scattering data[52, 78] on high-temperature $SrCl_2$, and high stability is found by the calculations.

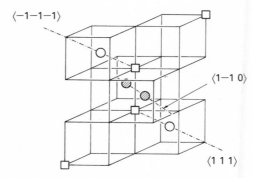

Fig. 7.22. Proposed interstitial–vacancy cluster in high-temperature fluorites.

The question of migration mechanisms in high-temperature fluorites seems to be more straightforward than that of the structural properties discussed above. At low temperatures, vacancies are known to be more mobile than interstitials, with activation energies of about 0.3–0.5 eV. Vacancies migrate by a straightforward jump mechanism whereas interstitials adopt a more

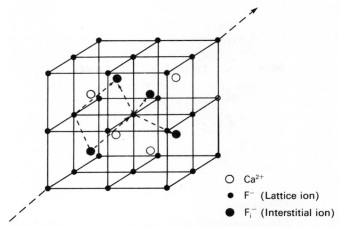

Fig. 7.23. Interstitial mechanism for anion transport in fluorites. The migrating interstitial displaces the central lattice F^- ion to one of three neighbouring interstitial sites.

complex migration route in which the migrating interstitial displaces a lattice anion into the neighbouring interstitial sites as shown in Fig. 7.23. Thermal expansion, however, lowers the interstitial activation energy; the larger lattice parameter leads to lower strain in the constrained saddle point shown in Fig. 7.23. In the superionic region both vacancies and interstitials contribute to the migration process; although if clusters of the type shown in Fig. 7.22 dominate, the anion transport is presumably more complex than a combination of interstitial and vacancy transport.

Further information on the superionic properties of the high-temperature fluorites has been provided by the application of molecular dynamics simulation techniques. These methods, discussed in Section 7.2.3, allow explicit representation of thermal effects. Application of the methods by Gillan and Dixon[53, 54] has clearly shown that the superionic behaviour is associated with only limited disorder as argued above; their work has also verified the validity of 'hopping' descriptions of the anion transport processes—an assumption which is necessarily made in applying the static simulation methods.

Dynamical simulation methods have also been applied to a second high-temperature superionic, AgI.[55] Static simulation techniques are inappropriate here, owing to the breakdown of the simple hopping model. For the fluorites, however, applications of the static simulations have yielded useful evidence on problems where the extraction of unambiguous information from experiment is difficult.

(b) β-Al_2O_3 type compounds

These layer-structured materials are possibly the most intensively studied superionics owing to the use of Na–β-Al$_2$O$_3$ in the commercially viable sodium–sulphur battery. The compounds are of general formula M$_2$O·11Al$_2$O$_3$ (M = alkali metal) when stoichiometric. As normally prepared, however, they are grossly non-stoichiometric with a typical composition being (M$_2$O)$_{1.3}$·11Al$_2$O$_3$. Their main structural features are illustrated in Fig. 7.24. They comprise 'spinel'-structured blocks of Al$_2$O$_3$† between which are 'sandwiched' the mobile alkali metal ions and the bridging oxygen ions. The conduction plane in which the Na$^+$ ions are located contains a hexagonal net of sites; two types of site alternate, known as *Beevers–Ross* (BR) and *anti-Beevers–Ross* (aBR).

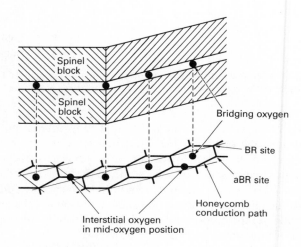

Fig. 7.24. The structure of Na–β Al$_2$ O$_3$.

The main problems investigated by the simulation methods concern the nature of the Na$^+$ migration mechanism, and the effect of deviation from stoichiometry on Na$^+$ mobility. Information on the latter point is provided from experimental studies of Hayes *et al.*,[56] who showed that the conductivity of near-stoichiometric Na–β-Al$_2$O$_3$ was considerably less than that of the grossly non-stoichiometric compound; thus they found that the activation energy changed from about 0.6 eV for near-stoichiometric β-Al$_2$O$_3$, to about 0.17 eV for the non-stoichiometric crystal. Excess Na$^+$—present as some form of interstitial in the conduction plane—is clearly therefore essential for high sodium mobility. The first question we must ask concerns the structure of the Na$^+$ interstitials. There is good evidence[57] that in the stoichiometric compound, Na$^+$ ions occupy the Beevers–Ross sites. Early calculations of Wang *et al.*[58] suggested that Na$^+$ interstitials had a split interstitial configu-

†It is interesting to note that the metastable phase, γ-Al$_2$O$_3$, has a cation-deficient spinel structure.

Computational techniques and simulation of crystal structures

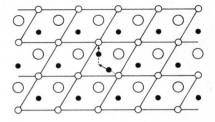

Fig. 7.25. Split interstitial configuration in β–Al$_2$O$_3$ (proposed by Wang et al.[58]).

ration, as shown in Fig. 7.25, in which the excess Na$^+$ ions displace a lattice cation from its site, with the ions in the resulting Na$^+$–Na$^+$ pair occupying mid-oxygen positions. More recent work by Walker and Catlow[59] which used large, explicity relaxed regions and shell model potentials, found that the most stable configuration of an isolated interstitial was at the anti-Beevers–Ross site, with the split interstitial configuration shown in Fig. 7.25 being the saddle point for an interstitialcy type of migration mechanism. The calculated activation energy is 0.57 eV—a result which accords well with the data of Hayes et al.[56] if we assume that the near-stoichiometric Na-β-Al$_2$O$_3$ contained a small excess of Na$^+$, presumably present as interstitials.

We are left, however, with two problems. The first concerns the structural properties of the phase. Diffraction data suggests high occupancy by Na$^+$ ions of mid-oxygen sites in non-stoichiometric Na–β-Al$_2$O$_3$—a result which is compatible with the models of Wang et al.[58] A plausible explanation of the occupancy of these sites can, however, be provided by the occurrence of extensive clustering between the Na$^+$ ions and the excess oxide ions which are of course also present in the conduction plane of non-stoichiometric β-Al$_2$O$_3$. The latter species themselves have a complex structure, being stabilized by the displacement of neighbouring spinel block Al^{3+} ions as illustrated in Fig. 7.26. The calculations of Walker and Catlow[59] suggest that clustering between these interstitials and excess Na$^+$ ions is important, and leads to displacements of the Na$^+$ ions towards the mid-oxygen sites.

The second problem concerns the activation energy for grossly non-stoichiometric Na–β-Al$_2$O$_3$; the measured value of about 0.17 eV is considerably lower than our calculated interstitialcy activation energy. A tentative explanation of this observation was advanced by Walker and Catlow.[59] They suggest that in highly non-stoichiometric β-Al$_2$O$_3$, there are 'conduction paths' as shown in Fig. 7.27, in which both BR and aBR sites are occupied. Migration of Na$^+$ ions within these paths will take place essentially by a vacancy mechanism. Separate sets of calculations on the related compound Na–β″-Al$_2$O$_3$ (Na$_2$O·MgO·5Al$_2$O$_3$), in which the conduction plane has fully

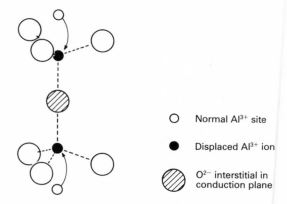

Fig. 7.26. Oxygen interstitial in Na–β-Al$_2$O$_3$. Note the stabilization by displacement of two neighbouring Al^{3+} lattice ions into interstitial positions.

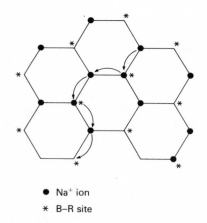

Fig. 7.27. Proposed conduction mechanism in grossly non-stoichiometric Na–β-Al$_2$O$_3$. Note Na$^+$ migration is effected essentially by a vacancy mechanism through 'paths' of occupied Beevers–Ross and anti-Beevers–Ross sites.

occupied BR and aBR sites,† found a calculated Na$^+$ vacancy migration energy of about 0.16 eV. This result clearly supports the type of model proposed by Walker and Catlow.

7.3.4 Crystal structure prediction

As discussed in Section 7.2.1, the availability of reliable potentials, together

† β''-Al$_2$O$_3$ also differs from β-Al$_2$O$_3$ by having thinner spinel blocks, and by containing Mg^{2+} in addition to Al^{3+}.

Computational techniques and simulation of crystal structures

with efficient minimization methods, opens up the possibility of predicting the minimum energy structures of crystals. Such techniques have already been applied with success to molecular crystals,[60] but until recently had received little attention in inorganic crystal chemistry.

The value and reliability of such methods can be shown by a relatively simple example, namely the recently discovered phase of VO_2 known as $VO_2(B)$.[61] This structure is based on a mode of linkage of the VO_6 octahedra that differs from that found in the rutile, anatase, and brookite structures. An idealized model based on regular octahedra is shown in Fig. 7.28(a). This was used as the starting point of a minimization study.[62] The final, minimum energy configuration is displayed in Fig. 7.28(b), where it is compared with the experimentally determined structure. The distortion of the observed from the ideal structure is seen to be well reproduced by the calculations; and indeed the discrepancies between predicted and observed structures are generally within the magnitudes of the amplitudes of their thermal vibrations, obtained from the Debye–Waller factor in the crystallographic study of this phase—discrepancies of at least this magnitude would be expected for a 'static' simulation. Energy minimization studies[62] of a number of titanates have also been undertaken. Again, good agreement between predicted and observed structures is obtained.

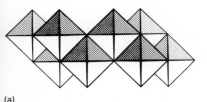 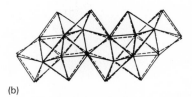

(a) (b)

Fig. 7.28. VO_2 (B) structure. (a) Ideal structure. (b) Observed structure (dotted line), which is compared with prediction of energy minimization studies (full line).

The most rewarding applications of these techniques are probably in the field of structural mineralogy. A particular incentive is provided here for theoretical studies, as difficulties in obtaining adequate single crystals often hamper accurate experimental structure determinations. In addition, in many geochemical areas one is often concerned with the structure and properties of minerals under extreme conditions of temperature and pressure where experimental studies are difficult; simulations may therefore provide information that is inaccessible to experimental methods.

Work of Parker[63,64] has shown that energy minimization methods can accurately predict the structures of a large number of minerals. The potentials used are essentially based on the ionic model with parameterized two-body short-range potentials. An important feature of the work is the use of certain, accurately known structures in parameterizing the potentials by the 'zero-strain' conditions described in Section 7.2.2. The subsequent applications of the techniques are, of course, to structures not included in the parameterization process.

A good example of this work is provided by olivine (Mg_2SiO_4), the structure of which is illustrated in Fig. 7.29. The feature to which we draw attention is the 'rumpling' of the planes of oxygen ions from the close-packed, hexagonal structure. This complex distortion is accurately reproduced by the calculations which yield a minimum energy structure which is close to the experimentally determined one. Energy minimization studies of other ortho-silicates (i.e. silicates which contain isolated SiO_4^{4-} groups) include the rare minerals $ZrSiO_4$ and $ThSiO_4$. Again the calculations accurately reproduce the most recent experimental structures.

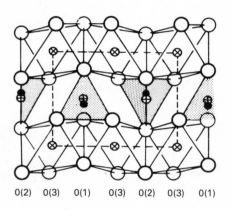

O(2) O(3) O(1) O(3) O(2) O(3) O(1)

Fig. 7.29. The structure of olivine (Mg_2SiO_4).

The other main applications reported to date are to ring and chain structured silicates. The systems studied are listed in Table 7.12. For all these structures, good agreement between theory and experiment have been obtained. This is of especial interest in view of the suggestions by a number of authors[65, 66] that there are considerable differences between the bonding of bridging and non-bridging oxygen atoms in these structures. The potentials used in the energy minimization studies do not include such differences, and their success in predicting the structures of these systems must throw doubt on the validity of these arguments. Indeed one of the main points to emerge from this work is the surprising degree of success which may be enjoyed in simulations of silicates by relatively simple potential models. We should note that the methods have proved less successful when applied to framework-structured silicates, although for zeolites, which are framework-structured

Table 7.12 *Some silicates whose structures have been successfully predicted by energy minimization techniques. (See Parker et al.[64])*

Formula	Name	Structural type
Mg_2SiO_4	Forsterite	Ortho-silicate
$ZrSiO_4$	Zircon	Ortho-silicate
$Na_2Be_2Si_3O_9$	—	Ring
a-$Sr_3Si_3O_9$	—	Ring
$Al_2Be_3Si_6O_9$	Beryl	Ring
$MgSiO_3$	Enstatite	Chain
$FeSiO_3$	Ortho-ferrosilite	Chain
Na_2SiO_3	Sodium meta-silicate	Chain

tion of the low-valence cations that are located in the channels in these structures. Accurate computer modelling of framework minerals, however, requires extension of the potential models to include bond-bending forces which arise from the covalent contribution to the bonding. Recent work suggests that such extensions are possible.

7.3.5 Surface simulations

Both perfect and defective surfaces have become amenable to simulation studies. Perfect-surface simulations generally describe a region within a few layers of the surface by an explicit atomistic method; the perturbation provided by the surface on the interior of the crystals very rapidly approaches zero; and it is not, in general, necessary to treat relaxation of the inner regions of the crystal.

Work of Tasker[68] has yielded interesting results on the energies and structures of the surfaces of ionic crystals. Thus the calculations are able to determine the surface orientation of lowest energy. They reveal interesting structural details, in particular the 'rumpling' effect illustrated in Fig. 7.30; this involves relative displacements of cations and anions in the surface layer. Analysis of the calculations shows that rumpling is attributable to ionic polarization. The anions and cations are polarized at the surface in opposite directions, but, to avoid increasing overlap with the ions on the plane below, the anions move outwards while the cations are able to move inwards. Surface structure may be investigated by the technique of low energy electron diffraction (LEED). Analysis of LEED data still, however, presents problems; and guidance from reliable simulations could be of considerable value in such studies.

Surface simulation techniques may be used on other two-dimensional structures, e.g. shear planes (discussed in Section 7.3.2) and grain boundaries. The latter are the junctions between crystallites in polycrystalline materials; and they are of considerable interest in contemporary materials

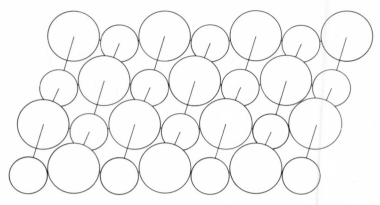

Fig. 7.30. Surface-rumpling calculated for NaCl structure by Tasker.[68]

science research as grain boundary processes greatly influence transport and reactivity in polycrystalline solids.

The techniques employed by *surface defect* simulations employ a hemispherical region I around the surface defect (e.g. a surface vacancy or surface impurity atom) is relaxed explicitly; more distant regions are treated by a Mott–Littleton type of method. Work of Mackrodt[69] and Tasker[70] has shown that surface defect energies may be considerably different from the energies of the same type of defect in the bulk. This may lead to segregation or depletion of impurity atoms from the surface region—a process which again may affect the reactivity of the solid.

Possibly the most exciting and far-reaching studies are those reported recently by Colbourn and Mackrodt.[71] They have studied the chemisorption on oxide surfaces by a combination of quantum mechanical calculations and surface simulations of the type described above. Their calculations show the initial rôle played by surface defects in leading to binding of the molecule to the surface; they could have major consequences for our understanding of catalysis.

7.4 References

1. Gillan, M. J., and Jacobs, P. W. M. *Phys. Rev.* **B28**, 759 (1983).
2. Ewald, R. P. *Ann. Physik* **64**, 253 (1921).
3. Tosi, M. In *Solid state physics* (eds F. Seitz and Turnbull) Vol. 16, p. 1. Academic Press, New York (1964).
4. Waddington, T. C. *Adv. Inorg. Radiochem.* **1**, 158 (1959).
5. Catlow, C. R. A., and Mackrodt, W. C. (eds) *Computer simulation of solids* Lecture Notes in Physics, **166** (1982).
6. Dick, B. G., and Overhauser, A. W. *Phys. Rev.* **112**, 90 (1958).
7. Catlow, C. R. A., and Norgett, M. J. *UKAEA report* **AERE M2936** (1976).
8. Norgett, M. J., and Fletcher, R. *J. Phys. C.* **3**, L190 (1970).

9. Mott, N. F., and Littleton, M. J. *Trans Farad. Soc.* **34**, 485 (1938).
10. Norgett, M. J., *UKAEA report* **AERE R7650** (1974).
11. Sangster, M. J., and Dixon, M. *Adv. Phys.* **25**, 247 (1976).
12. Mackrodt, W. C., and Kendrick, J. *Solid State Ionics* **8**, 247 (1983).
13. Henderson, B. *Defects in crystalline solids.* Edward Arnold, London (1972).
14. Jacobs, P. W. M. In *Mass transport in solids—Proc. NATO ASI* (eds F. Béniere and C. R. A. Catlow). Plenum Press, New York (1983).
15. Corish, J., and Jacobs, P. W. M. In *Surface and defect properties of solids* (eds J. M. Thomas and M. W. Roberts) Vol. 2, Chap. 7. Chemical Society, London (1973).
16. *Idem*, Vol. **11**, p. 218 (1977).
17. Vineyard, G. *J. phys. Chem. Solids* **3**, 157 (1957).
18. Lidiard, A. B. In *Handbuch der Physik*, Vol. **20** (ed. S. Flugge). Springer-Verlag, Berlin (1958).
19. Catlow, C. R. A. In *Mass transport in solids—Proc. NATO ASI* (eds F. Béniere and C. R. A. Catlow). Plenum Press, New York (1983).
20. Gerhardt-Anderson, R., and Nowick, A. S. *Solid State Ionics* **5**, 547 (1981).
21. Nowick, A. S., Wang, Da Yu, Park, D. S., and Griffith, J. In *Fast ion transport in solids* (eds P. Vashishta et al.). North-Holland Publishing Co., Amsterdam (1979).
22. Butler, V., Catlow, C. R. A., Fender, B. E. F., and Harding, J. H. *Solid State Ionics* **8**, 109 (1983).
23. Wuensch, B. J. In *Mass transport phenomena in ceramics* (eds A. R. Cooper and A. H. Heuer). *Mater. Sci. Res.* **9**, 211 (1975).
24. Wuensch, B. J. In *Mass transport in solids—Proc. NATO ASI* (eds F. Béniere and C. R. A. Catlow). Plenum Press, New York (1983).
25. Mackrodt, W. C. In *Computer simulation of solids* (eds C. R. A. Catlow and W. C. Mackrodt), *Lecture Notes in Physics* Vol. **166**. Springer-Verlag, Berlin (1982).
26. Mackrodt, W. C. In *Transport in non-stoichiometric compounds* (eds G. Petot-Ervas, Hj. Matzke, and C. Monty), p. 175. North Holland, Amsterdam (1984).
27. Catlow, C. R. A. In *Non-stoichiometric oxides* (ed. O. T. Sorensen). Academic Press, New York (1982).
28. Willis, B. T. M. *Proc. Brit. ceram. Soc.* **1**, 9 (1964).
29. Willis, B. T. M. *J. de Physique* **25**, 431 (1964).
30. Cheetham, A. K., Fender, B. E. F., and Cooper, M. J. *J. Phys.* C, **4**, 3107 (1971).
31. Catlow, C. R. A. *J. Phys.* C, **6**, L64 (1973).
32. Greiss, O. *Z. anorg. allg. Chem.* **430**, 175 (1977).
33. Gettmann, P. and Greiss, O. *J. solid-state Chem.* **26**, 255 (1978).
34. Bendall, P. J., Catlow, C. R. A., Corish, J., and Jacobs, P. W. M. *J. solid-state Chem.* **51**, 159 (1984).
35. Bendall, P. J., Catlow, C. R. A., and Fender, B. E. F. *J. Phys.* C, **17**, 797 (1984).
36. Catlow, C. R. A., Corish, J., and Chadwick, A. V. *J. solid-state Chem.* **48**, 65 (1983).
37. Fontanella, J. J., Treacy, D. J., and Andeen, C. G. *J. chem. Phys.* **73**, 2235 (1980).
38. Koch, F., and Cohen, J. B. *Acta Cryst.* **B25**, 275 (1969).
39. Cheetham, A. K., Fender, B. E. F., and Taylor, R. I. *J. Phys.* C, **4**, 2160 (1971).
40. Battle, P. D., and Cheetham, A. K. *J. Phys.* C, **12**, 337 (1979).
41. Lebreton, C., and Hobbs, L. M. *Radiation Effects* **74**, 227 (1983).
42. Dieckmann, R. In *Transport in non-stoichiometric compounds* (eds G. Petot-Ervas, Hj. Matzke, and C. Monty), p. 1. North Holland, Amsterdam (1984).
43. Mackrodt, W. C., and Colbourn, E. (Unpublished work).

44. Bursill, L. A., and Hyde, B. G. *Prog. solid-state Chem.* **7**, 177 (1972).
45. See, e.g. Tilley, R. J. D. In *MTP Int. Rev. Sci., Inorganic Chemistry* (ed. L. E. J. Roberts), Series (1), Vol. **10**, 279 (1972).
46. Catlow, C. R. A., and James, R. *Nature* **272**, 603 (1978).
47. Catlow, C. R. A., and James, R. *Proc. R. Soc.* **A384**, 157 (1982).
48. Cormack, A. N., Jones, R., Tasker, P. W., and Catlow, C. R. A. *J. solid-state Chem.* **44**, 174 (1982).
49. Schroter, W., and Nolting, J. *J. Phys. (Paris)* **41**, C, 6–20 (1980).
50. Carr, V. M., Chadwick, A. V., and Saghafian, R. *J. Phys.* C, **11**, L637 (1978).
51. Dickens, M. H., Hayes, W., Hutchings, M. T., and Smith, C. *J. Phys.* **C15**, 4043 (1982).
52. Dickens, M. H., Hutchings, M. T., Kjems, J. K., and Lechner, R. E. *J. Phys.* **C11**, L583 (1978).
53. Gillan, M. J., and Dixon, M. J. *J. Phys.* C, **13**, 1901 (1980).
54. Dixon, M. and Gillan, M. J. *J. Phys.* C, **13**, 1919 (1980).
55. Vashishta, P., and Rahman, A. In *Fast ion transport in solids* (eds P. Vashishta *et al.*), p. 527. North-Holland Publishing Co., Amsterdam (1982).
56. Hayes, W., Holden, L., and Tofield, B. C. *Solid State Ionics* **1**, 373 (1980).
57. Colomban, P., and Lucazeau, G. *J. chem. Phys.* **72**, 1213 (1980).
58. Wang, J. C., Gaffari, M., and Choi, S. I. *J. chem. Phys.* **63**, 772 (1975).
59. Walker, J. R., and Catlow, C. R. A. *J. Phys.* C, **15**, 1651 (1981).
60. Busing, W. R. Oak Ridge Laboratory Report URNL-5747 (1981).
61. Theobald, F., Cabala, R., and Bernard, J. *J. Solid State Chem.* **17**, 431 (1976).
62. Catlow, C. R. A., Cormack, A. N., and Theobald, F. *Acta Cryst.* **B40**, 195 (1984).
63. Parker, S. C. *UKAEA report* **AERE TP963** (1982).
64. Parker, S. C., Cormack, A. N., and Catlow, C. R. A. *Acta Cryst.* **B40**, 200 (1984).
65. Cruickshank, D. W. J. *J. chem. Soc.* 5486 (1961).
66. McDonald, W. S., and Cruickshank, D. W. J. *Acta Cryst.* **22**, 37 (1967).
67. Sanders, M. J., and Catlow, C. R. A. *Proc. Int. Zeolite Conf.* (eds A. Bisio and D. M. Olson), p. 131. Butterworths, London (1983).
68. Tasker, P. W. *Phil. Mag.* **A39**, 119 (1979).
69. Mackrodt, W. C., and Stewart, R. F. *J. Phys.* C, **12**, 5015 (1979).
70. Duffy, D. M., and Tasker, P. W. *Phil. Mag.* **A48**, 155 (1983).
71. Colbourn, E., and Mackrodt, W. C. *Surf. Sci.* **117**, 571 (1982).
72. Catlow, C. R. A. *Proc. Roy. Soc.* **A353**, 533 (1977).
73. Catlow, C. R. A. and Norgett, M. J. *J. Phys. C*, **6**, 1325 (1973).
74. Catlow, C. R. A., and Fender, B. E. F. *J. Phys. C*, **8**, 3267 (1975).
75. Cormack, A. N., Catlow, C. R. A., and Tasker, P. W. *Radiation Effects* **74**, 237 (1984).
76. Catlow, C. R. A. *Comments Solid State Phys.* **9**, 157 (1980).
77. Catlow, C. R. A., and Hayns, M. R. *J. Phys. C,* **5**, L237 (1972).
78. Dickens, M. H., Hayes, W., Kjems, J. K., Schnabel, P. G., and Smith, C. *J. Phys. C,* **17**, 3903 (1984).
79. Sanders, M. J., Leslie, M., and Catlow, C. R. A. *J. Chem. Soc. Chem. Commun.* p 1271 (1984).
80. Jackson, R. A. and Catlow, C. R. A. *Molecular Simulations* (In press).

8 Transport measurements

A. Hamnett

8.1 Introduction

Determining transport properties is one of the simplest and most informative macroscopic measurements that can be performed on a solid material, and the information obtained can be related both to the nature of the bonding and the dynamical properties of the lattice. However, as with all macroscopic measurements, the results do not directly lead to microscopic parameters; instead, transport data may suggest a model of the electronic structure. The predictions of this model may then be checked by spectroscopic, magnetic, and thermal data, and by reference to the crystal structure. It cannot be overemphasized in this chapter that transport measurements on their own will not yield an unambiguous picture of the electronic structure; the study of inorganic solids is quintessentially a multidisciplinary science.

Most of this chapter is devoted to inorganic semiconductors. Some, such as elemental Si and Ge, have been intensively studied because of their commercial importance. Others, such as VO_2, have extraordinary physical properties the study of which has greatly enhanced our understanding of electronic structure. For many, however, data are still scarce, and much remains to be explored.

8.2 Conductivity

The conductivity of any material is given by the product of the *number of carriers* and their *mobility*:

$$\sigma = ne_0\mu \qquad (8.1)$$

where e_0 is the charge on the carrier. Throughout this chapter the carriers are assumed to be electrons with charge 1.602×10^{-19} C. Until recently the identity of the carriers was rarely checked since ions were assumed to be essentially immobile, but the advent of fast-ion conductors has led to some increased caution in this regard. The units used in this chapter will be SI; n is thus expressed in m^{-3}, μ in $m^2V^{-1}s^{-1}$, and σ then has units $ohm^{-1}m^{-1}$.

8.2.1 Concentration of carriers

The electronic structure of semiconductors is described in detail elsewhere;[1,2] the salient features are summarized in Fig. 8.1. Overlap between orbitals of neighbouring atoms or ions gives rise to a series of narrowly spaced energy levels, termed bands. The highest occupied band is called the *valence band* (VB) and the lowest unoccupied band the *conduction band* (CB). As an example of this, the valence band of TiO_2 is formed primarily from the O $2p^6$ orbitals which overlap to form a band about 4 eV broad.[3] The rather narrower conduction band is formed primarily from the Ti $3d^0$ orbitals. In between the CB and VB is an energy gap called the *bandgap* which, in TiO_2, is about 3 eV wide; within this energy range, there are no extended electronic states.

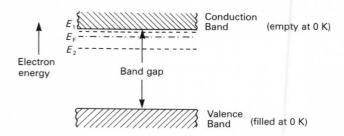

Fig. 8.1. Energy levels in a semiconductor. There are two broad bands of energy levels separated by a bandgap; the Fermi level is denoted by E_F and E_1 and E_2 denote the energy levels of localized states.

This picture is reasonably adequate at 0K but, as the temperature is raised, some thermal excitation of the electrons can occur from the VB to the CB. Statistical calculations, beyond the scope of this chapter,[1,3] lead to the conclusion that the number density of electrons at energy E in the CB, $n(E)$, is given by

$$n(E) = 2N_c(E)P_e(E) \qquad (8.2)$$

where $N_c(E)$ is the number density of states in the conduction band (i.e. the number of available energy levels for the electrons between energies E and $(E+dE)$ is $N_c(E)dE$), and $P_e(E)$ is a thermal distribution function related to the Boltzmann law but modified to take account of the quantum mechanical properties of the electrons. It is termed the Fermi function and has the form

$$P_e(E) = 1/[1 + \exp\{(E - E_F)/kT\}]. \qquad (8.3)$$

The reference energy E_F is called the Fermi energy and corresponds to the chemical potential of the electrons in the solid. The factor 2 in eqn (8.2) is to take account of the fact that each level may be occupied by two electrons of opposite spins.

Excitation of the electron from the VB leaves behind a vacancy or 'hole'. Electrons in the VB can move to fill these holes leaving, in turn, further holes. Conceptually, it is easier to consider the holes moving in a sea of immobile electrons, and we may develop expressions for the number density of these holes in close analogy to the formulae above for the electrons. Thus, the number density of holes in the VB at energy E', $p(E')$, is given by

$$p(E') = 2N_v(E')P_h(E') \tag{8.4}$$

where

$$P_h(E') = 1 - P_e(E') = 1/[1 + \exp\{(E_F - E')/kT\}] \tag{8.5}$$

and $N_v(E)$ is the number density of states in the valence band at energy E.

The total number density of electrons excited n_i can be obtained by integrating over the energy range $E = E_c$ to the top of the CB, and, in a similar way, the total number density of holes p_i is obtained by integrating eqn (8.4) from the bottom of the VB to E_v. Clearly, these integrals cannot be performed unless the functional form of N_v and N_c are available. In practice, precise functional forms are rarely available, but it is often found that the top of the VB and the bottom of the CB have simple parabolic shapes of the form

$$N_c(E) = 2\pi(2m_e)^{\frac{3}{2}}h^{-3}(E - E_c)^{\frac{1}{2}} \tag{8.6}$$

$$N_v(E) = 2\pi(2m_h)^{\frac{3}{2}}h^{-3}(E_v - E)^{\frac{1}{2}} \tag{8.7}$$

Even though the bands may deviate substantially from this form at energies more removed from the band edges, the functional form of eqns (8.3) and (8.5) ensures that little contribution to the total numbers of electrons or holes will come from these regions. The deviation from free-electron-like behaviour is accounted for by the two *effective masses*: m_e and m_h. For a free-electron band, m_e and m_h would both have the value of the free-electron mass. As is shown elsewhere, we can take into account the periodic potential in a solid in first-order by incorporating into the theory an *effective mass* for the electron or holes.[5] In terms of the overlap model we can say very approximately that the larger the overlap between different atoms or ions, the wider the band of states becomes; the wider this band, the smaller the effective mass will be. Substituting (8.6) and (8.7) into (8.2) and (8.4) and integrating, we have, after some approximations,

$$n_i = N_c \exp\{(E_F - E_c)/kT\}; \quad N_c = 2(2\pi m_e kT/h^2)^{\frac{3}{2}} \tag{8.8}$$

$$p_i = N_v \exp\{(E_v - E_F)/kT\}; \quad N_v = 2(2\pi m_h kT/h^2)^{\frac{3}{2}}. \tag{8.9}$$

Since the system is neutral, $n_i = p_i$ and

$$E_F = \frac{E_v + E_c}{2} + \tfrac{3}{4}kT\log_e(m_h/m_e). \tag{8.10}$$

The condition that $n_i = p_i$ defines the system as being *intrinsic*. It may be helpful to insert some numbers into the above equations: for normal values of m_e and m_h ($0.1m_0$–$10m_0$, where m_0 is the free-electron mass), N_V and N_C take the values 10^{25}–10^{26} m^{-3}. Thus, for Ge at 300 K, $n_i = p_i \cong 2.3 \times 10^{19}$ m^{-3} assuming a bandgap of 0.6 eV. By contrast, for TiO$_2$, with a bandgap of 3.0 eV, $n_i = p_i \cong 1$ m^{-3}.

The very small intrinsic carrier concentrations for TiO$_2$ imply that the total concentration of carriers in this material is likely to be dominated by impurities. In this case, the semiconductor is described as *extrinsic*. Truly intrinsic semiconductors are rather rare unless the bandgap becomes very small ($\lesssim 0.4$ eV), or strenuous efforts are made to remove all impurities. These impurities may be neutral or, more interestingly from the point of view of the number of carriers, they may ionize, releasing electrons into the conduction band, or they may capture electrons, releasing holes into the VB. The former type of impurity leads to an '*n*-type' semiconductor, the latter to a '*p*-type' material. Examples are legion; the basis of the modern electronics industry is silicon, but Si has a bandgap of 1.1 eV and the number of intrinsic carriers is therefore very low (1.0×10^6 m^{-3}) at room temperature.[6] To enhance the conductivity, very small amounts of phosphorus or boron can be annealed into the sample. The former has one extra electron which it can release into the CB forming *n*-Si whereas the latter has one too few valence electrons and can capture an electron from the VB to give *p*-type conduction. The amount of P or B required to give a conductivity of 1 $(\Omega\,\text{m})^{-1}$ can be calculated knowing the mobility of electrons in Si is 0.15 m^2 V^{-1}s^{-1}; since $\sigma = ne_0\mu$, $n \simeq 4.2 \times 10^{19}$ m^{-3}. This represents about one part of P in 10^9 parts Si. Since P must be the dominant impurity, this in turn implies that we must start with extremely pure Si—hence the need for clean-room facilities in semiconductor technology.

With binary compounds an additional source of impurity levels is non-stoichiometry. Thus, many oxides are either cation- or anion-deficient; the former, such as Ni$_{1-x}$O ($x \lesssim 0.01$) are *p*-type, whereas the latter, such at TiO$_{2-x}$ ($x \lesssim 0.01$) are *n*-type. The extent of non-stoichiometry varies with the partial pressure of oxygen in a manner that can usually be calculated from point-defect theory provided that x remains small.

Quantitative treatment of extrinsic semiconductors is obviously more difficult than the intrinsic case discussed above since n_i and p_i are no longer equal. However, we still have

$$n = N_c \exp\{(E_F - E_c)/kT\} \quad (8.11)$$

$$p = N_v \exp\{(E_v - E_F)/kT\} \quad (8.12)$$

and so
$$np = N_c N_v \exp\{-(E_c - E_v)/kT\} = N_c N_v \exp\{-E_g/kT\} \quad (8.13)$$

where E_g is the bandgap. Thus n and p are always related by the expression

$$np = n_i p_i = n_i^2. \quad (8.14)$$

We now suppose that there are N_d donor levels of energy E_d below the conduction band. Then the number of levels n_d occupied is given by

$$n_d = \frac{N_d}{1 + g_d \exp\{(E_d - E_F)/kT\}} \quad (8.15)$$

where g_d is a statistical number which takes the value $\frac{1}{2}$ for the common situation of a one-electron impurity level such as P in Si.

From the electroneutrality principle

$$n + n_d = N_d + p \quad (8.16)$$

where we have assumed that the number of acceptor traps N_a is negligible. Then

$$N_c \exp\{(E_F - E_c)/kT\} - N_v \exp\{(E_F - E_v)/kT\} = \frac{N_d g_d \exp\{(E_d - E_F)/kT\}}{1 + g_d \exp\{(E_d - E_F)/kT\}}. \quad (8.17)$$

In principle this may be solved as a cubic in $\exp(-E_F/kT)$. Rather than do this however we will examine two extreme solutions. At high temperatures (region (a) in Fig. 8.2), the left-hand side of eqn (8.17) is dominant and there is a return to the intrinsic situation described above. At lower temperatures (region (b) in Fig. 8.2), the second term on the left-hand side, representing the concentration of holes, becomes negligible, and we have

$$N_c \exp\{(E_F - E_c)/kT\} = \frac{N_d g_d \exp\{(E_d - E_F)/kT\}}{1 + g_d \exp\{(E_d - E_F)/kT\}} \quad (8.18)$$

Commonly $(E_c - E_d)$ is very small (about 0.05 eV) and $(E_d - E_F) \gg kT$, so the right-hand side of eqn (8.18) is approximately equal to N_d. Hence

$$E_f \equiv E_F - E_c = kT \log_e(N_d/N_c). \quad (8.19)$$

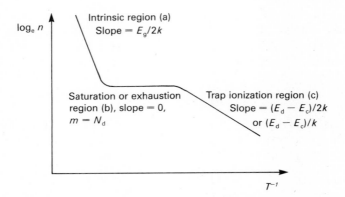

Fig. 8.2. Concentration of electrons in the conduction band for an n-type semiconductor, showing the three regions referred to in the text.

At very low temperatures (region (c) in Fig. 8.2), the assumption that E_d and E_F are well separated, i.e. that all the donor centres are ionized, breaks down and we must solve the quadratic equation (8.18) explicitly. If $N_a \ll N_d$

$$n = \left(\frac{N_a N_c}{2}\right)^{\frac{1}{2}} \exp\{-(E_d - E_c)/2kT\}. \tag{8.20}$$

If, in eqn (8.16), we cannot neglect the number of acceptor centres N_a, then N_a should be added to the left-hand side. Solving at low temperatures we find

$$n = \left(\frac{N_d - N_a}{2N_a}\right) N_c \exp\{-(E_d - E_F)/kT\}. \tag{8.21}$$

The number of carriers in an n-type extrinsic semiconductor is shown in Fig. 8.2 as a function of (inverse) temperature.

8.2.2 Mobility of carriers

Hitherto we have considered the *number* of carriers n, and how it might vary with temperature. However, as we saw above, the conductivity is determined by the *product* of n and the mobility μ, and it is to a consideration of the latter that we now turn.

At a microscopic level, the electrons do not move uniformly in the direction of the applied electric field in a conductor. Rather they progress in a complex three-dimensional random-walk upon which is superimposed a steady drift. Moreover, the actual velocity of motion of an electron is far greater that its drift velocity. This originates from the fact that the electrons are continually deflected or scattered by imperfections in the crystalline lattice. These imperfections are of two main types: impurities of the sort discussed above and, quite universally, lattice vibrations. Suppose the electron has a (real, thermal) velocity v, and the mean time between scattering events is τ. Then we can define a mean-free-path $l = v\tau$. If we start with n_0 electrons of velocity v at time $t=0$, the number that have *not* undergone a collision at time t is $n = n_0 \exp(-t/\tau)$. Consider an electron that has just made a collision at time $t=0$. Provided it has not suffered another deflection, its velocity in the x-direction (in which the field is assumed to be applied) is, after time t,

$$v_x = v_{x,0} - \mathscr{E}e_0 t/m_e \tag{8.22}$$

where ε is the field strength, m_e the effective mass, and Newton's laws of motion are assumed. If we average v_x over all times t, using the expression $\exp(-t/\tau)\mathrm{d}t/\tau$ for probability of a collision in time $\mathrm{d}t$

$$\bar{v}_x = \bar{v}_{x,0} - \frac{\mathscr{E}e_0}{\tau m_e}\int_0^\infty t\exp(-t/\tau)\mathrm{d}t = \bar{v}_{x,0} - \frac{\mathscr{E}e_0 \tau}{m_e}. \tag{8.23}$$

If the scattering is isotropic, $\bar{v}_{x,0}=0$. The current $J=ne_0\bar{v}_x=ne_0\,(\mathscr{E}e_0\tau/m_e=\sigma\mathscr{E}$ by definition of the conductivity σ. The *mobility* of the electron is then:

$$\mu_e = e_0\tau/m_e. \tag{8.24}$$

For a mobility of $0.01\,\text{m}^2\,\text{V}^{-1}\text{s}^{-1}$, and $m_e = m_0$, the free-electron rest mass, we have $\tau \cong 6 \times 10^{-14}\,\text{s}$. The thermal velocity of the electrons is given, in elementary kinetic theory, by

$$\tfrac{1}{2}m_e\bar{v}^2 = 3kT/2 \tag{8.25}$$

and writing $v = \sqrt{\bar{v}^2}$, we have

$$v = \left(\frac{3kT}{e_0}\right)^{\frac{1}{2}}\left(\frac{e_0}{m_e}\right)^{\frac{1}{2}}. \tag{8.26}$$

At 300 K, for $m_e = m_0$, $v = 7.15 \times 10^5\,\text{m s}^{-1}$ and l, the mean-free-path, is 7×10^{-9}m. Clearly for normal field strengths, $\mathscr{E}\mu_e$ is some orders-of-magnitude below v.

If we increase the scattering probability, τ and l decrease. However, we cannot decrease l indefinitely for, if l becomes comparable to the lattice dimensions (i.e. to the distance apart of the atoms in the crystal) and we assume for the moment that the impurity level has remained low, then the interaction between the electron and the lattice vibrations or 'phonons' has become so large that we must abandon altogether the theory developed above. In fact the lattice now has time to polarize around the electron, which becomes trapped in a potential well. Movement from site to site is now by hopping which is actually assisted by co-operative lattice vibrations.

In addition to the assumption that $l \gg a_0$ in the theory above, we have also implicitly assumed that τ is the same for all electrons. This is a very poor approximation since, depending on the dominant scattering mechanism, τ will vary rather strongly with the energy $E_e = E - E_c$. The first consequence of this more realistic assumption is that the expression for the flux J must be modified since we must clearly weight the more energetic electrons; in fact

$$J = \sum n(E_e)e_0^2\tau(E_e)/m_e = n(E_e)e_0^2\langle\tau\rangle/m_e \tag{8.27}$$

where the expectation value $\langle\tau\rangle$ is given by eqn (8.44) below. The proof of this equation is rather involved and is given elsewhere.[1]

To evaluate the expression for $\langle\tau\rangle$ we need to know the functional form of the dependence of τ on E_e. The calculation of this dependence is beyond the scope of this chapter, but it may be helpful to quote some results.[7] Scattering by lattice vibrations may be divided into two categories according to the character of the vibrations. Those in which the atoms within each unit cell move essentially in phase are termed 'acoustic phonons', whereas those in which the atoms within each unit cell vibrate with respect to each other are termed 'optic phonons' since it is these modes that may be optically active (see Chapter 9). If the major scattering is by acoustic modes, the mean displacement of the atoms from their equilibrium positions is about kT, and the mean-free-path will be proportional to T^{-1}. Since the thermal velocity $v \sim T^{\frac{1}{2}}$ and $\tau \sim l/v \sim T^{-\frac{3}{2}}$, we have

$$\mu_l = e_0\tau/m = AT^{-\frac{3}{2}} = BE_e^{-\frac{1}{2}}T^{-1}. \tag{8.28}$$

There is more controversy over the scattering by optic modes, but a commonly quoted formula, valid for $E_e \gg h\nu_e$, for the mobility is

$$\mu_0 = \chi(x)\{e^x - 1\}/x^{\frac{1}{2}} \tag{8.29}$$

where $x = h\nu_e/kT$, ν_e is the optic mode frequency, $\chi(x)$ is a slowly varying function of x, and the formula is valid provided interaction is weak. For impurity scattering by charged centres

$$\mu_1 = CT^{\frac{3}{2}} \tag{8.30}$$

whereas for neutral centres, μ_1 is independent of T. In any real material all scattering modes will be operative, though one will tend to dominate, and the overall mobility will be given by

$$1/\mu_T = 1/\mu_l + 1/\mu_0 + 1/\mu_1. \tag{8.31}$$

As the temperature changes, the dominant scattering mode may also change; an example of this is seen in Fig. 8.3. From the formulae above, the dependence of μ on E_e may be obtained; generally we find $\mu \approx E_e^{-s}$ where $s < \frac{5}{2}$.

In the case that l becomes sufficiently small that the electron is localized, and conduction is by hopping, the temperature-dependence of the mobility is now quite different. Since hopping is aided by excitation of suitable lattice modes, we expect the mobility to increase with temperature and it is generally found that

$$\mu = \mu_0 \exp(-\varepsilon_p/kT) \tag{8.32}$$

where ε_p is called the 'polaron' energy and is typically a few tenths of an electron-volt.

8.3 Measurement of conductivity

8.3.1 Contacts

The simplest possible experiment that can be performed on an unknown sample is to press the two Cd-plated contact needles of a multimeter on to opposite ends of the sample and measure the current passed on applying a known voltage. Unfortunately, this rarely, if ever, yields an accurate measure

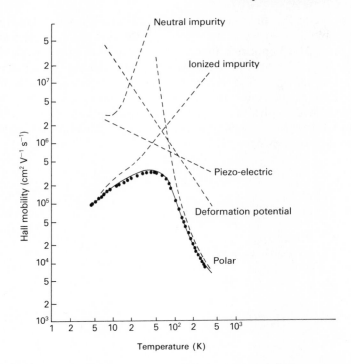

Fig. 8.3. The Hall mobility of GaAs (● ●) and the theoretical contributions to it from impurity, acoustic, and polar mode scattering. At low temperatures, the dominant scattering mode is from ionized impurities, but above 50 K the major scattering is from polar vibrational modes.

of sample resistance since for most semiconductors the contact resistance to Cd is very high. The reason for this is quite subtle. We stated above, that the Fermi level in a semiconductor or metal was a measure of the *chemical potential* of the electrons. Thus, if two conductors are placed in electrical contact, charge will flow from one to the other until the Fermi levels are equalized. This charge flux will, of course, give rise to a potential difference between the two conductors called the *contact potential*. Now, for metals, this contact potential is concentrated in an extremely narrow region at the point of contact and offers little barrier to electron tunnelling. However, for a semiconductor–metal junction, the relatively small number of carriers in the semiconductor does not permit the development of such large potential gradients, as a consequence of the relationship between charge and potential:

$$\partial^2 V / \partial x^2 = -\rho / \varepsilon \varepsilon_0 \tag{8.33}$$

where ε is the relative permittivity of the semiconductor, ε_0 the permittivity of free space ($= 8.8541 \times 10^{-12}$ F m^{-1}), ρ the charge density, and V the potential. If the maximum charge density in the semiconductor is N_d, the number density of ionizable impurities, there is clearly a limit to the rate at which the potential can change with distance into the semiconductor. On placing the semiconductor and metal in contact, as in Fig. 8.4(b), electrons from the surface region of the semiconductor flow into the metal giving rise to a spatially extended 'depletion region'.[8] Electrons in the conduction band must now tunnel through this region in order to complete the circuit, so a large contact resistance may develop.

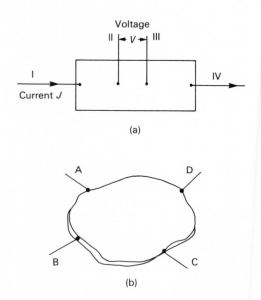

Fig. 8.4. (a) Simple arrangement for four-probe conductivity measurements: the current is driven between probes I and IV, and the voltage measured across II and III. (b) The arrangement for van der Pauw measurements; contacts A ... D are discussed in the text.

To avoid this problem, considerable effort must be devoted at the beginning of the study to finding a metallic contact material that will form an 'ohmic' or non-depleted contact with the semiconductor. The main variable is the work-function of the metal. To make contact to an n-type semiconductor, a metal of low work-function (such as Ga–In eutectic) would be used, whereas for a p-type material, a high work-function metal (such as gold) might be tried. If a metal dopant has been used, such as Zn in p-type GaP, diffusing excess dopant into the surface region can allow an extremely narrow depletion region of very low resistance to form.[9] In practice, determination of a suitable ohmic-contact metal is something of a 'black art' and may represent a major heuristic hurdle.

8.3.2 Electrode arrangements

Provided the junction is ohmic, the resistance measured with the two-probe technique described above should reflect the true resistance of the sample provided it is not too low ($\gtrsim 100\,\Omega$). For small values of resistance, residual contact resistance, impedance of the leads, and other effects start to become significant, and a different technique using four probes is employed. A rather straightforward experimental set-up is shown in Fig. 8.4(a). Current is forced through probes I and IV, and the potential dropped between probes II and III is measured using a high-impedance voltmeter. In principle, none of the probes need be ohmic, though in practice, more reliable results seem to be obtained if care is taken with the contacts as described above.

An important variant on the arrangement of Fig. 8.4(a) is the van der Pauw system illustrated in Fig. 8.4(b). This is based on a theorem, proved by van der Pauw,[10] for a sample of uniform thickness d, but arbitrary shape, with four arbitrarily positioned contacts (A,B,C,D) on the edges. If we define

$$R_{AB,CD} = \frac{V_D - V_C}{I_{A-B}} \quad \text{and} \quad R_{BC,DA} = \frac{V_A - V_D}{I_{B-C}} \tag{8.34}$$

where $V_D - V_C$ is the potential difference between contacts D and C caused by the passage of current I from A to B, etc., van der Pauw's theorem takes the form

$$\exp(-\pi R_{AD,CD} d\sigma) + \exp(-\pi R_{BC,DA} d\sigma) = 1 \tag{8.35}$$

where σ is the conductivity. This equation can be solved for σ; we find

$$\sigma = \frac{\log_e 2}{\pi d}\left(\frac{2}{R_{AB,CD} + R_{BC,DA}}\right) F\left(\frac{R_{AB,CD}}{R_{BC,DA}}\right) \tag{8.36}$$

where F is a slowly varying function of its argument and is defined elsewhere.[10] The main limitations on the van der Pauw technique are that the sample must be of uniform thickness, and that the contacts must be of minimal area and positioned on the edges. For irregularly shaped single-crystal samples the conductivity may prove difficult to measure owing to uncertainties in the geometrical correction. Great effort often has to be expended to ensure that the correct relationship between the measured resistance and the actual conductivity is determined.

One final class of materials for which conductivity data are often difficult to obtain is polycrystalline pellets; in such materials, as discussed below, contact between the individual grains comprising the sample is through narrow 'necks' that form during the process of 'sintering' or thermal hardening.[11] Such necks may have rather high resistances and the resultant d.c. measurement will reflect these rather than the true bulk resistance. Techniques that can disentangle bulk and 'grain-boundary' effects are described below, and it is of the utmost importance that they be used if the investigator suspects his data to contain a contribution from this effect.

8.4 Some results

Although the results below convey some information about the electronic properties of the solid, it cannot be overemphasized that further investigation, using the techniques described in later sections, is always desirable, and frequently essential.

8.4.1 Metallic compounds

As indicated above, the resistivity of most metals will *rise* with temperature since the carrier concentration is effectively constant and the mobility, dominated by interactions with lattice vibrations or phonons, will decrease as T^{-n}. However, the value of n will differ from that found in semiconductors since, although the mean-free-path l still varies as T^{-1}, the thermal velocity v is effectively constant for electrons in a metal, as only those electrons at the Fermi level are able to contribute to the conduction. Hence, at high temperature, $\sigma \sim T^{-1}$. At lower temperatures, the lattice vibrations must be treated more carefully. A general formula, due to Grueneisen, takes the form

$$\rho \equiv 1/\sigma \approx 4 \left(\frac{T}{\theta_D}\right)^5 \int_0^{\theta_D/T} \frac{e^x x^5}{(e^x - 1)^2} dx \tag{8.37}$$

where θ_D is the Debye temperature of the crystal, a measure of the vibrational 'stiffness' of the crystal. At very low temperatures, eqn (8.37) reduces to

$$\rho \approx T^5 \tag{8.38}$$

and the general behaviour is shown for metallic sodium in Fig. 8.5.[12]

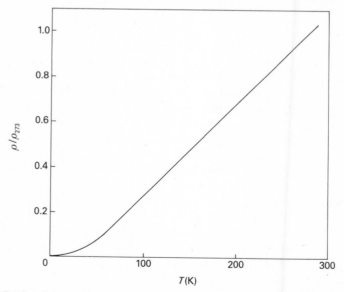

Fig. 8.5. Ratio of the resistivity ρ of metallic sodium at T to that at 273 K vs. T. The essentially linear variation at temperatures above 100 K is typical of simple metals.

8.4.2 Intermediate compounds

As might be expected, there are several classes of compound that cannot easily be described as either metals or semiconductors. The critical factor is overlap in energy of the conduction and valence bands. Where strong overlap occurs, the result is a classical metal. If the overlap is small, and indirect, the material is termed a *semimetal*, a familiar example being graphite. In practice, such materials are often difficult to distinguish from semiconductors with very small bandgaps. The latter class, which includes a very large number of binary and ternary chalcogenides and pnictides have considerable technical importance in low-temperature devices. Finally, there is a small category of compounds, exemplified in Fig. 8.6 by some ruthenate oxides,[13,14] which show very little change in resistance with temperature but which, on other criteria, are undoubtedly metallic. These materials find substantial use in resistors, though the fundamental reason for their behaviour is not known.

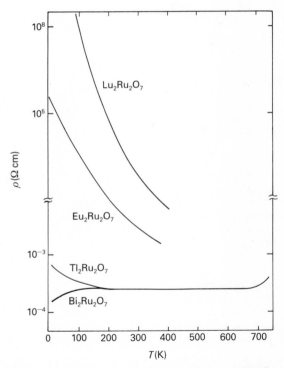

Fig. 8.6. The resistivities ρ of some ruthenium oxides; $Lu_2Ru_2O_7$ and $Eu_2Ru_2O_7$ behave as semiconductors with small bandgaps and room-temperature resistivities of the order of 10^2 ohm-cm. However, $Tl_2Ru_2O_7$ and $Bi_2Ru_2O_7$ have room-temperature resistivities typical of narrow-band metals, but their resistances remain constant from 200–700 K. Below 200 K, the resistance decreases for the Bi compound as expected for a metal, but the resistance of the Tl compound increases; the interpretation of this result is still controversial.

8.4.3 Semiconductors

The essential feature of semiconductors is that, over a large temperature range, the conductivity is dominated by the number of carriers. Typically, at temperatures above 10 K, three regions are found, corresponding to the three regions shown in Fig. 8.2. An example of this behaviour is shown by n-InP, as illustrated in Fig. 8.7.[15]

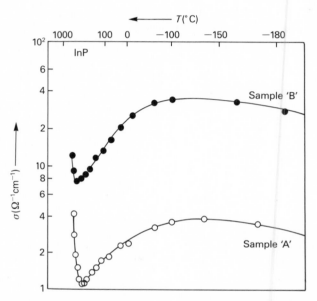

Fig. 8.7. Conductivity σ of various InP samples vs. T. At low temperatures the conductivity rises extremely slowly as expected for carriers ionized from shallow traps; at higher temperatures, the conductivity falls again due to exhaustion of these traps and the concomitant fall in mobility due to polar mode scattering. At very high temperatures, the conductivity rises sharply as intrinsic generation starts to play a more important role.

8.4.4 Organic conductors

In recent years, low-dimensional semiconductors and metals have been discovered. These compounds conduct essentially through spatially extended π and π^* orbitals, the prototype of course being graphite. Some examples of simple polymeric derivatives are illustrated in Fig. 8.8 where their conductivity is compared to the more familiar silicon and copper. It is noteworthy[16] that $(TMTSF)_2^+PF_6^-$ shows a transition to semiconducting behaviour below about 10 K, whereas $(TMTSF)_2^+ReO_4^-$ actually shows a superconducting transition at about 1 K.

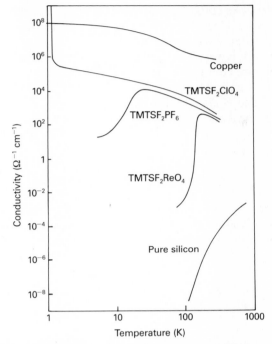

Fig. 8.8. Conductivities of some organic compounds compared to copper (as a typical metal) and silicon (as a typical semiconductor). All the organic materials are based on tetramethyltetraselenafulvene (TMTSF).

Changing the counter-ion can cause a dramatic change in electrical properties: the perchlorate salt is a superconductor at low temperatures whereas the other two are semiconductors.

8.5 Thermopower

8.5.1 The Seebeck effect

To deconvolute the conductivity into its two components, n and μ, an additional experimental probe is needed. The first of these that will be considered is the Seebeck effect. If a thermal gradient is maintained along a uniform bar of material there will be a flow of carriers to one end of the bar. This is because the differing temperatures lead to differing *electrochemical potentials* for electrons at the two ends, and charge flows until an electric field builds up that just compensates this difference. This electric field, measured as an e.m.f., provides a measure of the electron concentration.

Experimentally, it is not possible to perform the experiment in precisely the form described since contacts must be made to the material in order for the e.m.f. to be measured. The arrangement actually used is shown in Fig. 8.9(a).[17] The voltage V_{ab} is found to be a function only of the substances 'a' and 'b', and the *temperature difference* $(T_h - T_c)$. The Seebeck coefficient S_{ab} is defined as

$$S_{ab}(T) = \partial V_{ab}/\partial T_h \qquad (8.39)$$

where the sign convention is adopted that if V_{ab} is such as to induce a current to flow from 'a' and 'b' at the cold junction, $V_{ab} > 0$. Empirically and theoretically, $S_{ab}(T)$ can be written as the difference of two quantities, one referring only to material 'a' and the other to 'b'. Thus it is possible to define standards such as Pb, Pt, or Cu to which all other materials can be referred.

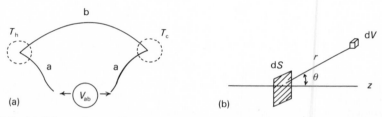

Fig. 8.9. (a) Schematic experimental set-up for the Seebeck experiment; T_h and T_c are the hot and cold junctions between two conductors 'a' and 'b'. The sample is represented by 'b', and the voltage is measured across 'a'. (b) Geometry for the calculation of the number of electrons in volume dV at (r, θ) that can reach elemental area dS.

To calculate the Seebeck coefficient explicitly, it is necessary to take a specific example. Consider an *n*-type semiconductor subject to a temperature gradient in the *z*-direction, the gradient being sufficiently small that the Fermi distribution law holds throughout the sample. Figure 8.9(b) shows a small element of area dS in the x–y plane. If $n(E,z)$ is the number of electrons at z of energy E and a fraction $dS|\cos\theta|/4\pi r^2$ of these electrons move from volume element dV at (r,θ) toward, dS, the number of electrons of energy E that make a collision in dV and pass through dS without a further collision is

$$\iint n(E,z)|\cos\theta|\exp(-r/l)\,dS\,dV/4\pi r^2 \qquad (8.40)$$

where l is the mean-free-path $= v\tau$. Hence the total flux of electrons of energy E through dS towards negative z is obtained by integrating over $dV = 2\pi r^2 \sin\theta\, d\theta\, dr$ for $z > 0$, and the number flowing to positive z is similarly obtained from an integration over $z<0$. We obtain, for the net flux in the $+z$ direction

$$J_e(E) = \frac{-\tau(E)v^2}{3} \cdot \frac{dn(E)}{dz} \quad (8.41)$$

$$\equiv -D(E) \cdot dn(E)/dz \quad (8.42)$$

where $D(E)$ is an equivalent diffusion coefficient for electrons of energy E, and eqn (8.42) is a statement of Fick's first law of diffusion.

The net flux of all electrons can be obtained by integrating over all energies and the overall current flow is then

$$J_T = -e_0 J_e + ne_0^2 \langle \tau \rangle \mathscr{E}/m_e \quad (8.43)$$

where $\mathscr{E}$ is the electric field, and $\langle \tau \rangle$ is a suitably weighted average value of τ as discussed above. Explicitly

$$\langle \tau \rangle = \int_0^\infty W(E) N(E) \tau(E) dE / \int_0^\infty W(E) N(E) dE \quad (8.44)$$

where $W(E)$, the weighting factor, has the form $Ef(E)(1-f(E))$, $f(E)$ is the Fermi function defined above, and $W(E)$ takes account of the fact that the more energetic electrons contribute more to $\langle \tau \rangle$.

It follows that if the net particle flux is zero, we have

$$-e_0(dT/dz) \int_0^\infty \{dn(E_e)/dT\} D(E_e) dE_e = ne_0^2 \langle \tau \rangle \mathscr{E}/m_e = ne_0 \mu \mathscr{E}. \quad (8.45)$$

Integrating $\mathscr{E}(\equiv dV/dz)$ around the circuit of Fig. 8.9(a) will then yield the Seebeck coefficient. To perform the integrals, an expression for $dn(E_e)/dT$ is necessary. This can be obtained from eqn (8.8), and on substitution we find, for the Seebeck coefficient[17]

$$S_a = -k/e_0 \left(\frac{-E_f \langle \tau \rangle + \langle E\tau \rangle}{kT \langle \tau \rangle} \right). \quad (8.46)$$

Physically, the term $\langle E\tau \rangle / \langle \tau \rangle$ represents the average kinetic energy of the moving electrons in the semiconductor and is called the *kinetic term*. If $\tau(E) \sim E^{-s}$, its value is easily found to be $(\frac{5}{2} - s)$. The term E_f/kT has the value $-\log_e(N_c/n)$ for an extrinsic *n*-type semiconductor, and so the final result is the remarkably simple formula

$$S_a = -k/e_0 \{(\tfrac{5}{2} - s) + \log_e(N_c/n)\}. \quad (8.47)$$

Since $s < \tfrac{5}{2}$, this result is always negative. Numerically, the value of k/e_0 is 88 μV K^{-1}, and so values of S_a between 100 and 1000 μV K^{-1} are expected.

The theory for an extrinsic *p*-type semiconductor is very similar, save that the total hole flux is now

$$J_T = +e_0 J_e + pe_0^2 \langle \tau \rangle \mathscr{E}/m_h \quad (8.48)$$

and

$$S_a = +k/e_0 \{(\tfrac{5}{2} - s) + \log_e(N_v/p)\}. \quad (8.49)$$

Two important factors emerge from this discussion. First, S_a is expected to be positive for a p-type semiconductor and negative for an n-type semiconductor. Thus a measurement of the sign of S_a can straightforwardly distinguish the signs of the two carrier types. Second, S_a clearly varies with the number of carriers n. Thus, if the concentration of carriers is given by eqn (8.21), then

$$\log_e(N_c/n) \sim C + (E_d - E_c)/kT \tag{8.50}$$

and a plot of S_a vs. $1/T$ should be linear with a slope of $(E_d - E_c)/k$, provided the kinetic term does not change significantly with T.

The theory developed so far applies to wide-band semiconductors where $\langle \tau \rangle$ has some meaning. If l is reduced to a value comparable to the lattice dimensions, the semiconductor will show 'small-polaron' behaviour as discussed above, and the electron can be considered as localized, that is, if we place a charge carrier on a site, the lattice will relax about that site to give a new interatomic distance, and the energy of relaxation is ΔH_R. Since it is this energy that is carried by the electron or hole when it moves, we can replace the term $\langle E\tau \rangle / \langle \tau \rangle$ in eqn (8.46) by ΔH_R.[17] If the *free energy* of deformation is ΔG_R, the number of mobile carriers at a given time is $c \cdot \exp(-\Delta G_R/kT)$, and the number of unoccupied sites is $\sim (1-c)$, where c is the fractional occupancy of lattice sites by carriers. We can then replace n/N_c by $c \cdot \exp(-\Delta G_R/kT)/(1-c)$ and obtain, for the Seebeck coefficient of a small-polaron semiconductor:

$$S_a = -k/e_0[-\Delta H_R/kT - \log_e\{c \cdot \exp(-\Delta G_R/kT)/(1-c)\}] \tag{8.51}$$

$$= -k/e_0[\Delta S_R/k - \log_e\{c/(1-c)\}]. \tag{8.52}$$

Application of this formula is not entirely straightforward, as shall be illustrated below.

8.5.2 Measurement of Seebeck coefficients[18-23]

In essence the experiment consists of generating a thermal gradient along a bar of the sample and measuring the resultant e.m.f. Since $S_a \sim 100-1000\, \mu\text{V K}^{-1}$ and thermal gradients of $5\, \text{K cm}^{-1}$ are desirable, the thermoelectric e.m.f. will be $\sim 1\, \text{mV}$ in samples whose impedance may be very large. Electrical noise problems therefore dominate the constructional principles, and well-screened leads and amplifiers are essential. In principle, samples of almost any shape and size could be used, but non-uniform samples have the particular problem that, unless the contact area with the heat sink and source are negligible, complex thermal gradients are likely to exist in the neighbourhood of the contacts, making reliable and reproducible results difficult to obtain. However, one advantage over normal d.c. conductivity measurements is that polycrystalline samples usually give good results since grain-boundary effects seem to be much less important.

The determination of a temperature difference using thermocouples and the measurement of the Seebeck voltage can be done simultaneously using the elegant subtraction circuit shown in block form in Fig. 8.10.[21] The outputs are directly proportional to the thermal e.m.f. V_{ab}, and the temperature difference ΔT between the ends of the sample. The thermoelectric power is then obtained by measuring the gradient of the X–Y recorder signal V_{ab}/T.

8.5.3 Results

The elementary theory discussed above is illustrated by results on n-GaAs, a wide-band semiconductor.[24] In the donor range 10^{15}–10^{18} cm^{-3}, m^* is approximately constant; and Fig. 8.11 shows a linear relationship between $|S|$ and

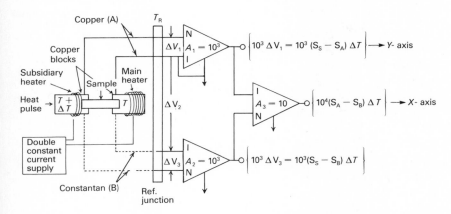

Fig. 8.10. Block diagram of the Seebeck apparatus used in Oxford to measure thermopowers in the range 5–1000 µV K^{-1}.

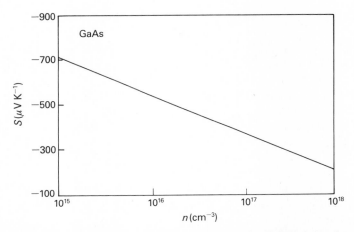

Fig. 8.11. Seebeck coefficient vs. electron concentration for n-GaAs. On the apparatus of Fig. 8.10, a sample of n-GaAs with $n = 5.9 \times 10^{17}$ cm^{-3} gave a Seebeck coefficient of -244 µV K^{-1} in good agreement with the data reported here.

the logarithm of the donor density, as predicted by eqn (8.47). As expected, S varies only slightly with temperature in the region 100–300 K. At lower temperatures the predominant scattering mechanism changes, and $|S|$ alters.

Some Seebeck coefficient measurements for p-NiO are shown in Fig. 8.12.[25] NiO can be made an extrinsic p-type semiconductor by doping with small quantities (<0.5%) of Li, which enters the lattice substitutionally. For every Li^+ entering the lattice, a hole must be generated in the Ni^{2+} levels. A dramatic increase in the conductivity is observed on adding Li, from which it can be deduced that the holes in the Ni^{2+} levels are mobile, at least in part. The conductivity results showed that near 300 K, σ was activated, with an activation energy of 0.3 eV. The results of Fig. 8.12 show conclusively that this activation must be associated with an exponential increase in the number

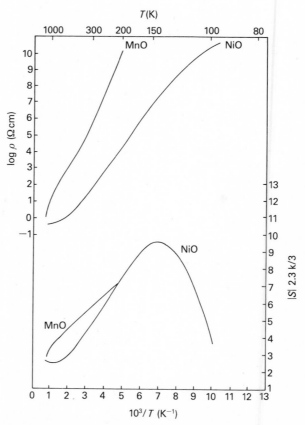

Fig. 8.12. Resistivity and Seebeck coefficient for four first-row transition-metal binary oxides vs. $1000/T$. Note that in the temperature range covered in this figure, only Fe_2O_3 is n-type.

of carriers with temperature and *not* an increasing mobility. In other words, for NiO the mean free path $l \gg a_0$. Therefore it appears that the positive holes introduced on doping with Li are trapped at Ni^{3+} sites localized by Coulombic attraction near the Li^+ ions. Thus the activation energy arises from the thermal excitation of the holes away from the Li^+ sites to parts of the crystal where they can move freely.

If $l \sim a_0$, we must use eqn (8.52). If we assume that $\Delta S_R/kT$ is small (it is expected to be less than 10 µV K^{-1}) then, for *p*-type materials

$$S \simeq 198 \log_{10}\{(N-p)/p\} \, \mu V K^{-1} \qquad (8.53)$$

where p is the polaron concentration. Equation (8.53) has been extensively tested [26] for the spinel series $A^{2+}[A_x^{2+}V_{2-2x}^{3+}V_x^{4+}]O_4$, where substitution of a 2+ ion (A) on the octahedral sites of the $A[V_2]O_4$ spinel causes oxidation of some V^{3+} to V^{4+}. Unlike the Ni^{3+} in NiO, the V^{4+} are present in AV_2O_4 as discrete ions, even on sites remote from the substitutional A cations, and the mobility of holes on the V sub-lattice is activated. The simplest assumption is that $p = x$ and $N = 2$ and (as can be seen from Fig. 8.13)[27] the solid solution with A = Mn follows eqn (8.53) for $x > 0.2$. For $x < 0.2$, the Seebeck coefficient becomes temperature-dependent and higher than that predicted from eqn (8.53). This seems to be associated with the same type of behaviour as in NiO, with some Mg^{2+}–V^{4+} units so firmly bound as not to participate in the small polaron conduction.[27]

In contrast to this rather straightforward example, there are many cases where the simple association of stoichiometry and p does not describe the

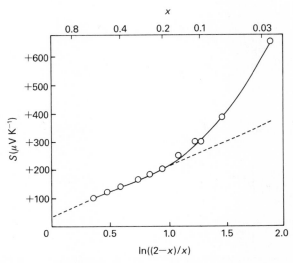

Fig. 8.13. Thermopower vs. $\ln((2-x)/x)$ in the system $Mn_xV_{(2-x)}O_4$ at 300 K.

experimental situation adequately, but nevertheless S alters in a well-defined way with x. Two reasons may be cited for such behaviour. First the crystal structure may be such as to lead to *pairs* of ions rather than *single* ions acting as single sites. Second, the presence of an ionized donor or acceptor site may polarize the lattice in its neighbourhood, and reduce the number of *available* sites to which the carrier can hop. An elegant example of the latter behaviour arises in the interpretation of the Seebeck effect in $M^+_x V_2 O_5$ bronzes, where M is a monovalent cation.[28] In the complex structures adopted by these oxides, there are *three* distinguishable *types* of V ion: a V_1 type, on which electrons donated by M reside, a V_2 type which forms V_2–O–V_2' clusters, and a V_3 type. The presence of V_2–O–V_2' clusters means that N is reduced from the expected value of 2 to $\frac{5}{3}$. However, the function

$$S = -198 \log_{10}\{(\tfrac{5}{3} - x)/x\} \tag{8.54}$$

still does not describe the observed behaviour. In fact, we need to make the assumption that each M^+ ion can bind to a V_1 or V_3 site, thus removing that site from the sum of sites available for hopping. Hence N becomes $(\tfrac{5}{3} - \tfrac{2}{3}x)$ and

$$S = -198 \log_{10}\{(\tfrac{5}{3} - \tfrac{2}{3}x)/x\} \tag{8.55}$$

fits the data well, as shown in Fig. 8.14.

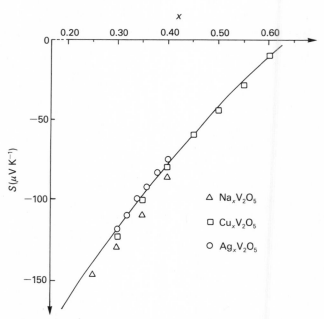

Fig. 8.14. Thermopower vs. monovalent metal concentration x in three vanadium bronzes, $Na_x V_2 O_5 (\triangle)$, $Cu_x V_2 O_5 (\square)$, and $Ag_x V_2 O_5 (\bigcirc)$ at 300 K. The full line is the theoretical curve discussed in the text.

8.5.4 Mixed conduction

If both electrons and holes contribute to conduction, the Seebeck coefficient is given by

$$S_t = (S_e \sigma_e + S_h \sigma_h)/(\sigma_e + \sigma_h) \qquad (8.56)$$

where S_e and S_h are the Seebeck coefficients for electrons and holes calculated from the formulae above, and σ_e and σ_h are the corresponding conductivities ($n_e e_0 \mu_e$; $p_h e_0 \mu_h$). Of course, S_e and S_h have opposite signs, and for intrinsic semiconductors, for which μ_e and μ_h are comparable, S_t will be small and its sign determined by rather subtle factors. In general, therefore, thermopower investigations on intrinsic semiconductors are much less informative, though there is one important exception to this statement; if the material has an extrinsic dopant which dominates the conductivity at low temperatures, but has a sufficiently small bandgap that intrinsic conduction becomes significant at high temperatures, the Seebeck coefficient may offer a very valuable probe. A well-known example is MnO. At ambient temperature, MnO as normally prepared is slightly deficient in Mn.[25] The excess holes are localized as Mn^{3+} ions and the compound is a small polaron semiconductor.[25] This is demonstrated by the fact that S decreases with T up to 1200 K as shown in Fig. 8.12, though the activation energy is lower than that found for the conductivity. This suggests that, as in p-NiO, the holes are trapped near Mn-vacancies (which appear to the crystal as negative wells, just like Li^+ ions in NiO), but that even when freed from these traps, the holes cannot move freely in the lattice but form small polarons whose mobility is activated. Fig. 8.15[29] shows that at high temperature, the Seebeck coefficient changes sign. The reason is that thermal excitation from the Mn $3d^5(e_g^2)$ levels to the Mn $4s$ conduction band becomes significant. Since the $4s$ band is much broader than the band formed from the e_g orbitals, $\mu_e \gg \mu_h$ (observed values at 1000°C being 10 cm²(Vs)⁻¹ and 0.2 cm²(Vs)⁻¹ (respectively) so conductivity in the conduction band becomes dominant, and S changes sign.

Another cause of a sign change in S arises in such solid solutions as $Ti_{1-x}Al_xO_{2-y}$ ($x \sim 10^{-3}$). S varies considerably with oxygen pressure; at low p_{O_2} the material is oxygen-deficient, leading to an excess of negative carriers, and giving $S < 0$. However, at high temperatures and pressures, S becomes positive due to holes forming in the O $2p$ band, as the material becomes 'superstoichiometric' (i.e. y becomes smaller than $x/2$).[30]

8.6 The Hall effect

8.6.1 Theory

We have seen that thermopower measurements can give some indication of the density of charge carriers. However, a more direct measure of n and μ can be made using the Hall effect.[1] The principles of the measurement are

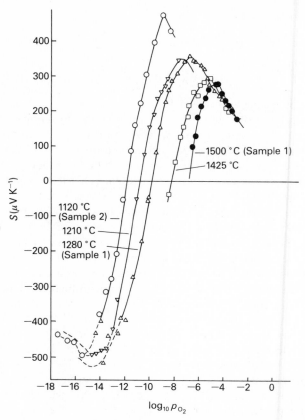

Fig. 8.15. Thermopower of MnO vs. $\log_{10} p_{O_2}$ at various temperatures. An inversion clearly occurs at sufficiently low oxygen pressure for all temperatures reported. Furthermore, the higher the temperature, the higher the pressure at which inversion is seen.

illustrated in Fig. 8.16, which shows a rectangular slab of material of width w and depth t. Let the current density in the $+x$-direction be J and we initially consider an extrinsic n-type semiconductor. A magnetic induction B is now applied in the $+z$-direction and the electrons experience a force $-e_0 vB = +e_0 v_x B$ in the $+y$-direction. Since the flux of electrons remains in the $(-)x$-direction, a field $\mathscr{E}_y$ forms such that

$$-e_0 \mathscr{E}_y + e_0 v_x B = 0; \quad \mathscr{E}_y = v_x B = -BJ/ne_0 \tag{8.57}$$

where $v_x = -J/ne_0$ and $\mathscr{E}_x = J/ne_0 \mu_n$. Clearly also

$$\mathscr{E}_y/\mathscr{E}_x \equiv \tan\theta = -B\mu_n \tag{8.58}$$

where θ is called the Hall angle.

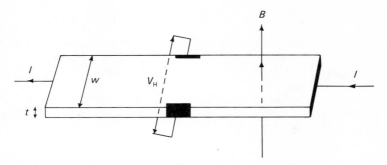

Fig. 8.16. Schematic diagram for the measurement of the Hall coefficient in a rectangular slab of width w and thickness t.

The Hall voltage $V_H = w\varepsilon_y = wRJB$, where R is termed the Hall coefficient. For electrons, $R = -1/ne_0$ and, by a straightforward extension, for holes, $R = +1/ne_0$.† Thus, the sign of the Hall voltage immediately indicates whether electrons or holes are the majority carriers. The size of the effect is easy to calculate: for a slab of width $w = 1$ cm, thickness $t = 1$ mm, and having a dopant density $n = 10^{16}$ cm^{-3}, $R \sim 10^{-3}$ m^3 C^{-1}. If the current passing is 10 mA, $J = 10^3$ A m^{-2}, $V_H = B/100$ volts $= 1$ mV for a field of 0.1 T ($= 10^3$ gauss).

Implicit in our theory is that all the electrons have the same relaxation time. In practice, this is not so and the formulae must be modified. In fact the modification is very simple (though the proof is not and we shall not give it).[1] The modified Hall coefficient can be written

$$R = \pm r/ne_0 \quad \text{where } r = \langle \tau^2 \rangle / \langle \tau \rangle^2. \tag{8.59}$$

For lattice scattering

$$\tau \sim aE^{-\frac{1}{2}}; \; r = 3\pi/8 = 1.18 \tag{8.60}$$

and for impurity scattering

$$\tau \sim aE^{\frac{3}{2}}; \; r = 315\pi/512 = 1.93. \tag{8.61}$$

Thus, r lies between 1 and 2 and represents a fairly small correction. To avoid having to calculate r, it is customary to *define* the Hall mobility;

$$\mu_H = |\sigma R|. \tag{8.62}$$

As in the theory of the Seebeck effect developed in Section 8.5.1, the presence of two types of carrier leads to more complex formulae. If there are both electrons and holes, we can no longer assume that there is no flux of electrons in the y-direction, only that the *net* flux of electrons and holes is zero. For the electrons, we may then write

† The units in this section are SI. The units of B are therefore Vs m^{-2} or T (Tesla), and hence $B\mu_e$ is dimensionless. The units of R are m^3 C^{-1}.

$$m_e \cdot v_{x,e} = -e_0 \mathscr{E}_x - e_0 v_{y,e} B;\ m_e \cdot v_{y,e} = -e_0 \mathscr{E}_y + e_0 v_{x,e} B. \tag{8.63}$$

Now, if $B\mu_e \ll 1$, $Be_0 v_{y,e} \simeq Be_0\mu_e\mathscr{E}_y = (B\mu_e)^2 e_0 \mathscr{E}_x$. Hence, $e_0 v_{y,e} B$ can be neglected and we have $\bar{v}_{x,e} = -\mu_e \mathscr{E}_x$. Hence, $v_{y,e} = -\mu_e \mathscr{E}_y - \mu_e^2 B \mathscr{E}_x$. For holes, we have $v_{y,h} = \mu_h \mathscr{E}_y - \mu_h^2 B \mathscr{E}_x$, and the total flux density is given by

$$J_y = -ne_0 \bar{v}_{y,e} + pe_0 \bar{v}_{y,h} = 0 \text{ at steady state} \tag{8.64}$$

and R, the Hall constant, is given by $\mathscr{E}_y/B\mathscr{E}_x$. Using (8.64) to evaluate $\mathscr{E}_y/\mathscr{E}_x$, we obtain:

$$R = \frac{p\mu_h^2 - n\mu_e^2}{e_0(p\mu_h + n\mu_e)^2} = \frac{1}{e_0}\frac{p - b^2 n}{(p + nb)^2};\ b = \mu_e/\mu_h. \tag{8.65}$$

This clearly has the correct behaviour if either p or $n \to 0$. In the event that τ is not constant, but varies with E in the same way for holes and electrons we simply multiply eqn (8.65) by the factor r defined above. Fairly commonly, $b > 1$ and hence $R < 0$ for an intrinsic semiconductor ($p = n$). For a p-type semiconductor, we therefore expect r to be positive at low temperatures, but to show a sign reversal at high temperatures.

8.6.2 Measurement of Hall coefficients[31-35]

Measurement of the Hall voltage, particularly on high-impedance samples, is bedevilled by problems. The simplest technique, using a d.c. current and d.c. magnetic induction, and measuring the static Hall voltage has several major drawbacks. The first is that the three probes, V_H, I, and B must be accurately orthogonal, or a misalignment voltage will develop. In principle, such a voltage is constant, but low-frequency noise due to temperature and field fluctuations, and $1/f$ noise from the contacts cause a low-frequency drift to appear, making static measurements difficult. Second, is that the coupling of thermal and magnetic field gradients across the sample gives rise to a number of effects, the most important being the Ettinghausen–Nernst and Righi–Leduc effects which will not be discussed here, but whose magnitude can easily exceed the Hall voltage unless very accurate temperature control is used. Finally, if the contacts are non-ohmic, space charge can build up which may reduce the measured Hall voltage substantially in high-impedance samples.

To eliminate these effects, it is now customary to use a.c. methods. However, modulating the current I or field B separately effects only a limited improvement, since in the first case the misalignment voltage remains, and in the second case space-charge effects cannot be eliminated, and pick-up by current loops becomes a major headache. The favoured technique is then to modulate I and B simultaneously; since V_H depends on the *product* of B and I, it is clear that V_H will have frequency components $\omega_B \pm \omega_I$ (since $2\cos A \cos B = \cos(A - B) + \cos(a + B)$). Picking out the signal at the cross-frequency eliminates misalignment and most thermogalvanomagnetic effects, and eliminates the inductive effects that plague normal a.c. magnetic measurements. The main limitations arise from intermodulation of the misalignment and magnetic induction voltages by non-linear circuit elements such as non-ohmic contacts, and from wide-spectrum noise such as Johnson noise. This latter can be reduced by using narrow-band filters. Misalignment may also be reduced by using an adaptation of the van der Pauw technique described earlier. If we define $R_{BD,AC}$ by analogy with Fig. 8.9, the Hall coefficient

$$R_H = V_H/wJB = tV_H/IB \text{ and } V_H/I = RB/d = \Delta R_{BD,AC} \quad (8.66)$$

where $\Delta R_{BD,AC}$ is the change in $R_{BD,AC}$ on application of the magnetic field. Hence

$$R = \Delta R_{BD,AC} d/B. \quad (8.67)$$

In practice, this result should be extrapolated to zero field.

8.6.3 Results

For non-magnetically ordered broad-band semiconductors, the Hall experiment is the most important single measurement as an adjunct to the conductivity. For these materials, Hall mobilities usually lie in the range 10–1000 cm^2 V^{-1}s^{-1}, and some typical results for InP are shown in Fig. 8.17.[36] Since μ_H lies within a factor of two of the drift mobility, Hall measurements can be used to estimate the number of carriers in a semiconductor without making any assumptions about their effective masses such as are necessary in interpreting the Seebeck data. This is, in fact, the standard way of assessing n in low-doped samples where chemical analysis is very difficult. (Note that 10^{14} carriers, which have a marked effect on the conductivity in GaAs, corresponds to only 1 part in 10^8 dopant.)

A more intriguing example is TiO_2. When stoichiometric, this compound is an insulator, but on slight reduction it becomes an n-type semiconductor, electrons being donated from the oxygen vacancy donors to a conduction band which is primarily of Ti^{4+} orbital parentage. The Hall mobility of the latter electrons has been determined by several groups with the results shown in Fig. 8.18.[37] Whilst there are significant differences in detail, the most important fact found by all workers is that μ_H decreases with temperature, showing that the electrons in the Ti-3d band are truly itinerant and not small-polaron in type.

In several materials the Hall voltage reverses at high temperatures, often associated with reversal of the Seebeck coefficient. A good example is MnO, referred to in Section 8.5.4,[25] where independent Hall-effect data, shown in Fig. 8.19, confirms that the hole mobility is significantly smaller than the

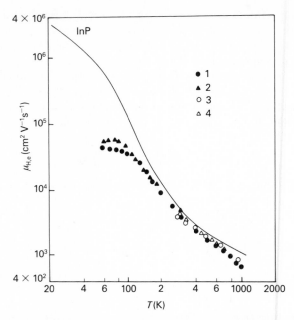

Fig. 8.17. Hall mobility for InP as a function of temperature. The four samples measured are all *n*-type with donor densities in the range $1-6 \times 10^{15}$ cm^{-3}. The full curve is a theoretical fit involving primarily optic mode scattering.

electron mobility. However, the theory of the Hall effect has not been fully worked out in detail for magnetically ordered materials, and the Hall reversal found in Fe_2O_3 is still not fully understood.

8.7 Photoconductivity[38]

8.7.1 Background

It is clear[39] that the techniques referred to in Sections 8.5 and 8.6 give information only on one type of carrier. Save in rather fortuitous circumstances, such as MnO, it is impossible to determine the transport properties of *minority* carriers in wide-bandgap extrinsic semiconductors by straightforward application of Hall or Seebeck measurements. There are, in addition, a wide range of materials which either cannot be doped, or where the dopant forms a deep trap, preventing the generation of itinerant carriers.

For such materials, non-equilibrium techniques must be employed. The most important of these use light energy; the CB is populated by optical pumping, and the properties of the carriers then explored. In the simplest

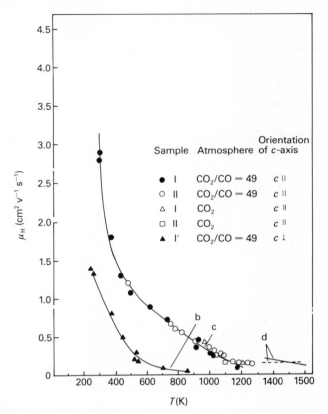

Fig. 8.18. Hall mobility for TiO_2 at high temperatures showing (i) that the mobility drops with temperature (ii) that there is some anisotropy in the mobility with that parallel to the c-axis (which contains strings of Ti ions) about 2–15 times larger than that perpendicular to the c-axis. Curve 'd' shows results obtained from the analysis of a simple point defect model in which the concentration of defects (in this case oxygen vacancies) was calculated from thermogravimetric data. Curves 'b' and 'c' are results at rather lower temperatures respectively perpendicular and parallel to the c-axis.

experiment, photoconductivity, a large potential gradient is maintained across the sample, which should have a high dark-resistivity. The small current that flows in the dark is considerably enhanced on illumination with light of energy comparable to, or greater than, the bandgap.

Experimentally, given that the impedance of the sample, even on illumination, will be very high (typically > 1 MΩ), great care must be taken to avoid loading the system. Response times are also usually rather slow, and a.c. transient techniques therefore difficult to employ; for samples of rather lower resistivity, where the conductivity enhancement will be much smaller, the reduced response times permit the use of light-chopping and phase-

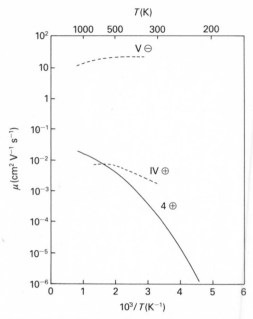

Fig. 8.19. MnO. Hole (IV ⊕) Hall mobility, hole (4 ⊕) drift mobility from the Seebeck measurement, and electron (V ⊖) Hall mobility vs. $1000/T$.

sensitive detection to improve the sensitivity. Another advantage of light-chopping is that thermal effects, which are a considerable nuisance in d.c. measurement, can be effectively diminished.

As we have seen, conductivity measurements can give information only about the product $ne_0\mu$. In the photoconductivity experiment, we are, in fact, measuring the combined hole and electron conductivities, and the situation is complicated. Only in the (albeit common) case that the mobility of one type of carrier greatly exceeds that of the other can even information about a single product be obtained. To deconvolute this product into mobility and concentration, the photo-Hall experiment has been devised; the a.c. current is stimulated by chopping the light, and an a.c. magnetic induction is also employed. A Hall voltage will then appear at the frequencies $\omega_L \pm \omega_B$.

8.7.2 Examples

Since the photocurrent can be measured as a function of the incident wavelength, information about the bandgap may be obtained. It is clear that this information will complement the *thermal* bandgap, calculated from the variation of σ with T, and the optical spectrum, in which localized excitations are also seen. As an example, the optical spectrum of NiO shows localized d–d excitations throughout the visible region whereas the photoconductivity onset, which has only clearly been seen in Li-doped samples, is found in the ultraviolet.[40,41] Even more striking is the case of PdO; the absorption threshold is about 0.8 eV, whereas the photoconductivity threshold is about 2.5 eV.[42] Again the *optical* threshold is dominated by d–d transitions.

Materials with deep traps can also be 'activated'; indeed they are ideal for photo-Hall measurements, as the resultant hole in the trap is quite immobile. Good examples are provided by colour centres in the alkali halide crystals,[32] whose excitation to the conduction band gives a carrier mobility of 9.0 cm^2 V^{-1}s^{-1}.

8.8 Conduction in amorphous and imperfectly crystalline materials

Implicit in most of the treatment so far has been the assumption that the lattice is perfectly periodic. For most samples that the chemist or materials scientist encounters in practice, this is not so; the presence of impurities and defects, often in substantial concentration, leads to a radically different *type* of behaviour.[43] Consider a spatially periodic lattice; if the potential is periodic, a well-defined band-structure will develop, of energy width B. We now introduce into the potential a random element which can vary from site to site in the lattice with a total energy spread V_0. Then it can be shown that the mean-free-path of the electrons in the lattice is reduced to the value

$$l = \frac{8a}{z^2}/(V_0/B)^2 \qquad (8.68)$$

where a is the distance between potential wells, and V_0 is assumed to be much smaller than B ($B = 2zb$ where z is the number of nearest neighbours and b the transfer integral). Clearly, as $V_0 \to B$, $l \to a$ and, as indicated above when thermal scattering was considered, a changeover to *localized* behaviour is expected. This is called the *Anderson transition* in disordered systems. The onset of localized behaviour depends in a rather interesting way on electron energy; the effect of a large V_0 is to broaden the spread of allowed electronic levels appreciably, from the value B to a value of approximately $(B^2 + V_0^2)^{\frac{1}{2}}$. More importantly, as V_0 is increased, the states at the energy extremes of the band become *localized* even though states in the middle of the band are still *delocalized*. The situation is sketched in Fig. 8.20(a) which shows that there must be a critical edge E_C dividing localized from delocalized states; this is called the *mobility edge*. Figure 8.20(b) illustrates the results of calculating the distance E_C from the middle of the band for a system where the critical value of V_0/B is 2. For values of V_0/B above this, the entire band will become localized. The maximum in E_C reflects the fact that the band is broadening faster in this region than the mobility edge is advancing; in this region, the

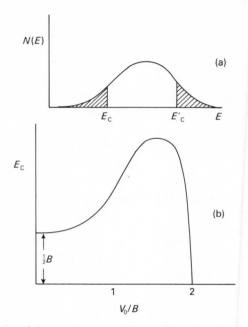

Fig. 8.20. (a) Density of electronic states $N(E)$ vs. E for a band in a disordered crystal; the hatched areas below E_C and above E'_C represent Anderson localized states in the band tails as discussed in the text. (b) Variation of E_C with V_0/B with respect to the middle of the band.

conductivity at $T = 0$ K, which falls approximately as l, levels off slightly before plummeting to zero at $V_0 = 2B$.

Experimentally, the Anderson transition has proved rather elusive to observe. Ideally, we would like to fill up the band from the bottom in a controlled way. If we have only a few electrons, they will occupy *localized* states, whereas above a certain critical concentration, $E_F > E_C$, and a transition to metallic behaviour will be seen. Unfortunately it is far from easy to alter the number of electrons in the CB without altering something else. Thus, we can form the series Na_xWO_3 in which the number of electrons in the W $5d$ CB can be arbitrarily increased, but the cations introduced increase the degree of randomness in the lattice and also cause structural modifications to occur. The latter can be quenched in the solid solution $Na_xWO_{3-y}F_y$ ($y = 0.1$–0.2),[43] the transport data for this compound are shown in Fig. 8.21. Clearly, as $x \to 0.3$, a transition to metallic behaviour is observed.

It is clear from Fig. 8.21 that the conductivity in the localized region does not vary exponentially with $1/T$. To understand this we can consider Fig. 8.22 which shows schematically the two mechanisms for electron hopping. One is the quantum route whose transition probability is given by

$$p_{i,j} = v_{ph} \exp\{-2aR - (W/kT)\} \qquad (8.69)$$

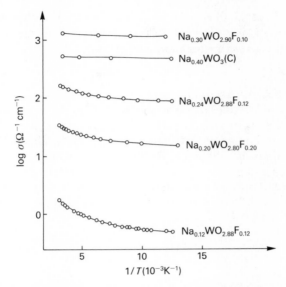

Fig. 8.21. The conductivity of $Na_xWO_{(3-y)}F_y$ vs. temperature showing a clear transition in behavioural type from a semiconductor at low sodium content to a quasi-metallic region as x increases.

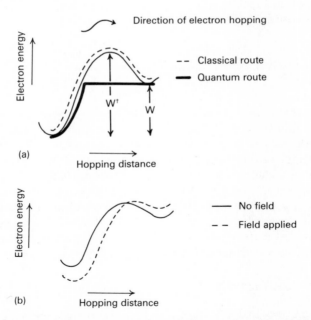

Fig. 8.22. (a) Activation energy vs. interatomic distance for electron hopping in a crystal; the classical and quantal routes for hopping are shown. (b) Effect of an electric field on the activation energy for hopping; note that the net effect is to lower the classical energy barrier.

where v_{ph} is a lattice distortion frequency, the term $2aR$ reflects the wavefunction overlap, and $\exp(-W/kT)$ is a Boltzmann factor. The other is the classical route, giving a transition probability

$$p_{i,j} = v_{ph}\exp(-W^{\dagger}/kT). \tag{8.70}$$

At some sufficiently low temperature, the first route must dominate and we need only consider states in the neighbourhood of the Fermi energy E_F. If the density of states is $N(E_F)$, and $V_0 \gg B$, then $N(E_F) \sim 1/a^3 V_0$ (from the bandwidth formula above, assuming all sites are localized). Further, if the distance that can be hopped is R, there are $4\pi(R/a)^3/3$ sites available to a given electron about to hop to another site. Let the energies of these states be distributed randomly over the energy range V_0 in this sphere of radius R. The smallest energy difference between states is then $3V_0/(4\pi(R/a)^3)$. From the expression for $N(E_F)$

$$W_{min} = \frac{3}{4\pi R^3 N(E_F)} \tag{8.71}$$

and

$$p = v_{ph}\exp\{-2aR - (W_{min}/kT)\} = v_{ph}\exp\left(-2aR - \frac{3}{4\pi R^3 N(E_F)}\right). \tag{8.72}$$

The function in brackets has a minimum value when $R = 9^{\frac{1}{4}}/(8\pi a N(E_F)kT)^{\frac{1}{4}}$ and the maximum hopping probability is then

$$p_{max} = v_{ph}\exp(-C/T^{\frac{1}{4}}). \tag{8.73}$$

The conductivity σ clearly has the functional form

$$\sigma \sim \sigma_0 \exp(-(T_0/T)^{\frac{1}{4}}). \tag{8.74}$$

The treatment above is termed the random-range hopping model and its main prediction is the $T^{-\frac{1}{4}}$ dependence for log σ.[44] This law is well-established for a number of materials at very low temperatures, e.g. P-doped Si shows $T^{-\frac{1}{4}}$ behaviour below 10 K. However, this behaviour may also be found at considerably higher temperatures. Thus Fe-doped Ge shows good $T^{\frac{1}{4}}$ behaviour to 300 K, and both VO_2 and Fe_3O_4 have recently been shown to exhibit similar conductivity.[45] In the case of VO_2, other evidence also points to the existence of a large number of localized traps in the bandgap, particularly the a.c. conductivity behaviour discussed below, and random-range hopping is a plausible hypothesis—see Fig. 8.23.

8.9 A.c. conductivity[46,47]

If a small a.c. modulated voltage $V_1\exp(i\omega t)$ is applied across the sample, the resultant current $I_1\exp(i\omega t - i\varphi)$ can be measured using phase-sensitive techniques. The ratio $(V_1/I_1)\exp(i\varphi)$ is called the *impedance Z* of the system; Z

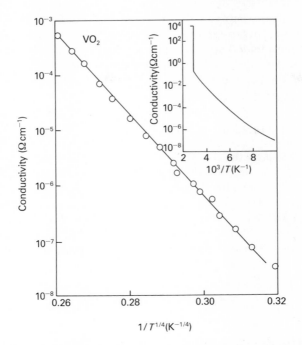

Fig. 8.23. VO_2. Conductivity vs. $T^{-\frac{1}{4}}$. The insert shows the conductivity of VO_2 vs. $1000/T$ over a wider temperature range that includes the semiconductor–metal transition at 340 K. The apparently variable activation energy at lower temperatures in the insert can obviously be understood in terms of the linearity of the $T^{-\frac{1}{4}}$ plot.

has both *magnitude* (V_1/I_1) and *phase* φ. The impedance of a pure dielectric can be obtained without difficulty for if the relative permittivity of the dielectric is ε, the area Γ, and the width d, the capacitance C is $\varepsilon\varepsilon_0\Gamma/d$, where $\varepsilon_0 = 8.854 \times 10^{-12}$ Fm^{-1}. The impedance of the sample is then $Z = 1/i\omega C$ and the *admittance* $Y = 1/Z = i\omega C = i\omega\varepsilon\varepsilon_0\Gamma/d$. If now the system has a finite conductivity $\sigma_a(\omega)$, then the expression for the admittance must be modified. The relative permittivity ε is now a complex number, $\varepsilon = \varepsilon' - i\varepsilon''$, where ε' and ε'' are both *real*, and the admittance becomes

$$Y = i\omega\varepsilon_0\Gamma(\varepsilon' - i\varepsilon'')/d = i\omega\varepsilon'\varepsilon_0\Gamma/d + \omega\varepsilon''\Gamma/d = i\omega\varepsilon'\varepsilon_0\Gamma/d + \sigma_a(\omega)\Gamma/d \quad (8.75)$$

where the *a.c. conductivity* $\sigma_a(\omega) = \omega\varepsilon''\varepsilon_0$. It is important to note that ε' and ε'' are not independent. In fact

$$\varepsilon'(\omega) = \varepsilon_\infty + 2/\pi \int_0^\infty \frac{\omega'\varepsilon''(\omega')d\omega'}{\omega'^2 - \omega^2} \quad \text{and} \quad \varepsilon''(\omega) = 2/\pi \int_0^\infty \frac{\varepsilon'(\omega') - \varepsilon_\infty}{\omega'^2 - \omega^2} d\omega'. \quad (8.76)$$

These equations are called the Kramers–Kronig relationships.

The relationship between the d.c. and a.c. conductivities is far from straightforward. In general $\text{Lt}(\omega \to 0)\sigma_a(\omega) \neq \sigma_{dc}$ since the two may well be probing different underlying physical processes. Where there is a substantial a.c. component to the conductivity, possible reasons are that the contacts may be non-ohmic, the material may be polycrystalline, or it may be amorphous or glassy.

8.9.1 Non-ohmic contacts

We saw above that if the electric contact between the metal and semiconductor is non-ohmic, a depletion layer will be set up in the semiconductor. This depletion layer has a width W given by $(2e_0 V/kT)^{\frac{1}{2}} L_D$ where V is the potential dropped across the layer and L_D is the Debye length of the semiconductor, defined as $(\varepsilon\varepsilon_0 kT/2e_0^2 N_D)^{\frac{1}{2}}$ with N_D the donor density. The depletion layer acts as a capacitor of width W and area Γ, so that $C = \varepsilon\varepsilon_0 \Gamma/W$. The bulk resistance of the semiconductor is in series with this capacitance giving an equivalent circuit

The impedance Z of this current is evidently $R + 1/i\omega C$. The admittance Y is given by

$$Y = 1/Z = i\omega C/(1 + i\omega CR) = (i\omega C + \omega^2 C^2 R^2)/(1 + \omega^2 C^2 R^2) = i\omega \varepsilon_0 \Gamma (\varepsilon' - i\varepsilon'')/d \tag{8.77}$$

whence

$$\sigma_a(\omega) = \omega \varepsilon'' \varepsilon_0 = \frac{d}{\Gamma} \cdot \left(\frac{\omega^2 C^2 R}{1 + \omega^2 C^2 R^2} \right). \tag{8.78}$$

At low ω values, $\sigma_a(\omega) \to 0$ and at high ω values, $\sigma_a(\omega) \to d/\Gamma R = \sigma_{dc}$. The frequency at which the change in $\sigma_a(\omega)$ is most pronounced is that for which $\omega^2 C^2 R^2 = 1$, i.e.

$$\omega^2 = 1/C^2 R^2 = W^2/(\varepsilon\varepsilon_0 \Gamma R)^2 = W^2 \sigma_{dc}^2/(\varepsilon\varepsilon_0 d)^2. \tag{8.79}$$

Typically, $W = 10^{-6}$ cm, $R = 1$ kΩ, $\Gamma = 1$ cm^2, and $\varepsilon = 10$, we have $\omega \approx 1$ kHz (see Fig. 8.24(a)).

8.9.2 Polycrystalline materials

In the course of sintering powdered compacts, the 'necks' joining crystallites of materials together in the compact may become highly resistive, either because donor centres migrate out of the necks, or because gases adsorbed on the surface become charged and give rise to a depletion layer. A quantitatively very similar treatment for such 'grain-boundary' effects can be given to that described above for the non-ohmic case, save that there is now a

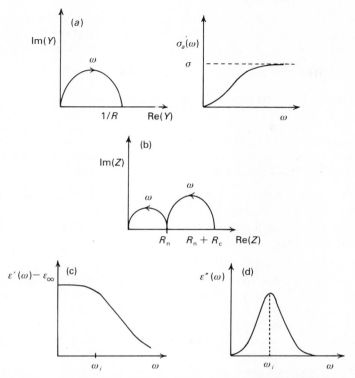

Fig. 8.24. (a) Imaginary part of the admittance Y vs. the real part for the equivalent circuit commonly found for non-ohmic contacts. Also shown is the a.c. conductivity vs. frequency which asymptotically approaches the true d.c. value for the sample. (b) Typical plot of the real vs. imaginary part of the impedance for a Koopsian equivalent circuit commonly found in samples showing pronounced grain-boundary effects. (c) and (d): Plots of $\varepsilon(\omega) - \varepsilon\infty$ vs. frequency ω and $\varepsilon'(\omega)$ vs. ω.

complex distribution of resistors and capacitors in the sample, and the capacitance of the crystallites themselves cannot be neglected in comparison with the capacitance of the 'necks'. The main effect of this distribution is to broaden the frequency range over which the a.c. conductivity shows marked changes, though we still have the qualitative result that $\sigma_a(\omega) \to 0$ as $\omega \to 0$, and $\sigma_a(\omega) \to d/\Gamma R$ as $\omega \to \infty$. If the main physical cause of the grain-boundary effects is migration of donors out the necks, the conductivity of the necks must also be included; this leads to an equivalent circuit first proposed by Koops[48]

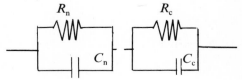

where the subscript 'c' refers to the crystallite, and 'n' to the neck. Provided the time constants $R_n C_n$ and $R_c C_c$ are sufficiently different, it is possible to evaluate all four parameters by plotting the imaginary part of Z versus the real part (see the references for further details); see Fig. 8.24(b).

8.9.3 Amorphous and imperfectly crystalline materials[46]

If a defect exists in a semiconductor at which an electron is trapped and constrained to hop between allowed energy states, the effect is that of a fluctuating dipole which contributes to the a.c. conductivity but not to the d.c. response. Two types of trap can be distinguished:

(1) The electron may hop between centres of fixed relative *energy* which are distributed in space (random range hopping).
(2) The centres are fixed in spatial orientation, but have a distribution of relative *energies* (random energy hopping).

In compounds with both acceptor and donor centres (compensated semiconductors) electron transfer may take place between two such neighbouring species to give a strongly bound $D^+ A^-$ pair. Hopping between adjacent sites of the acceptor A round D^+, or between D sites around A^- can then provide a simple model of the underlying physical process. As a specific example, NiO is usually compensated; there are both Ni vacancies and O vacancies, with the former present in larger concentration. Ionized Ni vacancies will be negatively charged, and will attract positively charged oxygen vacancies. The result is that the two centres 'cancel each other out' as far as the d.c. conductivity is concerned, but if there is a nearby unionized Ni vacancy, hopping between V'_{Ni} and V_{Ni} will give an a.c. conductivity due to the fluctuating dipole.

Quantifying this model is straightforward; consider two sites 1 and 2, separated by a distance $r^i_{1,2}$, where the superscript 'i' labels this *pair* of sites, as in Fig. 8.25. The associated dipole *change* as the electron hops from 1 to 2 is $e_0 r^i_{1,2} \cos \theta^i_{1,2}$. To calculate the current associated with such a change of

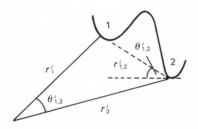

Fig. 8.25. Pair of sites 1 and 2 separated by a distance $r^i_{1,2}$ over which an electron must hop to create a dipole change.

we remember that the total polarization P in any given direction is just the sum of the resolved dipoles in that direction per unit volume. From the polarization we can, in turn, calculate the dielectric constant ε, since the polarization due to moving charges is given by $P = \varepsilon_0(\varepsilon - \varepsilon_\infty)E$, where, ε_∞ is the dielectric constant at infinite frequency. To calculate the polarization, we consider the rate at which electrons hop from 1 to 2. Let the fractional occupancies of these states be f_1^i, f_2^i, where obviously $f_1^i + f_2^i = 1$. The static polarization can then be written $P^i = f_1^i e_0 r_{1,2}^i \cos\theta_{1,2}^i$. Furthermore, if $p_{1,2}^i$ and $p_{2,1}^i$ are the probabilities of an electron hopping from 1 to 2 or vice versa, we find

$$\partial f_1^i / \partial t = -p_{1,2}^i f_1^i + p_{2,1}^i f_2^i \qquad (8.80)$$

$$= p_{2,1}^i - f_1^i(p_{1,2}^i + p_{2,1}^i). \qquad (8.81)$$

To calculate the probability with which a jump can occur, we can consider the potential distribution of Fig. 8.22(b) above. The presence of an electric field E alters the activation energy for a jump by an amount $\beta E r_{1,2}^i \cos\theta_{1,2}^i$, where β is a number (<1) representing the fraction of the potential difference between 1 and 2 at the maximum. The new values of the transition probabilities are:

$$p_{1,2}^{i\prime} = p_{1,2}^i \exp(-\beta E r_{1,2}^i \cos\theta_{1,2}^i e_0 / kT) \simeq p_{1,2}^i (1 - \beta E r_{1,2}^i \cos\theta_{1,2}^i e_0 / kT) \qquad (8.82)$$

with a similar expression for $p_{2,1}$. If we write $\omega_i = p_{1,2}^i + p_{2,1}^i$ and $\xi = e_0 E r_{1,2}^i \cos\theta_{1,2}^i / kT$ and E is an alternating field of frequency ω, i.e. $E = E_0 \exp(i\omega t)$, we can write $\xi \equiv \bar{\xi} \exp(i\omega t)$,

$$\partial f_1^i / \partial t = p_{1,2}^i (1 + \xi) - f_1^i \{\omega_i + \xi(p_{2,1}^i - p_{1,2}^i)\} \qquad (8.83)$$

and, on solving

$$f_1^i = \{\bar{\xi}_0 p_{1,2}^i p_{2,1}^i / \omega_i\} \{e^{i\omega t} / (i\omega + \omega_i)\} \qquad (8.84)$$

whence the polarization is

$$P^i = \frac{e_0^2 (r_{1,2}^i)^2 \cos^2\theta_{1,2}^i}{V} \cdot \frac{p_{1,2}^i p_{2,1}^i}{\omega_i} \cdot \frac{e^{i\omega t}}{(i\omega + \omega_i)} \cdot E_0 \qquad (8.85)$$

$$= \varepsilon_0(\varepsilon_i(\omega) - \varepsilon_{i,\infty}) e^{i\omega t} E_0, \qquad (8.86)$$

assuming $\beta = \frac{1}{2}$. Writing $\gamma^i = e_0^2 r_{1,2}^2 \cos^2\theta_{1,2}^i / e_0 V k T$ and remembering that $\varepsilon(\omega) = \varepsilon'(\omega) - i\varepsilon''(\omega)$, we find finally

$$\varepsilon_i(\omega) - \varepsilon_{i\infty} = \gamma^i p_{1,2}^i p_{2,1}^i / (\omega^2 + \omega_i^2); \quad \varepsilon_i(\omega) = \gamma^i p_{1,2}^i p_{2,1}^i / \{\omega_i(\omega^2 + \omega_i^2)\}. \qquad (8.87)$$

These functions are plotted in Fig. 8.24(c) and (d), and the overall dielectric response can be obtained by summing over all pairs. This summation is tedious but not difficult, and the shape of the function allows considerable simplification. We shall not explicitly do the summation but merely quote the results. If $p^i_{1,2}$ be written as in eqn (8.69) above, we find, for $v_{ph}\omega$, in the case of random range hopping

$$\varepsilon''(\omega) = \frac{\pi^2 e_0^2 N^2 a^{-5}}{48kT\varepsilon_0}\{\log_e(2v_{ph}e^{-W/kT}/\omega)\}^4 \tag{8.88}$$

$$\sigma(\omega) = \frac{N^2 a^{-5}}{kT}\omega\{\log_e(2v_{ph}e^{-W/kT}/\omega)\}^4 \tag{8.89}$$

where W and a are defined in eqn (8.69). The behaviour of eqn (8.89) can be closely approximated by a power law, $\sigma(\omega) \simeq \omega^{0.86}$, over a wide range of frequencies.

In the other case, of random-energy activated hopping, we assume that there are pairs of trapping states of energy E below the mobility edge separated by a distance r. To be specific, we suppose that these traps constitute the Anderson-localized states of an imperfectly crystalline semiconductor as discussed in Section 8.8. The energy density of such traps is $g(E) = G\exp[-f\{(E_C - E), T\}]$ where $f\{(E_C - E), T\}$ is some as yet undetermined function, and E_C is the mobility edge as described above. It is reasonable to suppose that the activation energy for hopping is $(E_C - E) = W$. Provided that E is at least $3kT$ removed from the Fermi level, we can write, for the probability that the *pair* is *singly* occupied, $\exp\{-(E - E_F)/kT\}$.

Inserting these formulae into the polarization equation, we obtain

$$\sigma(\omega) = N^2 r^2 (2v_{ph})^{(1-s)}\omega^s / kT \tag{8.90}$$

where s is defined as $2f(W',T)kT/W'$, and W' has the value $kT\log_e(2v_{ph}/\omega)$. If now $f(W,T) \sim W$, then the distribution of localized states constitutes an exponential tail; at constant T, moreover, s is a *constant*. Since there is considerable independent optical evidence for such an exponential distribution, the constancy of s in eqn (8.90) in a wide variety of materials can be understood. The ubiquity of this power law variation in $\sigma(\omega)$ can be gauged from Fig. 8.26.[46]

8.10 Conclusions

Transport data from both a.c. and d.c. conductivity, thermopower, photocurrent, and Hall measurements can be obtained fairly easily. By varying the temperature and frequency, it is possible to build up quite a detailed picture of the following features of a solid:

(a) the mobilities and effective masses of carriers in valence and conduction bands, from which the range of likely bandwidths can be deduced,
(b) the energy separation of the major bands,
(c) the energy of acceptor and donor states near the edges of the conduction and valence bands,

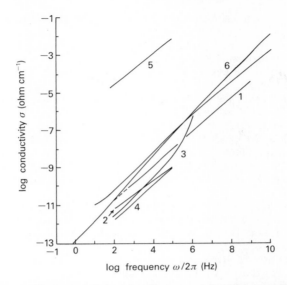

Fig. 8.26. Plots of log (conductivity) vs. log (frequency) for (1) a vanadium phosphate glass, (2) doped and compensated silicon, (3) aluminium oxide, (4) amorphous selenium, (5) scandium oxide, and (6) trans-β-carotene.

(d) the existence of significant disorder in the material, together with some indication of whether electron hopping in localized traps is through random range or random energy hopping.

Transport measurements are now applied routinely to newly synthesized materials in much the same way as more traditional types of spectroscopy might be applied to molecular species to obtain a quick 'spectrum', and to establish a provisional model for the material. More detailed studies can then be carried out if properties of interest are revealed.

8.11 References

1. Smith, R. A. *Semiconductors*. Cambridge University Press (1978).
2. Hannay, N. B. (ed.) *Treatise on solid-state chemistry*. Plenum Press, New York (1976).
3. Riga, J., Teuret-Noel, C., Pireaux, J. J., Caudano, P., Verbist, J. J., and Gobillon, Y. *Phys. Scripta* **16**, 351 (1977).
4. Blakemore, J. S. *Semiconductor statistics*. Pergamon Press, New York (1962).
5. Smith, R. A. *Wave mechanics of crystalline solids*. Chapman and Hall, London (1969).
6. Landolt-Bornstein, New Series: Vol. III/17a–f *Electronic and transport properties of semiconductors*. Springer-Verlag, Berlin (1982).
7. Butcher, P. N. *Electrons in crystalline solids* (Proc. int. course in solid state

physics). International Centre for Theoretical Physics, Trieste, IAEA, Vienna (1973).
8. Many, A., Goldstein, Y., and Grover, N. B. *Semiconductor surfaces*. North-Holland, Amsterdam (1965).
9. Sze, S. M. *Physics of semiconductor devices* (2nd edn), p. 307. Wiley-Interscience, New York (1969).
10. van der Pauw, L. J. *Philips Research Reports* **13**, 1 (1958).
11. Burke, J. E. and Rosolowski, J. H. *Treatise in solid state chemistry* (ed. N. B. Hannay) vol. 4, p. 621. Plenum Press, New York (1976).
12. Bleaney, B. and Bleaney B. *Electricity and magnetism* (2nd edn) p. 87. Oxford University Press (1965).
13. Sleight, A. W. and Bouchard, R. J. *NBS Spec. Publ.* **364**, 227 (1972).
14. Subramanian, M. A., Aravamudan, G., and Subba Rao, G. V. *Prog. solid state chem.* **15** 55 (1983).
15. Folberth, O. G. and Weiss, H. *Z. Naturf.* **10a**, 615 (1955).
16. Bechgaard, K. and Jerome, D. *Scient. Am.* **247**(7), 50 (1982).
17. Heikes, R. R. and Ure, R. W. (eds) *Thermoelectricity*. Wiley-Interscience, New York (1961).
18. Zrudsky, D. R. and Strowalter, A. B. *Rev. sci. Instr.* **44**, 497 (1973).
19. Subha, V. and Ramesh, T. G. *J. Phys.* **E9**, 435 (1976).
20. Freeman, R. H. and Bass, J. *Rev. sci. Instr.* **41**, 1171 (1970).
21. Eklund, P. C. and Mabatah, A. K. *Rev. sci. Instr.* **48**, 775 (1977).
22. Berglund, C. N. and Beairsto, R. C. *Rev. sci. Instr.* **38**, 66 (1967).
23. Caskey, G. R., Sellmayer, D. J. and Rubin, L. G. *Rev. sci. Instr.* **40**, 1280 (1969).
24. Sutradhan, S. K. and Chattopadhyay, D. *J. phys.* **C12**, 1693 (1979).
25. Bosman, A. J. and Van Daal, H. J. *Adv. Phys.* **19**, 1 (1970).
26. Goodenough, J. B. *Prog. solid-state Chem.* **5**, 145 (1971).
27. Reuter, B. and Riedel, E. *Ber. Buns. Phys. Chem.* **71**, 189 (1970).
28. Galy, J., Darriet, J., Casalot, A., and Goodenough, J. B. *J. solid-state chem.* **1**, 339 (1970).
29. Hed, A. Z. and Tannhauser, D. S. *J. chem. Phys.* **47**, 2090 (1967).
30. Kofstad, P. *Non-stoichiometry, diffusion and electrical conductivity in binary oxides*. Wiley, New York (1972).
31. Olson, E. E. and Wertz, J. E. *Rev. sci. Instr.* **41**, 419 (1970).
32. Eisele, I. and Kevan, L. *Rev. sci. Instr.* **43**, 189 (1972).
33. Lupu, N. Z., Tallan, N. M., and Tannhauser, D. S. *Rev. sci. Instr.* **38**, 1658 (1967).
34. Altweiz, M., Finkewrath, H., and Stöckel, T. *J. Phys.* **E6**, 623 (1983).
35. Guthrie, G. L. *Rev. sci. Instr.* **36**, 1177 (1965).
36. Rode, D. L. *Semiconductors and Semimetals*, vol. 10 (eds R. K. Willardson and A. C. Beer). Academic Press, New York (1975).
37. Bransky, I. and Tannhauser, D. S. *Solid State Comm.* **7**, 245 (1969).
38. Bube, R. H. *Photoconductivity of solids*. Wiley, New York (1960).
39. Mort, J. and Lai, C. M. (eds) *Photoconductivity and related phenomena*. Elsevier (1976).
40. Makerov, V. V., Ksendsov, Ya. M. and Kruglov, V. I. *Fiz. Tverd. Tela* **9**, 663 (1967).
41. Powell, R. J. and Spicer, W. E. *Phys. Rev.* **B2**, 2182 (1970).
42. Rey, E., Kamal, M. R., Miles, R. B., and Joyce, B. S. H. *J. Mat. Sci.* **13**, 812 (1978).

43. Mott, N. F. and Davis, E. A. *Electronic processes in non-crystalline materials* (2nd edn). Clarendon Press, Oxford (1979).
44. Mott, N. F. *Phil. Mag.* **19**, 835 (1969).
45. von Schulthess, G. and Wachter, P. *Solid State Comm.* **15**, 1645 (1974).
46. Lewis, T. J. *Dielectric behaviour of non-crystalline solids* vol. **3**, 186. Spec. Per. Rep. Chem. Soc. (1977).
47. Daniel, V. V. *Dielectric relaxation*. Academic Press, New York (1967).
48. Koops, C. G. *Phys. Rev.* **83**, 121 (1951).

9 Vibrational spectroscopy

D. M. Adams

9.1 Introduction

The study of the vibrational motions of the atoms of a solid is known as lattice dynamics. It is a large and complex field in which much progress has been made. Theoretical models have been derived which more or less accurately mirror the behaviour of particular types of material: metallic, ionic, molecular, etc. The models are caricatures of real solids; each is characterized by emphasis upon some particular idiosyncracy which reduces the representation to manageable proportions. The end-point of such calculations is comparison of the theoretically estimated values with experimental ones. The quantities of interest are of two sorts: spectroscopic and thermodynamic. From the normal modes of vibration of the lattice the partition function Z may be calculated.

$$Z = \sum_{\text{All modes}} \sum_{n_i=0}^{\infty} \exp\{-(n_i + \tfrac{1}{2})h\omega_i/kT\} \qquad (9.1)$$

and thence the Helmholtz free energy

$$A = -kT\ln Z + U \qquad (9.2)$$

where U is the energy of the static lattice in which each lattice particle occupies its mean position, ω_i the ith normal mode of the lattice, and n_i the vibrational quantum number. From A, other thermodynamic properties, such as specific heats or elastic constants, may be derived using standard relations.

An adequate introduction to even the most basic principles of the vibrational spectroscopy of solids cannot be attempted in the space of this chapter. The aim of this treatment is, therefore, to form a bridge between undergraduate studies and specialist works, to which reference is essential for those undertaking research in this field. A knowledge of basic spectroscopic practice is assumed. Likewise, aquaintance with the elementary aspects of group theory is presumed, to the extent of obtaining and reducing a representation using either Cartesian or internal coordinates.

9.2 Lattice dynamics: some basic concepts

To understand some of the features of the vibrational spectra of powdered solids, and most certainly those of oriented single-crystals, some insight into lattice dynamics is essential. Many of the key ideas can be introduced by considering a linear chain, Fig. 9.1(a).

9.2.1 The longitudinal modes of a monatomic linear chain

We consider the *longitudinal* vibrations: that is, those in which the atom displacements are along the chain axis. We assume that the forces between the atoms extend only to nearest neighbours, and take the crystal to be harmonic. The equation of motion is

$$m\ddot{x}_n = f(x_{n+1} - x_n + x_{n-1} - x_n) \tag{9.3}$$

where x_n is the displacement from equilibrium of the nth atom, and f the force constant. The equations of motion of *all* the atoms are of this form, differing only in the value of n. We consider a wave solution of the type

$$x_n = A\exp\{i(kx_n^0 - \omega t)\} \tag{9.4}$$
$$= A\cos(kx_n^0 - \omega t) + iA\sin(kx_n^0 - \omega t) \tag{9.5}$$

where k is the 'wave-vector' (although it is actually a scalar in one dimension), and x_n^0 is the undisplaced position of the nth atom. The real part of this complex exponential is interpreted physically as atomic displacement. Substitution of (9.5) in (9.3) yields

$$-\omega^2 m = f(e^{ika} + e^{-ika} + 2) \tag{9.6}$$
$$= 2f(\cos ka - 1) \tag{9.7}$$

and thus

$$\omega^2 m = 4f\sin^2(ka/2) \tag{9.8}$$

where ω is the circular frequency ($=2\pi\nu$, where $\nu=$ frequency (in Hertz), and $\nu/c = \tilde{\nu}$ where c is the velocity of light, and $\tilde{\nu}$ the wave number (in cm^{-1})). Hence

$$\omega = \pm(4f/m)^{\frac{1}{2}}\sin(ka/2). \tag{9.9}$$

Note that the maximum value of the circular frequency $\omega_{\text{max.}}$ is $2(f/m)^{\frac{1}{2}}$. Thus, the lattice acts as a band pass filter, passing frequencies from zero to $\omega_{\text{max.}}$ but blocking higher (imaginary) ones. Notice also that we began with equations representing the *coupled* harmonic oscillations of the atoms of the chain, but that our solutions are *uncoupled* vibrations termed 'normal modes'. Thus, if one atom starts vibrating, it will not continue to do so with constant amplitude but will transfer energy to others.

We now introduce a boundary condition requiring the solutions of (9.3) to be periodic over some long distance: i.e. $x_n = x_{n+N}$, where N is the number of masses in the chain. Hence

$$\cos kna = \cos(kna + kNa) \qquad (9.10)$$

and thus $kNa = 2\pi l$, where l is an integer. For $l = N/2$, $k = \pi/a$, and there are N possible k-values in the range $-\pi/a \leqslant k \leqslant \pi/a$, which is called the first Brillouin zone. Equation (9.9) is a *dispersion relation*, i.e. k varies with ω. The relation is shown in Fig. 9.1(b). The phase velocity, defined as ω/k, can be obtained by noting that when ka is very small, eqn (9.9) reduces to

$$\omega = (4f/m)^{\frac{1}{2}}(ka/2)$$

so that

$$\omega/k = a(f/m)^{\frac{1}{2}}. \qquad (9.11)$$

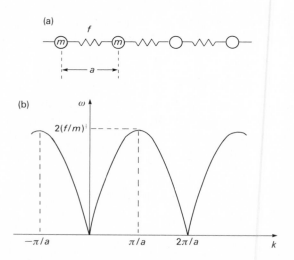

Fig. 9.1. (a) A linear chain or one-dimensional crystal. (b) Variation of the normal mode frequencies with wave-vector for the chain.

As ω varies from zero to $\omega_{\text{max.}}$, so also does the phase velocity. Vibrational waves of different frequency travel at different speeds. $k = 0$ corresponds to a rigid translation of the entire chain along its axis; by definition this has zero potential energy and zero frequency. For $k \neq 0$, the atoms move longitudinally but with phase differences with respect to each other.

9.2.2 Longitudinal and transverse modes of a linear chain

As a first step in the direction of a real crystal, we now allow the atoms of our linear chain to be displaced in two orthogonal directions normal to that of the chain, as well as longitudinally. The force constants will be equal for the two transverse displacements; that is, the modes are degenerate. In general $f_l > f_t$. At every 'wave-vector' k, there will be three independent vibrational modes: one longitudinal (l) and two transverse (t). The forms of their dispersion curves are shown in Fig. 9.2. The dispersion relation is said to have three 'branches'.

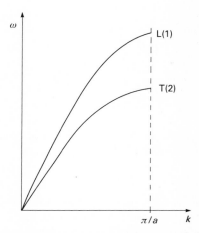

Fig. 9.2. Dispersion curves for the transverse (doubly degenerate) and longitudinal (non-degenerate) branches of a linear chain.

9.2.3 Long-range forces

The 'nearest-neighbour interactions only' restriction which lies behind eqn (9.3) is a serious one. If more distant neighbour interactions are included, (9.3) must be modified to

$$m\ddot{x}_n = \sum_p f_p(x_{n+p} - x_n + x_{n-p} - x_n) \tag{9.12}$$

where the interactions between atoms n and $(n+p)$ separated by distance pa are represented by a force constant f_p. The result of this improved analysis is that only the *shape* of the dispersion curve is affected. The more terms included in the summation (9.12), the more structure is in the dispersion curve. A curve for $f_2 = \frac{1}{2}f_1; f_3 = f_4 = \ldots = 0$ is given in Fig. 9.3.

9.2.4 The diatomic linear chain: acoustic and optic branches

Further essential concepts emerge from an analogous study of linear

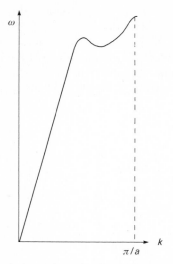

Fig. 9.3. The dispersion relation when the interatomic forces extend to second-nearest neighbours. (After Cochran.[1])

diatomic lattices, such as that of Fig. 9.4(a). Solutions to the equations of motion, for nearest-neighbour interactions only, are

$$\omega^2 = f(m^{-1} + M^{-1}) \pm f\{(m^{-1} + M^{-1})^2 - \frac{4}{mM}\sin^2 ka\}^{\frac{1}{2}}. \qquad (9.13)$$

Clearly, the corresponding dispersion relations split into *two* branches; see Fig. 9.4(b). At $k=0$ the upper value of ω is

$$\omega_0 = \{2f(m^{-1} + M^{-1})\}^{\frac{1}{2}}. \qquad (9.14)$$

Here M and m oscillate in antiphase with their centre of mass stationary. These vibrations are strongly dipolar for all k-values. This allows strong coupling to a radiation field; hence, the name *optical* is applied to it although, as will appear below, only values of $k \approx 0$ are sampled by infrared and Raman spectroscopy.

The second solution

$$\omega_A = ka\{2f/(M+m)\}^{\frac{1}{2}} \qquad (9.15)$$

represents the long-wavelength longitudinal elastic (or sound) waves in the region of the origin, with a frequency-independent velocity of ω_A/k. The atoms move together, as one unit, corresponding to a translational motion; this lower branch as a whole is termed *acoustic*. At $k=0$ the acoustic mode corresponds to a translation of the entire lattice.

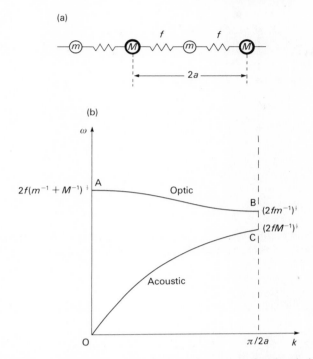

Fig. 9.4. (a) A linear diatomic chain. (b) Dispersion curves for the chain.

At the edge of the repeat of the chain (or unit cell), $k = \pi/2a$ and thus $\sin^2 ka = 1$. The two solutions are

$$\omega_O = (2fm^{-1})^{\frac{1}{2}} \quad \text{and} \quad \omega_A = (2fM^{-1})^{\frac{1}{2}}. \tag{9.16}$$

The modes correspond to standing waves with movement of one or other type of atom. All these modes are illustrated in Fig. 9.5.

9.2.5 Three-dimensional lattices

Most of the features of the one-dimensional model carry over into three dimensions. The equations of motion are found to have solutions of a form which represent plane waves. In certain directions of high symmetry the motion can be resolved into *purely* transverse and purely longitudinal contributions, although in general waves in lattices are neither transverse nor longitudinal. Since the force constants governing motion will generally differ with the direction of the motion, the dispersion relations will differ for t.o., l.o., t.a., and l.a. branches (t = transverse, l = longitudinal, o = optical, and a = acoustical). The point is well made by the dispersion relations for KBr (cubic), Fig. 9.6. In three-dimensional space, $\mathbf{k}$ is a vector.

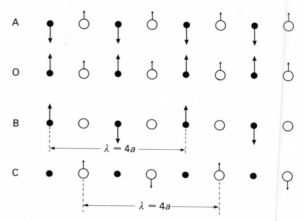

Fig. 9.5. Modes of vibration of a linear diatomic chain at $k=0$, (A,O), and at $k=\pi/2a$, (B,C); see Fig. 9.4(b). At A and O, $\lambda = 2\pi/k \to \infty$. At B and C, $\lambda = 2\pi/(\pi/2a) = 4a$.

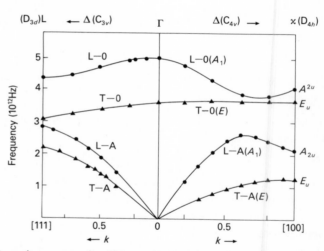

Fig. 9.6. Dispersion curves for KBr at 90 K (k in units of π/a). (After Woods, Brockhouse, Cowley, and Cochran.[2])

9.2.6 Phonons

A quantum of excitation of a lattice, $\hbar\omega$, is termed a *phonon*. Both phonons and photons are Bosons and are not conserved; they can be created and destroyed in collisions. We have seen that the normal modes are plane waves propagating throughout the crystal lattice. Consequently, phonons are not localized particles, although for most purposes we may consider them as fairly localized wave-packets. The momentum $\hbar q$ and the position are related by the Uncertainty Principle. Strictly speaking, a phonon in a harmonic crystal can have no momentum, because there is no interaction between the vibrations of neighbouring atoms. Real crystals are anharmonic: their phonons have momentum, their lattices exhibit thermal conductivity and are compressible. Hence, in general, phonon frequencies depend upon both temperature and pressure.

9.2.7 Sampling k-space

(a) Raman spectroscopy

In Raman scattering—an inelastic process—both energy and momentum are conserved. Thus, considering only Stokes scattering, if ω_1 is the circular frequency of the incident photons, and ω_s that of the scattered photons, the energy gained by the sample is $\hbar\omega$, where $\omega = \omega_1 - \omega_s$. Photon momenta $\hbar k$ are related by $q = k_1 - k_s$. This equation represents a momentum gain by the medium of $\hbar q$. A vector diagram illustrating this process is in Fig. 9.7(a). From it we see that

$$q^2 = k_1^2 + k_s^2 - 2k_1 k_s \cos\varphi. \tag{9.17}$$

For the common 90° scattering geometry it follows that

$$q^2 = k_1^2 + k_s^2 \tag{9.18}$$

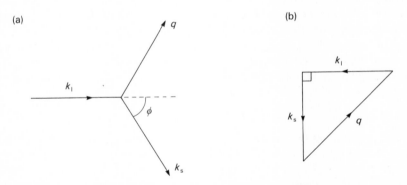

Fig. 9.7. Vector diagrams representing Stokes–Raman scattering at (a) $\varphi°$, (b) 90°.

for which Fig. 9.7(b) provides a convenient representation. Since $k_1 \approx k_s$ it follows that $q \approx k_1(2)^{\frac{1}{2}}$. Other useful scattering geometries, necessary in specific applications (see below), use $\varphi = 0$ or 180°.

For a vibrational wave of circular frequency $\omega = 2\pi\nu$ propagating with phase velocity $u = \nu\lambda$ in a medium of refractive index $n = c/u$, where c is the velocity of light, we have

$$q = 2\pi/\lambda = 2\pi v n/c = 2\pi \tilde{v} n. \qquad (9.19)$$

In a typical Raman experiment using blue laser light of approximately $2 \times 10^4 \text{ cm}^{-1}$ ($\tilde{v}_1$), for a solid of refractive index 1.5

$$q = 2\pi \times 2 \times 10^4 \times 1.5 \times 2^{\frac{1}{2}} \approx 3 \times 10^7 \text{ m}^{-1}.$$

This is to be compared with the maximum value q can take, which is at the Brillouin zone edge with value π/d, where d is the lattice constant. For $d = 10$ Å, $q_{max.} \approx 3 \times 10^{10} \text{ m}^{-1}$, three orders of magnitude less than the largest light-scattering excitation wave-vectors. Hence, in Raman scattering, only modes with q between 0 and $(3 \times 10^7) \text{ m}^{-1}$ are sampled—that is, modes with very long wavelengths.

(b) Infrared spectroscopy

In infrared experiments, even longer waves are sampled with $10^6 > q > 10^4 \text{ m}^{-1}$. Note, however, that $q \approx 0$ can arise in two distinct ways: (1) q is actually close to zero, or (2) from combinations of the type $q_1 - q_2 \approx 0$, where neither q_1 nor q_2 are near zero, but are from highly populated special symmetry points in the Brillouin zone. Correctly assigned (which is not easy) such combinations can give information on the shapes of dispersion curves.

(c) Inelastic neutron scattering

To obtain information on vibrational modes with q values anywhere in the Brillouin zone, two conditions must be satisfied: (1) the sampling radiation wavelength must be comparable with interatomic distances, and (2) it must have energy quanta $\hbar \omega$ similar in size to those of the lattice phonons. The first of these conditions is satisfied by X-rays, but not the second. However, thermal neutrons do comply with both. Neutron inelastic scattering has, accordingly, become an extremely important tool in lattice dynamics. It is beyond the scope of this chapter.

(d) Selection rules—'a first bite'

No change of dipole moment accompanies acoustic modes. Accordingly, they are not infrared-active. They may be Raman-active but, at $q \approx 0$ they are extremely low in frequency having, generally, $\tilde{v} < 2 \text{ cm}^{-1}$. Brillouin spectroscopy is the name given to their study, which is usually persued using different experimental procedures from Raman spectroscopy. In this chapter we shall only be concerned with optic branch modes.

For the optic branch, only t-o modes can interact with incident radiation. L.o. modes are not infrared-active, except under special conditions, although they can often be studied by Raman scattering with correct choice of geometry (see Section 9.4.4). Group theory makes no distinction between l-o and t-o modes. This is dramatically illustrated by the phonon dispersion curves of KBr (Fig. 9.6). It is easily shown using 'factor group analysis'

(Section 9.4) that the three optic branch modes of KBr all belong to the triply degenerate representation T_{1u} of O_h. In contrast, the dispersion diagram shows that this degeneracy is partially lifted, resulting in a non-degenerate l.o. and a doubly degenerate t.o. mode. The reason why group theory makes this incorrect prediction is examined in Section 9.4.

9.2.8 L.o.–t.o. mode splitting

We have seen that optic branch modes may be sampled at $k \approx 0$ using Raman and infrared spectroscopy, subject only to the usual selection rules. We now note an extremely important division of optic modes; it is made on the basis of whether or not the mode carries an electric dipole moment:

Non-polar modes do not generate an electric dipole moment and, thus, are not infrared active; they may be Raman active. Their frequencies are determined mainly by short-range forces in the lattice.

Polar modes carry an electric dipole moment and have radically different long-wavelength properties from those of non-polar modes. Their degeneracies are often lower than predicted by group theory. The oscillating dipole associated with a polar mode generates a lattice polarization; and the polarization in turn generates a macroscopic electric field, E (not to be confused with the externally applied radiation field). The long-range nature of dipole–dipole interaction couples similar transition moments throughout a crystal. As a result, polar mode frequencies depend upon the orientation and magnitude of the mode wave-vector q relative to E.

For polar modes the dielectric polarization of the lattice, P, is made up of two components which arise, respectively, from (i) the displacement of the ions during the lattice vibration, and (ii) from the effect of the field E. It will be shown that for *longitudinal* modes, $\partial \mu / \partial Q$, E, and P are all parallel to the direction of wave propogation q, and that $E_\parallel$ then reinforces the effect of ionic displacement, thus making $\omega_l > \omega_t$. In contrast, for *transverse* modes, $E\perp$ is smaller than $E_\parallel$ and, in the absence of incident electromagnetic radiation, is actually zero. The magnitude of this l.o.–t.o. splitting is given by the Lyddane–Sachs–Teller (LST) relation:

$$\frac{\omega_l}{\omega_t} = \left(\frac{\varepsilon_0}{\varepsilon_\infty}\right)^{\frac{1}{2}} \qquad (9.20)$$

where ε_0 is the dielectric constant at frequencies considerably lower than those of the lattice vibrations (usually at radio frequencies), and ε_∞ is the corresponding high-frequency dielectric constant ($=n^2$).

The origin of l.o.–t.o. splitting is thus seen to lie in the anisotropy created by the very existence of a propagation direction relative to E. This propagation direction is an axis of symmetry and therefore reduces the lattice symmetry. L.o.–t.o. splitting may be present even in a cubic crystal, provided that the lattice has a basis, that is it is *not* monatomic and thus can generate optic branch modes. Note that the effect is not dependent upon the presence of incident electromagnetic radiation, i.e. l.o.–t.o. splitting takes place even in the dark!

Let w be the mass-weighted displacement from equilibrium separation of the positive, relative to the negative, ions. This motion generates a macroscopic electric field E. Assuming that the restoring force obeys Hooke's law, and is proportional also to the long-range field E, the equation of motion of the ions is:

$$\ddot{w} = aw + bE \tag{9.21}$$

where a and b are scalars, a representing the short-range forces. The dielectric polarization of the lattice P is also proportional to E, but modified by the vibrations as indicated. Thus:

$$P = fw + gE, \tag{9.22}$$

and it can be shown from the conservation-of-energy principle that $b=f$. Equations (9.21) and (9.22) are coupled, and must therefore be solved together. Assuming a temporal variation of the form $X = X_0 e^{-i\omega t}$, (where $X = w, E, P$), and eliminating w,

$$P = E\left\{g - \frac{b^2}{\omega^2 + a}\right\} \tag{9.23}$$

$$= \frac{E}{4\pi}\left\{\varepsilon_\infty - 1 - \frac{\omega_0^2(\varepsilon_0 - \varepsilon_\infty)}{\omega^2 - \omega_0^2}\right\} \tag{9.24}$$

$$= \frac{E}{4\pi}\left\{f(\omega) - 1\right\} \tag{9.25}$$

where use has been made of the known equivalences: $a = -\omega_0^2$, $b = \{(\varepsilon_0 - \varepsilon_\infty)^{\frac{1}{2}}\omega_0\}/2\pi^{\frac{1}{2}}$, and $g = (\varepsilon_\infty - 1)/4\pi$.

When incident radiation couples with the lattice vibrations, the disturbance propagates with the velocity of light. Physically, this mixing of the mechanical and radiative (light) waves arises by virtue of the magnetic field and its derivatives of the light wave. The mixing is termed 'retardation'. It generates a new 'particle', the *polariton*, which is neither pure photon nor pure phonon but is an optical excitation containing a variable amount of both mechanical and radiative energy.

Figure 9.8 (broken line) shows the system in the absence of photon–phonon mixing. ω_t and ω_l are both independent of q. The dispersion curve of the photon is simply $\omega = cq$ (*in vacuo*), and $\omega = (c/n)q$ in a medium of refractive index $n = \varepsilon_\infty^{\frac{1}{2}}$. These are plotted in terms of the dimensionless

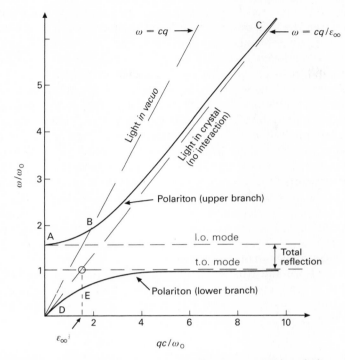

Fig. 9.8. Dispersion relations for an electromagnetic wave and optic branch modes travelling in a crystal, without coupling (broken lines); with coupling of the t.o. mode and the photon near $q=0$ (full lines).

quantities ω/ω_0 and qc/ω_0, taking the values of ε_∞ ($=2.25$) and ε_0 ($=5.62$) for NaCl.

With the mechanical-radiative interaction switched on, the situation becomes that shown by full lines in Fig. 9.8. ω_l is unaffected. Where the photon curve crosses ω_t.

$$\omega_t = \omega_{\text{photon}} = ck = 2\pi v_t \quad \text{or} \quad k = 2\pi \tilde{v}. \qquad (9.26)$$

For an infrared process with $\tilde{v}_T = 1000 \text{ cm}^{-1}$, k is therefore $(2\pi \times 10^5) \text{ m}^{-1}$. Compared with k at the Brillouin zone edge, this is very close to zero, and photon–phonon mixing is to be expected. Only the *transverse* lattice vibrations mix with the radiation field because an electromagnetic wave is a transverse wave. L.o. modes cannot appear in absorption. ω_t and the photon are replaced by the two branches of the polariton dispersion curve; for any given q, there are two values for ω, one between zero and ω_0, and one greater than ω_l. The behaviour of the photon–phonon system is analagous to the case of quantum mechanical resonance between two states of the same symmetry; the states repel each other (non-crossing rule). The region of q-space in which the interaction is significant is generally $(10^5 < |q| < 10^6)\,\text{m}^{-1}$, corresponding roughly to lattice vibrations in the 100–1000 cm^{-1} interval. As $|k|$ at the Brillouin zone boundary is typically about $10^{10}\,\text{m}^{-1}$, it is clear that the photon perturbs phonons only in the region very close to $q = 0$, that is, phonons with very long wavelengths. At the crossover point (near B and E) both branches represent excitations with equal phonon and photon character. At A the excitation is phonon-like (about 70 per cent of the energy is mechanical) although with sufficient retardation to replace the t.o. phonon with a polariton. This upper branch of the polariton dispersion curve meets the longitudinal excitation asymptotically at zero wave-vector, producing a threefold degeneracy in accord with the predictions of group theory. The splitting of the l.o. and t.o. modes seen in data such as those of Fig. 9.6 is a consequence of wave-vectors being to the right of the ω_t–photon crossover point. At D the excitation is largely photon-like with about 30 per cent mechanical energy. Near C the behaviour is photon-like and the lattice behaves like a collection of infinite-mass particles. More detailed descriptions are given by Born and Huang[3], Reissland[4], and Hayes and Loudon[5]. These very small wave-vector excitations can be studied directly by Raman scattering close to the forward direction. This follows from the vector diagram of Fig. 9.9.

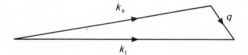

Fig. 9.9. Vector diagram representing conditions for the study of polaritons.

9.2.9 Dispersion of the dielectric constant

A valuable dispersion relation follows from eqn (9.25) and the definition of electric displacement,

$$D = E + 4\pi P = \varepsilon E. \tag{9.27}$$

Elimination of P gives the dielectric constant, $\varepsilon = f(\omega)$, conveniently written as

$$\varepsilon = \varepsilon_\infty + \frac{\varepsilon_0 - \varepsilon_\infty}{1 - (\omega/\omega_0)^2} \tag{9.28}$$

$$= \varepsilon_\infty \frac{\omega_l^2 - \omega^2}{\omega_t^2 - \omega^2}. \tag{9.29}$$

Plainly, ω_0 is the frequency (usually in the infrared) at which the dielectric constant exhibits a singularity: the crystal is then resonating by being driven in its normal mode of vibration. At $\omega=\omega_0$, $\varepsilon\to\infty$ and ω_0 is identified with ω_t. Combining the LST relation with (9.28) to give eqn (9.29) shows ω_l to be the frequency at which $\varepsilon=0$. This dielectric function is shown in Fig. 9.10. For a

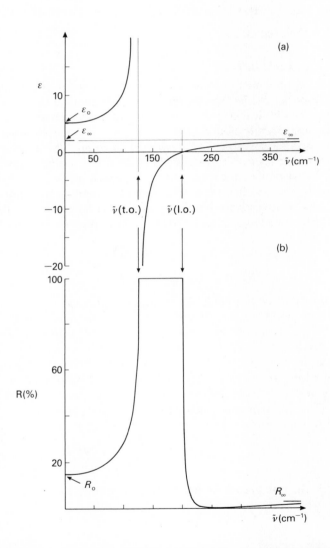

Fig. 9.10. (a) Dispersion of the dielectric constant for an infinite cubic crystal without phonon–photon interaction. Calculated for $\varepsilon_\infty=2.0$, $v_1=200\,\text{cm}^{-1}$, and $v_t=125\,\text{cm}^{-1}$, where $\omega=\tilde{v}\times 2\pi c$. (b) Reflectance of the same material, calculated from $R=(\varepsilon^{\frac{1}{2}}-1/\varepsilon^{\frac{1}{2}}+1)^2$. Note the total reflectance between $\tilde{v}_t$ and $\tilde{v}_1$.

cubic crystal with a single (undamped) infrared-active mode (9.28) may be rewritten as

$$\varepsilon = \varepsilon_\infty + \frac{S}{1-(\omega/\omega_t)^2} \tag{9.30}$$

where $S = 4\pi\rho$ is the transition strength, and can be shown to be proportional to $\frac{\partial \mu}{\partial Q}$. ρ is the oscillator strength.

9.2.10 Microscopic theories

The model upon which the above discussion is based is of a cubic diatomic lattice. Nevertheless, this *macroscopic* model has enabled us to recognize and appreciate the physical origin of the main features which are essential for an intelligent approach to obtaining and interpreting the vibrational spectra of solids. There also exist so-called *microscopic* theories. The same macroscopic dielectric properties may be calculated using them, though allowing further insights. In particular, the microscopic approach allows the analysis to be generalized more readily to include molecular and other complex crystal lattices, and to investigate how a crystal lattice affects molecular internal modes. Space precludes consideration of these theories, but we note some results.

The macroscopic field E must be replaced by

$$E = E_{\text{eff.}} - E_1 - E_2. \tag{9.31}$$

$E_{\text{eff.}}$ is the value of the field at the centre of the ion. It differs from the macroscopic field because E is not simply the total field; it is the total field averaged over the entire unit cell. $E_{\text{eff.}}$ is the total field with the contribution of the ion itself excluded. It can be shown that two terms make up this difference: $E_1 = -\frac{4}{3}\pi P$, and E_2, the field arising from the *displacements* of neighbour ions and their induced moments. E_2 depends upon the crystal structure and is actually zero for a cubic crystal. In the case of molecular crystals E_2 is the term which describes factor group splitting.

These considerations lead to a modified form of eqn (9.28)

$$\varepsilon = \varepsilon_\infty + \left(\frac{\varepsilon_\infty + 2}{3}\right)^2 \left(\frac{\partial \mu}{\partial Q}\right)^2 \cdot \frac{4\pi A}{\omega_t^2 - \omega^2} \tag{9.32}$$

where A is the number of ion pairs per unit volume and

$$\left(\frac{\varepsilon_\infty + 2}{3}\right)^2 = \left|\frac{E_{\text{eff.}}}{E}\right|. \tag{9.33}$$

Recalling that, at ω_1, the function (9.28) becomes zero, we have

$$\omega_1^2 - \omega_t^2 = \frac{4\pi A}{\varepsilon_\infty}\left(\frac{\varepsilon_\infty + 2}{3}\right)^2 \cdot \left(\frac{\partial \mu}{\partial Q}\right)^2 \tag{9.34}$$

which is generally known as the Haas–Hornig equation. It allows $(\partial\mu/\partial Q)$ to be obtained from the width of a reflectance band (see Section 9.3.4). Examples of its use have been given by its authors.[6] Details of other microscopic theories, particularly as applied to molecular crystals, are given by Decius and Hexter.[7]

9.2.11 Molecular crystals

The basic concepts of lattice dynamics were introduced in terms of models of atomic lattices. We now note further detail of importance in respect of molecular and complex ionic crystals.

The vibrational potential energy of a molecular crystal may be represented by

$$V = \sum_n V_n + \sum_n \sum_k V_{nk} + V_E + V_{En}. \qquad (9.35)$$

The first term is the sum of the potential energies due to all the internal coordinates of all the molecules. The potential energy of a molecule in a crystal differs from that of the same molecule in the gas phase due to the *static field effect*. This represents the influence on the molecule of the surrounding lattice in its *equilibrium* configuration. There is usually a shift in vibrational frequencies from gas to condensed-phases due to the static field. The symmetry of the molecular site in the lattice is simply that of the Wyckoff site on which its centre of gravity resides. Very commonly the *site symmetry group* is of lower order than that of the molecule; degeneracies may be lifted and selection rules affected.

The cross-terms $\sum_n \sum_k V_{nk}$ represent the *dynamical* effects arising from interactions with internal vibrations of other molecules in the same unit cell. This is called *correlation field* or *factor group splitting*. Clearly, it is absent if there is only one molecule per primitive cell.

Thus far the effects noted relate to modifications of the internal modes of the molecules. In a crystal further degrees of freedom are present; these may be regarded as arising from the three components of the translation vector of the free molecule, and the three rotational motions about the principal axes. They are termed *external* or *lattice modes*. Thus, for each molecule in the primitive cell, there are three translational modes in which the molecules oscillate about their Wyckoff sites; and three librational modes in which motion of a rotational nature is performed about the same origin. The potential energy due to both types is included in V_E. Libratory modes usually occur at higher frequencies than translatory ones. Librational mode frequencies can be estimated by $\omega_R(a) = (f_a/I_a)^{\frac{1}{2}}$, where f_a and I_a are the force constant and moment of inertia respectively about the ath axis.[8] Finally, we note existence of the cross-terms V_{En} between internal and external coordinates, the importance of which increases as internal and external mode frequencies approach each other.

9.3 How to obtain a spectrum

9.3.1 A preliminary: temperature

Careful attention must be given to the temperature(s) at which a spectrum is to be obtained. At room temperature $kT \approx 200 \text{ cm}^{-1}$. Hence a far-infrared spectrum (the region $< 200 \text{ cm}^{-1}$) will be seriously distorted by 'hot bands' arising from upper-level transitions unless the sample is cooled. In general, anharmonic effects ensure that $0 \rightarrow 1$, $1 \rightarrow 2$, etc. transitions are at slightly different frequencies; the net result is an envelope of bands, tailing off asymmetrically. For two levels 400 cm^{-1} apart, nearly 15 per cent of the vibrational energy is associated with the first excited vibrational state at room temperature, and the quality of a spectrum even at much higher frequencies is usually improved significantly by cooling the sample. Analagous considerations apply to Raman spectroscopy, as is clear from Fig. 9.11.

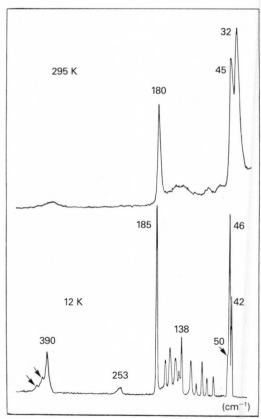

Fig. 9.11. The Raman spectrum of $Hg_2(NO_3)_2 \cdot 2D_2O$ at 295 K and 12 K.

The low-temperature spectrum shows almost all the bands required by theory, but little could have been deduced with certainty from the 295 K result. The literature is littered with reports in which 'dizzy' theoretical 'castles' have been erected on an unstable foundation of ambient temperature data, sometimes none too well-assigned. A variety of reliable cryostats is commercially available, while simple ones are easily constructed in the laboratory. For further details, consult Miller and Stace.[9]

9.3.2 Particle size and shape

Many solids can be obtained only in powder form. Moreover, even given crystalline material, the commonest procedure is to grind it into powder form and make a KBr disc. The spectra yielded by microcrystalline specimens may differ very significantly from those of the same material in the bulk, especially when strongly dipolar vibrations are involved. The force constants affecting atoms at a surface are quite different from those influencing atoms well within the solid. Moreover, the degree to which surface effects extend into the bulk depends upon the long-range coulomb forces, so the vibrational modes of a particle are dependent upon both size and shape. Vibrational modes associated with the surface may appear in a variety of guises. For our purposes the most important are the so-called 'optical surface modes' which are essentially localized vibrations at the surface with a wave-vector parallel to it. They can be excited by infrared radiation.

Even a summary of the main results in this complex field cannot be attempted here. The theoretical treatment used depends upon the dimensions of the particle relative to the wavelength of the sampling radiation. We shall look only briefly at a case in which the particle size is less than the wavelength of the infrared radiation used, namely a study of the infrared-active modes of three corundum-type oxides.[10] The shapes of the powder particles were determined by scanning electron microscopy, and the spectra (using KBr discs) were then compared with theoretical spectra computed on the basis of appropriate crystallite shape. Apart from shape, the essential data input consists of knowledge of the frequency-dependence of the complex refractive index, $\hat{n}$, obtained from single-crystal reflection spectra as outlined in section 9.3.4. From this, the polarizability, a, is given by

$$a = v/\{4\pi[g + (\hat{n}^2 - 1)^{-1}]\},$$

where g is a shape factor having a value between 0 and 1, and v is the particle volume. Finally, the extinction cross-section, c, is obtained:

$$c = \pi k \{\tfrac{8}{3} k^3 |a|^2 - \text{Im}(a)\},$$

where $k = 2\pi/\lambda$.

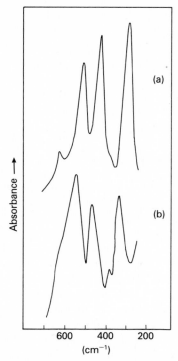

Fig. 9.12. Infrared spectra (using KBr discs) of a-Fe_2O_3: (a) lath-shaped particles, and (b) a mixture of morphologies.

Figure 9.12 shows the results for $a - Fe_2O_3$. Sample (a) consisted only of lath-shaped particles with dimensions of approximately $(0.4 \times 0.08 \times 0.08) \mu m^3$ and its spectrum was in good agreement with one calculated on the basis of this shape. Sample (b) was prepared at higher temperature than (a). Sintering had resulted in a range of morphologies which yielded a very different spectrum, thus dramatically illustrating the magnitude of the surface effects. A reasonable theoretical fit was obtained on the basis of a combination of three particle shapes with different statistical weights.

Raman spectra may also be affected by morphologically-dependent resonances which can generate sharp bands not present in the spectrum of the bulk material. Such peaks appear to arise from resonance-enhancement of the Raman emission, or from enhancement of a fluorescence background.[11]

9.3.3 Transmittance and reflectance

In general, when incident radiation interacts with a solid surface, both absorption and reflection occurs, with the relative proportions dependant upon the optical properties of the material. The optical properties are fully specified when the complex refractive index $\hat{n}(\tilde{v})$ at wave number $\tilde{v}$ is known.

$$\hat{n}(\tilde{v}) = n(\tilde{v})\{1 - i\kappa(\tilde{v})\} \tag{9.36}$$

$$= n(\tilde{v}) - ik(\tilde{v}) = |\hat{\varepsilon}(\tilde{v})|^{\frac{1}{2}} \tag{9.37}$$

where $n(\tilde{v})$ is the real refractive index, $\kappa(\tilde{v})$ the extinction coefficient, and $\varepsilon(\tilde{v})$ the complex dielectric constant. The absorption index, $k(\tilde{v}) = n(\tilde{v})\kappa(\tilde{v})$, may also be expressed in terms of $a(\tilde{v})$, the power absorption coefficient (neper m^{-1}), which measures the intensity attenuation of electromagnetic radiation per unit length within the solid: $k(\tilde{v}) = a(\tilde{v})/4\pi\tilde{v}$. In the following discussion, for simplicity, $n(\tilde{v})$ is written n as before, $\varepsilon(\tilde{v})$ as ε, etc. The complex dielectric constant, $\hat{\varepsilon} = \varepsilon' - i\varepsilon''$. Hence, from (9.37)

$$\varepsilon' = n^2 - k^2 \quad \text{and} \quad \varepsilon'' = 2nk. \tag{9.38}$$

The transmissivity, $\tau(\tilde{v})$, of a solid specimen of thickness d depends upon both absorption and reflection effects. It has values $0 \leqslant \tau(\tilde{v}) < 1$.

$$\tau(\tilde{v}) = e^{-a(\tilde{v})d} \tau_R(\tilde{v}, d) \tag{9.39}$$

where the first term represents absorption losses, and the second term those due to single and multiple reflection. The problems associated with estimation of $\tau_R(\tilde{v}, d)$ are severe. Moreover, for intensely absorbing solids, d may be as little as a few micrometres, thus creating generally insuperable difficulties in the way of transmission work with single-crystal slabs of strong absorbers. Clearly, for powdered specimens (as mulls or discs), the experiment is essentially that of determining the transmission spectrum although, as explained below, even then reflectance and particle-size effects can cause major distortions in appearance. For these reasons and, especially, when work with oriented single-crystals is indicated, infrared reflectance spectroscopy is a most important tool.

9.3.4 Infrared reflectance

The basic features of a reflectance spectrum follow simply from the Fresnel formula, which gives the reflectance R at normal incidence:

$$R = \left|\frac{\hat{n} - 1}{\hat{n} + 1}\right|^2 \tag{9.40}$$

Since $\hat{n} = \varepsilon^{\frac{1}{2}}$, and the behaviour of ε is known from eqn (9.28), R may be calculated (Fig. 9.10). When ε is negative, between ω_l and ω_t, $\hat{n}$ is imaginary: because electromagnetic waves with frequencies corresponding to imaginary values of $\hat{n}$ cannot propagate in the lattice, total reflection occurs between ω_l and ω_t, which thus define the width of the reflectance band.

These data refer to a harmonic cubic crystal without photon–phonon mixing. In reality this simple model must be modified to take account of such mixing, and also of existence of anharmonic forces, and departures from dielectric linearity. In practice this is most readily done by introduction of a frequency-independent damping factor γ in eqn (9.21), leading to the modified dispersion relation

$$\varepsilon = \varepsilon_\infty + \frac{\varepsilon_0 - \varepsilon_\infty}{1 - (\omega/\omega_0)^2 - i\gamma(\omega/\omega_0)}. \tag{9.41a}$$

Under these conditions, the width of the reflectance band is now a little greater than $(\omega_1 - \omega_t)$. Reflectance spectra are often fitted with a function of this type in which the summation is over all infrared-active t.o. modes, including second-order processes. Thus

$$\varepsilon = \varepsilon_\infty + \sum_j s_j / \{1 - (\omega/\omega_j)^2 - i\gamma_j(\omega/\omega_j)\}, \tag{9.41b}$$

where

$$\varepsilon_0 - \varepsilon_\infty = \sum_j s_j = \sum_j 4\pi \rho_j = \frac{2}{\pi} \int_0^\infty \frac{\varepsilon''(\omega)}{\omega} d\omega$$

and ω_j, γ_j, and ρ_j are the circular frequency, damping constant, and oscillator strength respectively of the jth mode. Values of ω_j, γ_j, and s_j are first estimated, a process in which a little experience is invaluable. $\varepsilon(\omega)$ and thence $R(\omega)$ are then calculated, compared with the experimental value of $R(\omega)$ (ie. the reflectance spectrum), and cycles of refinement continued until an acceptable fit is achieved. Worked examples of the effect of varying the damping γ, which corresponds to the gradient at the band edge, are given on p. 185 of Decius and Hexter.[7]

An alternative approach to analysis of reflectance spectra makes use of the Kramers–Krönig (K–K) relations for determination of the dielectric function. The complex reflectivity, $\hat{r}$, and the intensity of radiation R reflected normally from the surface are related by

$$\hat{r} = |r|e^{i\theta} \quad \text{and} \quad R = |r|^2. \tag{9.42}$$

Here θ is the phase difference between incident and reflected rays. The phase angle is generated from the experimental R versus $\tilde{\nu}$ spectra using the Kramers–Krönig inversion relationship

$$\theta(\tilde{\nu}_i) = \frac{2\tilde{\nu}_i}{\pi} \int_0^\infty \frac{\ln\{R(\tilde{\nu})\}^{\frac{1}{2}}}{\tilde{\nu}^2 - \tilde{\nu}_i^2} d\tilde{\nu}. \tag{9.43}$$

The physical basis of the K–K relations is simply that the electric displacement caused in a sample by the radiation field always lags behind that field. Then:

$$n = \frac{1 - R}{1 + R - 2R^{\frac{1}{2}}\cos\theta} \quad \text{and} \quad k = \frac{-2R^{\frac{1}{2}}\sin\theta}{1 + R - 2R^{\frac{1}{2}}\cos\theta}. \tag{9.44}$$

Although much used, in practice the K–K method is subject to a variety of 'ills' which, in general, render it less appealing than a well-parameterized classical dielectric dispersion function such as (9.41) and (9.45).

It should be appreciated that all methods of analysis used in this field are subject to specific conditions, and that some thought must be directed to them when selecting the best approach for a given problem. A particularly important situation in which the function (9.41) is troublesome, is when it is applied to very wide reflectance bands, such as are shown by many oxides, complex oxides, and complex fluorides. These are well fitted by use of the factorized form of the dielectric function, introduced by Berreman and Unterwald:[12]

$$\varepsilon = \varepsilon_\infty \prod_j \frac{\omega_{jl}^2 - \omega^2 \pm i\gamma_{jl}\omega}{\omega_{jt}^2 - \omega^2 \pm i\gamma_{jt}\omega} \qquad (9.45)$$

This model is successful with wide bands because it introduces independent variation of t.o. and l.o. mode damping constants and, as noted before, these correspond to the gradients at the band edges. Examples of the use of function (9.45) are given by Servoin and Gervais.[13]

Another difficulty arises when the reflectance spectrum contains both broad and narrow features: the narrow ones do not appear in $\varepsilon(\omega)$ unless a large number of $R(\omega)$ data points are sampled. Sceats and Morris[14] have reported a method which avoids such problems by means of a Fourier Transformation of $R(\omega)$ data to the time domain. The method is a general one.

Whichever method of analysis is used, the result is ε', ε'', n, and k as a function of frequency. The t.o. modes correspond with ε'' maxima, while l.o. modes occur where ε' becomes zero on changing from negative to positive values, as shown by Fig. 9.13, which represents the real and imaginary parts of (9.41). The same information may be obtained from n/k crossovers. Infrared reflectance methods are applied widely, particularly using polarized radiation to assist in determination of symmetry species of modes, and in the study of materials at high temperatures. Examples are given in Section 9.5.

9.3.5 Spectroscopy of single crystals: the indicatrix

Much symmetry-related information can be obtained from spectroscopy of oriented single-crystals, when the symmetry species to be determined are those of the space group of the crystal. Only phonons travelling in certain specified directions have the symmetry of these species, so the direction and polarization of wave propagation in the crystal must be specified carefully.

In a crystal subjected to an externally applied electromagnetic field the electric displacement produced is given by

$$\boldsymbol{D} = \varepsilon \boldsymbol{E} \qquad (9.46)$$

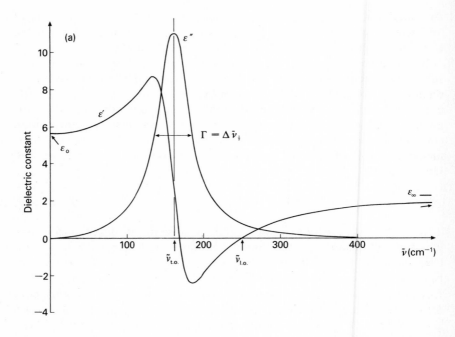

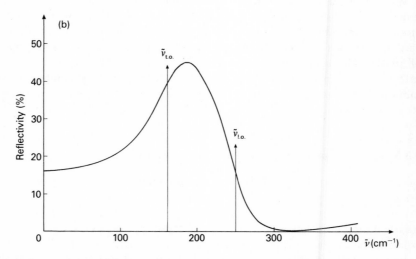

Fig. 9.13. (a) The real (ε') and imaginary (ε'') parts of the complex dielectric constant for NaCl. (b) The infrared reflectance spectrum of NaCl calculated from the data of (a) and the real part of eqn (9.41).

where E is the applied field, and the dielectric constant ε is a symmetric second-rank tensor. It is a standard result that such a tensor may be represented geometrically by a quadric, in this case an ellipse. It can be shown that, when referred to principal axes, the semi-axes of this ellipse are the principal refractive indices, $n_x = \varepsilon_{xx}^{\frac{1}{2}}$, etc. This ellipse is commonly termed the *indicatrix*. Since ε is a property of the crystal, symmetry requires the indicatrix axes to be related to the crystal axes;[15] see Table 9.1. Note that D, which represents the response of the crystal to E, is not constrained by symmetry; it depends upon the magnitude and orientation of E. The importance of the indicatrix is that, in the directions of its axes, the ordinary and extraordinary rays from a refracted beam travel together, though with orthogonal polarizations. It is the indicatrix axes which must be used when determining the symmetry species of vibrations.

Table 9.1 *Relations between crystal axes and the principal axes of the dielectric tensor*†

Optical classification	System	Symmetry relations
Isotropic	Cubic	Any orientation allowed. $\varepsilon_x = \varepsilon_y = \varepsilon_z$
Uniaxial	Tetragonal	$z\|C_4$ or S_4.
	Hexagonal	$z\|C_6$ or C_3. ε_x, ε_y in any directions in plane perpendicular to z.
	Rhombohedral (trigonal)	$z\|C_3$.
Biaxial	Orthorhombic	$x,y,z \|$ to the three axes in D_2 or D_{2h}. $z\|C_2$ and $x,y \perp \sigma_v$ in C_{2v}.
	Monoclinic	$z\|C_2$ or $\perp \sigma$. No symmetry restriction on x,y, which also show dispersion.
	Triclinic	No symmetry restrictions. All three axes may show dispersion.

†After Nye[15], and Decius and Hexter[7].

9.3.6 Spectroscopy at high pressures

Vibrational spectroscopy at high pressures is important both for study of phase relations, and for obtaining data basic to theories of lattice dynamics. In the harmonic approximation phonons have neither pressure nor temperature dependence. Accordingly, the variation of lattice vibration frequencies with P and T is a valuable source of information on anharmonic parameters.

Cochran[1] has given an excellent introduction to these topics; further food for thought is offered by Reissland.[4] The P and T dependencies are related by

$$\left(\frac{\partial v}{\partial T}\right)_P = \left(\frac{\partial v}{\partial T}\right)_V - \frac{a_T}{\chi_T}\left(\frac{\partial v}{\partial P}\right)_T \qquad (9.47)$$

where a_T is the volume thermal expansion coefficient, and χ_T the isothermal compressibility. The second term represents the 'thermal strain' contribution to the phonon shift with temperature at constant pressure, while the first term is the 'self-energy' shift. It arises from changes in phonon interactions with T which occur even at constant volume, and is the only term in the equation not readily susceptible to experimental determination. (The γ_j term of (9.41b) is the imaginary part of the phonon self-energy.)

Many solids undergo phase-transitions at elevated pressures, see, for example, Pistorius.[16] The transitions are commonly accompanied by changes in the numbers and intensities of the vibrational modes, or at least are associated with changes of slope in $\tilde{v}/P$ plots. Figure 9.14 illustrates the spectroscopic consequences of three successive pressure-induced phase changes in $Hg(CN)_2$. These spectra were obtained using a diamond anvil high pressure cell, the commonest and most versatile device for studying spectra and other properties to very high pressures. High-pressure spectroscopic equipment and results have been reviewed comprehensively.[17,18]

9.4 Group theory and analysis of the vibrational spectra of solids

In order to analyse the vibrational spectra of crystalline solids, only the most superficial knowledge of space group theory is required if attention is restricted to $k=0$ and to first-order spectra, as is the case in this chapter. Procedures for handling $k \neq 0$ calculations are not nearly as well-documented as those for zone-centre work. Those intending to work in that area are advised to begin with Burns and Glazer[19] and then proceed to Decius and Hexter (p.123).[7] Note also an interesting pictorial approach due to Pawley.[20]

Because the dimensions of a unit cell are invariably small compared with the wavelengths of the radiation used in observing vibrational spectra, all unit cells can be considered as experiencing the same electric field. Consequently, analysis is based upon one *primitive* unit cell which, by definition, has all the distinctive features of the structure. There are two particularly important situations in which this approximation is invalid: (i) when the linear dimensions of a solid particle approach those of the probing wavelengths (see Section 9.3.2), and (ii) for l.o.–t.o. splitting, which is not recognized in group theoretical analyses. The reason is indicated in Fig. 9.15, where transverse and longitudinal phonons are shown propagating in direction q in an infinite diatomic cubic lattice. The oscillating transverse electric field of an electromagnetic wave with $k=0$ can couple to the t.o.

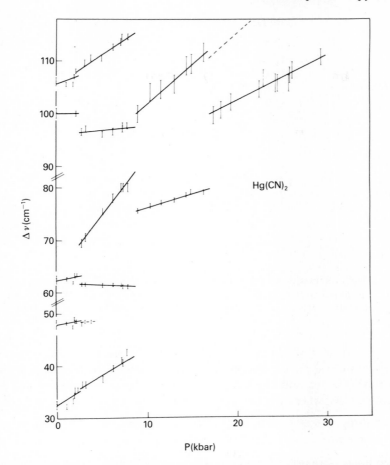

Fig. 9.14. Variation of the Raman-active lattice modes with pressure for Hg(CN)$_2$. The breaks indicate phase changes.

mode of *all* the unit cells. In contrast, when the incident electric vector is parallel to the oscillating dipole of the l.o. mode, it can only couple with it for *one* unit cell. Hence, only the two t.o. modes can be infrared-active. Nevertheless, the group theoretical analysis can make no such distinction as it is based upon the assumption that *all* unit cells experience identical electric fields, whatever the direction of observation.

Crystallographers generally describe structures in terms of the cell of highest possible symmetry which may be two, three, or four times the volume of the primitive cell. For example, zinc blende adopts a face-centred (F) cubic structure of symmetry $F\bar{4}3m$ with $z=4$. The primitive cell therefore contains only *one* formula unit. In contrast, $p-C_6H_4Cl_2$ has symmetry $P2_1/a$, $z=2$.

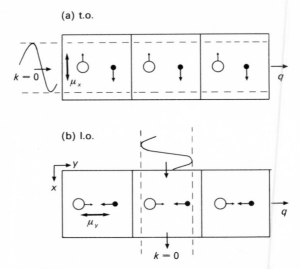

Fig. 9.15. (a) Interaction of a transverse optic phonon travelling in direction q with $q=0$, with the transverse electric field of an electromagnetic wave with $k=0$, and the amplitude represented by dashed lines. (b) The corresponding situation for the l.o. phonon.[21]

Since the label P shows that the cell is already primitive, the vibrational analysis must be carried out on the basis of the bimolecular cell ($z=z'=2$). For reference, we note the following relations between primitive and non-primitive cells, where z is the number of formula units in the larger cell:

Centering:	P	A, B, or C	I	R	F
z'(primitive):	z	$z/2$	$z/2$	$z/3$	$z/4$

There is no objection to working out a representation for a non-primitive cell, but the result must be divided by the appropriate factor as above. The method of analysis now described avoids this problem.

9.4.1 The factor group

In the Seitz notation a symmetry operation is denoted by $\{R|t\}$ where R indicates the rotational part of the operation (proper or improper), and t is the translational part. The identity operation for R is E, and that for t is O. Thus $\{E|t\}$ represents a pure translation, and $\{R|O\}$ a rotational operation without associated translation (i.e. a typical point group operation). $\{R|t\}$ corresponds to a coordinate transformation $x \to x' = Rx + t$, where

$$R = \begin{pmatrix} \pm 1 & 0 & 0 \\ 0 & \cos\theta & -\sin\theta \\ 0 & \sin\theta & \cos\theta \end{pmatrix}$$

The symmetry of a lattice is described by operations which, in general, have both point and translational symmetry; the lattice is invariant under operations of both kinds. A space group ($\mathscr{G}$) contains an infinite set of translational operations. However, since crystal vibrations are to be analysed on the basis of a single (primitive) cell, the translations between unit cells may be treated as identity operations. Thus, an operation $\{R|t\}$ is regarded as leaving a point invariant if it shifts that point into an adjacent cell, such that the initial and final positions are separated by a simple translation of one unit cell length. Other operations $\{R|t/n\}$ which describe screw axes or glide planes, are taken as equivalent to $\{R|O\}$. The new group ($\mathscr{F}$) is termed the *factor group* (or unit cell group) of the space group. Although ($\mathscr{G}$) contains all the operations of $\mathscr{F}$, the factor group is not a subgroup of $\mathscr{G}$. The factor group is always isomorphous with one of the thirty-two crystallographic point groups, although it should be noted that factor groups differ from point groups in several important respects in addition to those already indicated. 'Isomorphous' means that whilst the operations of the factor group and its associated point group are not identical, the operations of the factor group combine with each other in precisely the same way as do the rotations of the point group. The entries in their character tables are the same.

Space groups are described in two notations: Hermann–Mauguin (H–M), and Schoenflies. For example, these notations are respectively $P\,4/mnm$ and D_{4h}^{14} for the space group no. 136. The factor group is simply the Schoenflies label relieved of its superscript, D_{4h}. Space groups are given in both notations in *International tables for X-ray crystallography* (ITXR),[22] and in many other references.

9.4.2 Wyckoff sites and the site group

In all space groups there are sets of 'general' points or Wyckoff sites. Most space groups also have sets of 'special' sites; there is often more than one such set. For each set the environment of every member is identical. The symmetry operations associated with any one of these points defines the 'site group'. Each site group is necessarily a subgroup of the factor group. All the Wyckoff sites, and their symmetry labels, are listed for each space group in ITXR. For example, space group no. 14, $P2_1/c = C_{2h}^5$ has

$$\begin{array}{ll} 2(a-d) & C_i \\ 4\ e & C_1 \end{array}$$

where the H–M labels of ITXR have been replaced by their Schoenflies equivalents. Thus, there are two a sites of symmetry C_i, and also, independently, two b sites of this symmetry, etc. Finally, there is a set, e, of four 'general' sites. The coordinates of all these sites are listed in ITXR.

9.4.3 Factor group analysis

We now illustrate the procedures by which the $k=0$ vibrational modes of a crystalline solid may be analysed. In crystallographic papers a structure is described, *inter alia*, in terms of its space group and the atomic coordinates of all the constituent atoms. The latter are often given in terms of Wyckoff site occupancy. If the paper lists only the atom coordinates, these must be translated into Wyckoff sites by reference to ITXR.

Reduced representations for all sets of Wyckoff sites have been tabulated.[23] (This is the most compact and easily used compilation of such data published. It is out of print, but copies may be had at cost from the authors.) Some of these data have been published recently, in a different format, by Rousseau, Bauman, and Porto.[24] The following calculations are based upon use of either set of tables. Procedures for factor group analysis (FGA) without benefit of tables of reduced representations are described in many places, most conveniently by Decius and Hexter.[7]

An alternative approach to factor group analysis uses an *ascent-in-symmetry* method.[25,26] The three components of a vector on each occupied site are labelled according to the symmetry species of the site group. Using correlation tables, the group–supergroup relations are used to couple the motions of individual sets of sites, and to relabel them with the species of the factor group. This is a rapid and easily applied method.

9.4.4 L.o.–t.o. splitting and selection rules for Raman spectra

In Section 9.2.8 it was noted that l.o.–t.o. splitting occurs for polar vibrational modes. The l.o. modes do not usually appear in infrared spectra, although they determine the widths of reflectance bands. In contrast, l.o. modes may be observed directly in Raman spectra. Consider the case of a crystal with factor group D_{3h}. Raman activity is associated with the species A'_1, E', and E'' but E' also carries infrared-activity. Therefore l.o.–t.o. splitting will occur in E' spectra but *not* in those of A'_1 and E'' origin. To investigate the precise conditions under which the splitting can be observed, it is necessary to use the scattering tensor matrices. These are given by Hayes and Loudon,[4] and in a compact form by Decius and Hexter[7] (p. 312). Thus:

$$\begin{vmatrix} 0 & d & \\ d & 0 & \\ & & 0 \end{vmatrix} \quad \text{and} \quad \begin{vmatrix} d & & \\ & -d & \\ & & 0 \end{vmatrix}$$
$$E(x) \qquad\qquad\qquad E(y)$$

Consider the experiment $x(yx)y$, where x and y are in a plane perpendicular to the unique axis, $C_3(z)$. In this notation, due to Porto, the terms are:

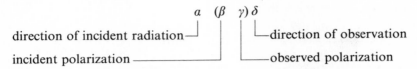

The polar mode excited in the experiment $x(yx)y$ has an electric moment parallel to x. Hence, it can couple with both longitudinal and transverse components of the phonon with wave-vector $\mathbf{q}$; see Fig. 9.16(a). Both E'(t.o.) and E'(l.o.) modes are, therefore, permitted to be Raman active. They will appear as two bands having the same frequencies as determined for the same modes by infrared reflectance spectroscopy.

In the scattering geometry $x(yx)z$, (Fig. 9.16(a)) both l.o. and t.o. components can, again, couple with $E(x)$. However, $\mathbf{q}$ now moves in a direction which is neither parallel nor perpendicular to the c-axis. Therefore it is neither pure A'_1 nor pure E' in type, but is a hybrid mode, E'(H). In practice, depending upon the nature of the bonding in the crystal, it may turn out to have more of one type than the other.

In identifying and assigning l.o. components it is most helpful to separate them from t.o. modes by an appropriate choice of scattering geometry. Sometimes this can be done with 90° scattering though details depend upon the factor group. More commonly, it is necessary to resort to back-scatter (or 180°) geometry. Again using a crystal of D_{3h} symmetry, and indicating back-scattering by a 'bar' over the coordinate in question, consider the experiment $y(xx)\bar{y}$, (Fig. 9.16(b)). The tensor component (xx) is associated with both A'_1 and E' species. No l.o.–t.o. splitting is possible for the A'_1 modes since they are not simultaneously infrared active. The component (xx) is associated specifically with $E(y)$. Thus, the local electric field oscillates with its vector parallel to y, thereby coupling to the l.o., but not the t.o. component. This scattering geometry therefore gives, uniquely, the E' (l.o.) modes, together with A'_1 modes. Note that use of 180° geometry does not restrict observations to l.o. components. Thus, for $z(xx)\bar{z}$ the electric field vector is still aligned parallel to y but this is now able to couple only to transverse components of $\mathbf{q}$, (Fig. 9.16(b)).

9.5 Worked examples

9.5.1 An atomic lattice: NiAs

This solid has the symmetry of space group no. 186, $P6mc = C_{6v}^4$, with $z = z' = 2$. The factor group is thus C_{6v}. There are two Ni atoms on sites a, and two As atoms on sites b. Sites a and b are both of C_{3v} symmetry.

(a) Cartesian method

As in point group calculations, the number of atoms unshifted by each

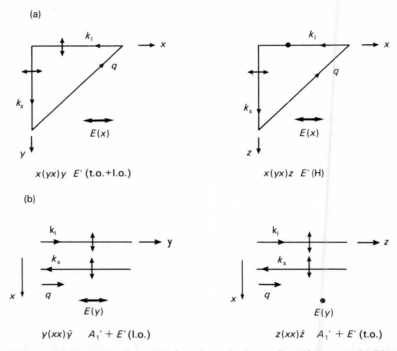

Fig. 9.16. (a) Phonon selection rules for E' species in D_{3h} for 90° scatter. (b) Phonon selection rules for E' species in D_{3h} for back scattering.

symmetry operation is determined and then multiplied by $(\pm 1 + 2\cos\theta)$. Consider the sites a. Since they are of C_{3v} symmetry the only operations of C_{6v} which leave them unshifted are: $E, 2C_3(z)$, and $3\sigma_v$. Thus:

	E	$2C_3$	$3\sigma_v$
Number of Ni atoms unshifted	2	2	2
$\times (\pm 1 + 2\cos\theta)$	3	0	1
$\chi'(R)$	6	0	2

Reduction of $\chi'(R)$ by the standard method, using the C_{6v} character table, yields

$$\Gamma(2a) = A_1 + B_1 + E_1 + E_2. \tag{9.48}$$

An identical relation applies for the $2b$ sites (As). Thus,

$$\Gamma(3N) = 2A_1 + 2B_1 + 2E_1 + 2E_2. \tag{9.49}$$

Note that the result contains $12 = 3N$ modes. Three of these modes are acoustic, transforming as x,y, and z; they are seen from the C_{6v} table to be $A_1(z)$ and $E_1(x, y)$. The remaining nine modes are from the optic branch:

$$\Gamma(\text{optic}) = A_1 + 2B_1 + E_1 + 2E_2. \quad (9.50)$$

Four bands are, therefore, predicted in the Raman spectrum (A_1, E_1, and $2E_2$), and two in the infrared spectrum (A_1 and E_1).

(b) Factor group analysis using tables

The method described above was used to generate the tables of Adams and Newton[23] (A&N), and of Rousseau, Bauman, and Porto[24] (RBP). Thus, with these tables, the factor group analysis for NiAs consists simply in adding the entries for the rows $2a$ and $2b$ to obtain ($3N$) as above.

Using the RBP method, first note from their Table 25A that sites a and b in C_{6v}^4 are labelled C_{3v}^d. Table 25B then shows that *each* set (a or b) yield a vector ($A_1 + B_1 + E_1 + E_2$) as before.

(c) Factor group analysis by ascent in symmetry

The three degrees of translational freedom possessed by an atom on a C_{3v} site transform as $A_1(z)$ and $E(x,y)$. In other words, the oscillations of the atom can be resolved into components parallel to z (i.e. the crystal c-axis), and two orthogonal (degenerate) components normal to z. In order to describe how the vibrations of the atoms on the two a sites are coupled to form modes labelled according to the irreducible representations of C_{6v} the correlation

$$\begin{array}{ccc} C_{3v} & \times 2 & C_{6v} \\ A_1 & \longrightarrow & A_1 + B_1 \\ E & & E_1 + E_2 \end{array}$$

is used which, of course, yields the same result as the other methods.

9.5.2 Lattice and internal modes of a molecular or complex-ionic crystal without factor group coupling

In the NiAs example, all the optic branch modes were vibrations of the entire unit cell. When a crystal is composed of molecules or complex ions, their internal modes will appear in the crystal spectrum at energies not dissimilar from those of the isolated species. Moreover, since the bonding between the polyatomic units is generally much weaker than that within them, there is a physically meaningful basis upon which to divide the optic branch modes into contributions which we can identify with internal and external modes respectively (see Section 9.2.11).

(a) K_2PtBr_4

This material is tetragonal, $P4/mmm = D_{4h}^1$ (space group no. 123), with

$z = z' = 1$. The factor group is D_{4h}. The atoms occupy the Wyckoff sites $1a$(Pt), $2e$(K), and $4j$(Br). Taking appropriate rows from A & N or RBP tables yields $\Gamma(3N) = 21$ as in Table 9.2.

The three acoustic translational modes $T_A = T(x,y,z)$ are then subtracted, leaving the vector of optic branch modes. This consists of contributions from: (1) the internal modes of $[PtBr_4]^{2-}$, Γ(vib.), (2) translatory lattice modes of K^+ and $[PtBr_4]^{2-}$, T, and (3) librational modes of the anion, R. In this case, the site $1a$ on which the anion is located has D_{4h} symmetry, as does the isolated anion. Hence, the translations of $[PtBr_4]^{2-}$ are simply the entries of row (i) in Table 9.2. To these must be added the translations of the two K^+ ions, which we identify as row (ii). Rows (i) and (ii) represent all the translations of the lattice, but three of them ($A_{2u} + E_u$) have already been identified as acoustic in type (T_A). Thus (i) + (ii) – $T_A = T$.

Table 9.2 Factor group analysis for K_2PtBr_4

D_{4h}	A_{1g}	A_{2g}	B_{1g}	B_{2g}	E_g	A_{1u}	A_{2u}	B_{1u}	B_{2u}	E_u	
Pt on $1a$							1		1	1	(i)
K on $2e$							1		1	2	(ii)
Br on $4j$	1	1	1	1	1		1	1		2	
$\Gamma(3N)$	1	1	1	1	1	0	3	1	1	5	$= 3N$
T_A							1			1	
Γ (optic)	1	1	1	1	1	0	2	1	1	4	

Table 9.3 Summary of factor group analysis for K_2PtBr_4

D_{4h}	N_T	T_A	T	R	Γ (vib)	Activities
A_{1g}	1				1	Raman ($xx+yy$, zz)
A_{2g}	1			1		
B_{1g}	1				1	Raman ($xx-yy$)
B_{2g}	1				1	Raman (xy)
E_g	1			1		Raman (xz, yz)
A_{2u}	3	1	1		1	Infrared (z)
B_{1u}	1				1	
B_{2u}	1		1			
E_u	5	1	2		2	Infrared (x, y)

The libratory modes R are simply $R(x,y,z)$ for D_{4h}, i.e. ($A_{2g} + E_g$) from the point group table. Γ(vib.) is then obtained from

$$\Gamma(3N) = \Gamma(\text{vib.}) + T + T_A + R \tag{9.51}$$

although it may also be obtained directly by a standard point group calculation. We note from Table 9.3 that no translatory modes are permitted in the Raman spectrum, and no libratory modes in the infrared spectrum. The results of a factor group analysis are usually summarized as in Table 9.3. Use of internal coordinates shows that the ν(Pt–Br) modes have symmetry A_{1g}, B_{1g}, and E_u. In-plane deformations δ(Br–Pt–Br) are B_{2g}, B_{1u}, and E_u, whilst the sole out-of-plane deformation, π(Br–Pt–Br) is of A_{2u} type.

(b) The single-crystal Raman spectrum of K_2PtBr_4

The electric moment induced in a crystal by an incident electric field E is given by $P = aE$, where $a =$ the polarizability tensor. Written in full

$$P_x = a_{xx}E_x + a_{xy}E_y + a_{xz}E_z \tag{9.52a}$$

$$P_y = a_{yx}E_x + a_{yy}E_y + a_{yz}E_z \tag{9.52b}$$

$$P_z = a_{zx}E_x + a_{zy}E_y + a_{zz}E_z. \tag{9.52c}$$

The subscripts on a_{ij} relate to the symmetry species of the modes associated with the term. Consider the experiment of Fig. 9.17. The tetragonal crystal of K_2PtBr_4 has been mounted such that its unique axis (c) is vertical. From Table 9.1 we see that the indicatrix has an axis parallel to c. Lasers emit plane-polarized light: we chose to have this plane parallel to z (i.e. E_z). In the normal Raman experiment light is collected at 90° to the laser input direction, and scattered radiation polarized parallel to x (i.e. P_x is selected using a Polaroid film). The experiment of Fig. 9.17(a) is thus $x(zx)y$. Equation (9.52a) reduces in this case to $P_x = a_{xz}E_z$, and the subscripts on a_{ij} indicate that the mode thus observed is of E_g symmetry. The factor group analysis shows that this mode is an anion libration. Further experiments $x(zz)y$ and $x(yx)y$ generate A_{1g} and B_{2g} spectra respectively. However, note that a_{xx} and a_{yy} are associated with both A_{1g} and B_{1g} species. Thus, experiments involving these tensor components always yield A_{1g} and B_{1g} modes together, although the A_{1g} set can be identified from the $x(zz)y$ result. Appropriate geometries for this are $z(xx)y$, in which the crystal has been rotated by 90° about the y-axis relative to the Fig. 9.17(a) arrangement; and $x(yy)\bar{x}$ in which back-scattering geometry has been used (Fig. 9.17(b)). This completes observation and assignment of the Raman-active modes.

(c) The single-crystal infrared spectrum of K_2PtBr_4

All six infrared-active modes (i.e. $2A_{2u} + 4E_u$) appear in mull or disc spectra of the microcrystalline material so the problem is to distinguish the two types. One approach would be to observe the transmission through a thin section of a crystal cut parallel to the c-axis. Light polarized parallel to $c(=z)$ would excite A_{2u} modes only. It is often simpler to study the infrared reflectance

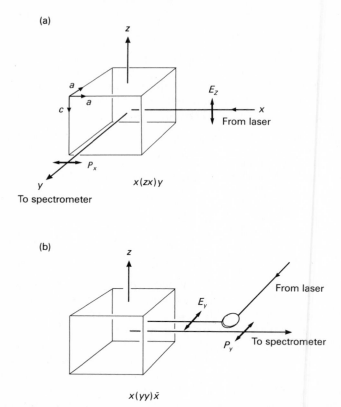

Fig. 9.17. Geometry for single-crystal Raman study showing experiments to observe (a) E_g and (b) $A_{1g} + B_{1g}$ species of tetragonal K_2PtBr_4.

from a crystal than to prepare very thin sections. Figure 9.18 shows the results from such an experiment, together with the optical constants derived from Kramers–Krönig analysis. In this material the $E_u\,\delta(\text{Br–Pt–Br})$ and $A_{2u}\,\pi(\text{Br–Pt–Br})$ modes were found to be in near coincidence at 129 and 128 cm^{-1} respectively,[27] a point which could not have been proved without recourse to single-crystal methods.

9.5.3 Factor group coupling: brucite

In K_2PtBr_4 there is only one complex ion per primitive cell, so we briefly examine the coupling which arises when more than one such component is present, looking at an extremely simple case.

Brucite, $Mg(OH)_2$, contains two hydroxyl ions per primitive (D_{3d}^3) unit cell, aligned with the crystallographic c-axis. The individual $\nu(\text{O–H})$ modes couple

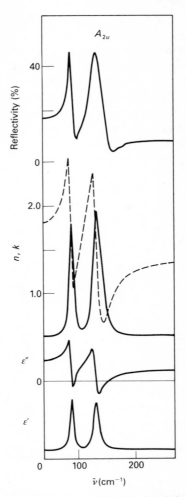

Fig. 9.18. Infrared reflectance spectrum of K_2PtBr_4 for light polarized parallel to the unique axis (A_{2u} species), and the derived dielectric parameters.

in-phase and out-of-phase to yield new vibrations which are a property of the unit cell, and are classified by means of its irreducible representations. Thus:

$$\begin{array}{cc} \leftarrow H-O \rightarrow & \leftarrow H-O \rightarrow \\ \leftarrow O-H \rightarrow & \text{and} \quad O \rightarrow \leftarrow H \\ A_{1g}(3653\ \text{cm}^{-1}) & A_{2u}(3698\ \text{cm}^{-1}) \end{array}$$

9.5.4 Line groups and polymers: orthorhombic lead oxide

The yellow form of PbO has an orthorhombic unit cell, $Pbcm = D_{2h}^{11}$, space group no. 57, with $z=4$. Chains run parallel to the b-axis and are stacked along c at $\frac{1}{4}$, $\frac{3}{4}$. There are two chains per unit cell. Since the distances between them suggest van der Waals bonding, we attempt a first-order interpretation of the spectra on the basis of one isolated chain, Fig. 9.19. The repeat unit of the chain is $(PbO)_2$. To obtain the internal modes of vibration, we subtract three translations from the $3n$ degrees of freedom but only *one* rotation, $R(z)$, since this is the only physically meaningful rotation for an infinite chain. Hence $3n-4=8$. (Note, however, that $R(z)$ would also be physically meaningless for a chain with all its atoms on the chain axis, and that, in such a case, the number of internal modes is $(3n-3)$.)

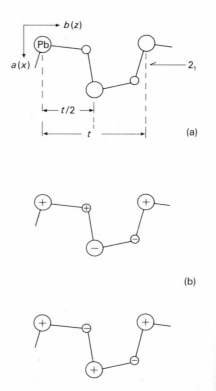

Fig. 9.19. (a) The chain in orthorhombic PbO (massicott). (b) Approximate forms of the A_2 and B_2 out-of-plane modes.

The symmetry elements of the chain are: $\{E|O\}$; $\{C_2^z|00\frac{1}{2}\}$, a two-fold screw axis along the chain direction, now relabelled z to accord with the standard form of the character table; $\{\sigma^{zx}|O\}$, the plane in which the chain lies; and $\{\sigma^{yz}|00\frac{1}{2}\}$, a glide plane. These elements define a group isomorphous with C_{2v}. Proceeding as in a point group calculation:

	E	$\{C_2^z\|00\tfrac{1}{2}\}$	$\{\sigma^{zx}\|O\}$	$\{\sigma^{yz}\|00\tfrac{1}{2}\}$
Number of atoms unshifted	4	0	4	0
$\{\pm 1 + 2\cos\theta\}$	3	-1	1	1
Γ(reducible)	12	0	4	0

Note that when counting the numbers of unshifted atoms, if an operation shifts an atom into a translationally equivalent one in another repeat unit, the atom is counted as unshifted. An equivalent statement applies to internal coordinates. Reduction of the above representation gives:

$$\Gamma(3n-4) = 4A_1 + 2A_2 + 4B_1 + 2B_2. \tag{9.53}$$

Subtracting $T(x,y,z)$ and $R(z)$ (this information is taken from the right-hand side of the C_{2v} character table):

$$\Gamma(3n-4) = 3A_1 + A_2 + 3B_1 + B_2. \tag{9.54}$$

The character for the A_2 and B_2 species under σ^{zx} is -1, indicating that these two modes are out-of-plane deformations. Their forms are easily written down by inspection (Fig. 9.19). The $3A_1 + 3B_1$ modes are all complex deformations within the plane of the chain, although the higher ones undoubtedly contain a high ν(Pb–O) contribution.

The lattice modes of orthorhombic lead oxide.

The chains lie in $ab = xy$ planes at $\tfrac{1}{4}, \tfrac{3}{4}$. The chain axis is y and the 'centre of gravity' of each chain lies on a line $(0,y,0)$. All the symmetry elements of the chain are (in this case) *also* elements of the unit cell space group. Hence, the centre of gravity is located on a site of C_{2v} symmetry. There must be two such sites per primitive cell, since there are two chains per cell. There are no C_{2v} sites in space group no. 57. However, all the space groups 47 to 74 inclusive have the same factor group, D_{2h}. Accordingly, we look in them for C_{2v} sites of the correct orientation. A convenient set occurs in $Pmmm = D_{2h}^1$, no. 47, as $2m:(0,y,0)$ and $(0,\bar{y},0)$. The translatory and rotatory motions associated with these sites yield the desired chain lattice modes. Using A & N tables, row $2m$ of no. 47 (D_{2h}^1) gives:

$$\Gamma(T+T_A) = A_g + B_{1g} + B_{3g} + B_{1u} + B_{2u} + B_{3u}. \tag{9.55}$$

From the standard D_{2h} character table, T_A is seen to span $B_{1u} + B_{2u} + B_{3u}$. Hence, the three g-modes are the desired optic branch translatory modes. The same result follows from RBP tables 8A and 8B using sites C_{2v}^y. However, the RBP tables do not permit determination of the librational modes in this instance because the librational degrees of freedom in their C-tables are not given in terms of cartesian components, as is the case in A & N. From the latter we have: Γ(librational) $= B_{2g} + A_u$. The same result follows from ascent-in-symmetry, from $R(y)$ on sites $2m$.

Note that *the method used* here to deduce the lattice modes of a polymeric chain *is a general one*, applicable both to chain and sheet (layer) structures. Thus the site(s) of the 'centre of gravity' is defined by the elements of symmetry of the chain or sheet, which are *also* elements of symmetry of the space group of the material.

9.5.5 L.o.–t.o. splitting in Raman spectra: benitoite

As a final example, we note the case of benitoite, $BaTiSi_3O_9$. This ring silicate has symmetry $P\bar{6}c2 = D_{3h}^2$, no. 188, with $z = 2$. The factor group is non-centric and hence allows the possibility of l.o.–t.o. splitting affecting the Raman spectrum. The E' species carries both infrared and Raman activity; it is this species which may show Raman bands additional to the predictions of factor group analysis. The selection rules applicable are those of Section 9.4.4. A full analysis of the selection rules, and of the single-crystal Raman and infrared-reflectance data, has been given.[28]

9.6 References

1. Cochran, W. *The dynamics of atoms in crystals.* Arnold, London (1973).
2. Woods, A. D. B., Brockhouse, B. N., Cowley, R. A., and Cochran, W. *Phys. Rev.* **131**, 1025 (1963).
3. Born, M. and Huang, K. *Dynamical theory of crystal lattices.* Clarendon Press, Oxford (1954).
4. Reissland, J. A. *The physics of phonons.* Wiley, London (1973).
5. Hayes, W. and Loudon, R. *Scattering of light by crystals.* Wiley, Chichester (1978).
6. Haas, C. and Hornig, D. F. *J. chem. Phys.* **26**, 707 (1957).
7. Decius, J. C. and Hexter, R. M. *Molecular vibrations in crystals.* McGraw-Hill, New York (1977).
8. Mitra, S. S. in *Optical properties of solids* (eds S. Nudelman and S. S. Mitra), Plenum Press, New York (1969).
9. Miller, R. G. J. and Stace, B. C. *Laboratory methods in infrared spectroscopy* (2nd edn). Heyden, London (1972).
10. Serna, C. J., Rendon, J. L., and Iglesias, J. E. *Spectrochim. Acta* **38A**, 797 (1982).
11. Owen, J. F., Barber, P. W., and Chang, R. K. *Microbeam Anal.* **17**, 255 (1982).
12. Berreman, D. W. and Unterwald, F. C. *Phys. Rev.* **174**, 791 (1968).
13. Servoin, J. L. and Gervais, F. *Appl. Opt.* **16**, 2952 (1977).
14. Sceats, M. G. and Morris, G. C. *Phys. Stat. Solidi (a)* **14**, 643 (1972).
15. Nye, J. F. *Physical properties of crystals.* Clarendon Press, Oxford (1957).
16. Pistorious, C. W. F. T. *Progr. in solid-state Chem.* **11**, 1 (1976).
17. Ferraro, J. R. *Coord. Chem. Rev.* **29**, 1 (1979).
18. Sherman, W. F. and Wilkinson, G. R. in *Advances in infrared and Raman spectroscopy* (eds R. J. H. Clark and R. E. Hester). Heyden, London (1980).
19. Burns, G. and Glazer, A. M. *Space groups for solid state scientists.* Academic Press, New York (1978).
20. Pawley, G. S. *Acta Cryst.* **A30**, 585 (1974).

21. Sherwood, P. M. A. *Vibrational spectroscopy of solids.* Cambridge University Press (1972).
22. *International Tables for X-ray crystallography* Vol. 1. Kynoch Press, Birmingham (1969).
23. Adams, D. M. and Newton, D. C. *Tables for factor group and point group analysis.* (Available from the authors.) (1970).
24. Rousseau, D. L., Bauman, R. P., and Porto, S. P. S. *J. Raman Spectrosc.* **10**, 253 (1981).
25. Boyle, L. L. *Acta Cryst.* **A28**, 172 (1972).
26. Boyle, L. L. *Spectrochim. Acta* **28A**, 1347 and 1355 (1972).
27. Adams, D. M. and Hills, D. J. *J. C. S. Dalton Trans.* 947 (1977).
28. Adams, D. M. and Gardner, I. R. *J. C. S. Dalton Trans.* 315 (1976).

10 Thermodynamic aspects of inorganic solid-state chemistry

A. Navrotsky[†]

10.1 Introduction

Classically, thermodynamics is a macroscopic science; its equations relate the observable bulk properties of a material (volume, enthalpy, heat capacity, free energy, etc.) to imposed external conditions (pressure, temperature, bulk compositions, etc.). Thus one can perform thermodynamic measurements, tabulate the data, and use the results to calculate the properties of a system under various conditions—all with minimal regard for details of structure and bonding in the phases involved. However, the thermodynamic properties of a material are related to interatomic interactions through the partition functions of statistical mechanics; so that, in principle, were interatomic potentials known accurately, and were computational procedures suitably sophisticated, macroscopic thermodynamic parameters could be calculated from first principles. The difficulty in practice arises both from the lack of sufficiently accurate potentials for complex systems, and from the complexity and cost of dealing with many-body-interactions in the partition functions. These difficulties are amplified in the solid state, where atoms are close together, interactions may be highly directional, and the unit cell may be large and of low symmetry. Thus the relation of thermodynamics to structure and bonding in solids must be made using a number of approaches—some relatively rigorous, and some largely empirical.

The purpose of this chapter is to describe some of the special features of the application of thermodynamics to problems in solid-state chemistry. Three inter-related aspects will be stressed. The first is experimental technique: how does one obtain thermodynamic data for crystalline materials which typically require high temperatures and long reaction times to reach equilibrium? The second is the question of level of sophistication: how does one go from the simple, 'ideal' approximations of textbook thermodynamics to equations complex enough to describe real solids, while retaining some

---

[†]Much of the work of A. Navrotsky's group referred to in this chapter was supported by the US National Science Foundation (Grants DMR 7810083 and 8106027).

theoretical rigour? The third point is the relation of thermodynamics to structure and bonding: can one predict the systematic variation of thermodynamic properties with chemical composition by considering the atomistic reasons for the stability of a given structure? Some specific examples, taken mainly from oxide and silicate chemistry, will be given to illustrate these points. References will be given to key papers and review articles, rather than to more detailed single studies, whenever possible. The specific examples will, to a large extent, relate to one group of materials—silicates, M_2SiO_4, with the olivine structure—because these are complex enough to show a variety of phenomena typical of complicated solids, yet sufficiently well-characterized to provide information on structure, thermodynamics, properties, and phase equilibria.

10.2 Heat capacities, entropies, and lattice vibrations

10.2.1 Basic relations and magnitudes

The heat capacity of a substance may be defined under conditions of constant pressure or constant volume:

$$C_V = (\partial E/\partial T)_V \tag{10.1}$$

$$C_P = (\partial H/\partial T)_P \tag{10.2}$$

where C_V and C_P are the heat capacities, E and H are internal energy and enthalpy, T is temperature, P is pressure, and V is volume. Because work must be done to expand a material during heating at constant pressure, C_P is larger than C_V, and

$$C_P - C_V = TV\alpha^2/\beta \tag{10.3}$$

where α and β are thermal expansion and compressibility respectively. Thus

$$\alpha = V^{-1}(\partial V/\partial T)_P \tag{10.4}$$

$$\beta = -V^{-1}(\partial V/\partial P)_T. \tag{10.5}$$

The reciprocal of the compressibility is the bulk modulus K used in the physics and geophysics literature.

For an ideal gas, $(C_P - C_V)$ is equal to the gas constant R ($= 8.314 \, \text{JK}^{-1}$), α is $1/T$ ($= 3.3 \times 10^{-3} \, \text{K}^{-1}$ at room temperature), and β is $1/P$ ($= 1 \, \text{bar}^{-1}$ at atmospheric pressure). For typical solids (see also Table 10.6), α is of the order of $10^{-5} \, \text{K}^{-1}$, β is of the order $10^{-6} \, \text{bar}^{-1}$, and $(C_P - C_V)$ becomes appreciable only at quite high temperatures ($> 1500 \, \text{K}$).

For a crystalline material, the thermal expansion and compressibility reflect the response of bond lengths and bond angles to changes in T and P. Although single parameters, the volume expansion and volume compressibility, can be defined as above, much insight can be gained by looking at the response along individual crystallographic axes, or of individual bond lengths. Thus, for example, the Mg–O bond in silicates such as Mg_2SiO_4 (olivine) expands with increasing temperature, and contracts with increasing pressure to a much greater extent than does the Si–O bond. The response of individual coordination polyhedra to pressure and temperature, and its relation to bond character has been developed extensively by Finger and Hazen.[1]

At sufficiently high temperatures, when vibrations are fully excited, the heat capacity C_V approaches its classical value of $3nR$ per mole, where n is the number of atoms per formula unit (e.g. $n=7$ for Mg_2SiO_4), and the factor of 3 arises because there are three independent vibrational degrees of freedom per atom. As an approximation to C_P at high temperature (since $(C_P - C_V)$ is small), this generalization is known as the 'law of Dulong and Petit'. At low temperature, as $T \to 0$ K, both C_P and C_V approach zero. For non-metallic materials at sufficiently low T, $C_V = aT^3$ (a is a constant), while for metals, $C_V = aT^3 + \gamma T$. The term linear in T, arising from electronic motion, predominates at the lowest temperatures.

10.2.2 Experimental techniques

To obtain the standard entropy (S^0_{298}) of a phase, its heat capacity must be measured down to temperatures well within the region where the T^3 law holds, and below any likely magnetic or other ordering phenomena. Adiabatic calorimeters to measure down to liquid-helium temperatures, and up to 350–400 K, exist in a number of laboratories. A relatively large amount of sample, 1–10 g, is required, especially for the lower temperature range. Because of this, and because measurements are painstaking and time-consuming, calorimetric entropies are not routinely available for many substances, especially for exotic and interesting materials that can only be synthesized in small amounts. In the range 200–1000 K, differential scanning calorimetry (DSC) has made great headway in obtaining heat capacities with an accuracy of about 1 per cent on samples as small as a few milligrams. Above 1000 K, heat contents are generally measured by drop calorimetry, and the heat capacities, with an accuracy of about ±5–10 per cent are obtained by taking the derivative $(\partial H / \partial T)_P$. These methods are summarized in Table 10.1.

10.2.3 Some examples

(a) Low-temperature heat capacities of silicates with the olivine structure

Figure 10.1 shows measured heat capacities of olivine silicates M_2SiO_4 (M = Mg, Mn, Fe, Co). Apart from the 'spikes' seen in the transition-metal silicates (discussed below), the heat capacity at a given temperature is smallest for Mg_2SiO_4 but fairly similar for the transition-metal silicates. Within a given structure type, lattice vibrations are usually most easily

Table 10.1 *Means of measuring heat capacities and heat contents*

Method	Basic measurement	Sample required	Approximate temperature range	Ref.
Adiabatic low-temperature calorimetry	Temperature change for known heat input under adiabatic conditions	1–10 g	5 – ~500 k	(2,3,4)
Differential scanning calorimetry	Temperature differences or power difference between sample and standard during controlled heating conditions	0.5–100 mg	200–1000 K	(5)
Drop calorimetry	Heat released when hot sample is dropped into calorimeter at room temperature	1–10 g	300–2500 K	(6,7)
Transposed temperature-drop calorimetry	Heat absorbed when sample is dropped from room temperature into calorimeter at high temperature	50–500 mg	300–1200 K	(8,9)

excited (C_P is largest, the Debye temperature (see below) is lowest) for the material with the largest unit cell, and with the longest and weakest M^{2+}–O bonds. This same trend is generally seen in S^0_{298}; see Table 10.2. The spikes in the heat capacity are due to magnetic transitions. At high temperature, the spins are generally completely disordered, so that there is an added entropy of $2R\ln N$, where N is the number of electrons involved. However this full entropy is not achieved until well above the actual transition temperature (Néel point) as the inset in Fig. 10.1 shows. This implies that a significant amount of residual short-range order initially exists in the high-temperature state, disappearing rather gradually with increasing temperature. Thus heat capacity measurements, especially when combined with structural and magnetic studies, are very useful in elucidating the magnetic ordering of solids.

(b) Heat content and ionic conductivity in PbF_2

Lead difluoride, which adopts the fluorite structure, undergoes a gradual transition (higher than first-order) near 700 K to a state of high ionic conductivity. Figure 10.2 shows $(H_T - H_{298})$, measured by drop calorimetry[10] as a function of temperature. Up to about 600 K the curve is linear with a slope (heat capacity at constant pressure) $C_P = 82 \pm 2\,\text{JK}^{-1}\,\text{mol}^{-1}$. Between about 625 and 750 K the curve is sigmoid with a maximum slope of

Table 10.2 Thermal properties of some silicate olivines

	V^0_{298} (ml/mol)	S^0_{298} (JK^{-1}mol^{-1})	$S^0_{298, \text{lattice}}$ (JK^{-1}mol^{-1})	$C_{p,298}$ (JK^{-1}mol^{-1})	S^0_{1000} (JK^{-1}mol^{-1})	$C_{P,1000}$ (JK^{-1}mol^{-1})	$H^0_{1000} - H^0_{298}$ (kJ^{-1}mol^{-1})
Mg$_2$SiO$_4$	43.79	94.11†	94.11	118.60	276.22	175.14	109.28
Mn$_2$SiO$_4$	48.60	155.90	129.14	128.50	345.76	177.33	113.40
Fe$_2$SiO$_4$	46.39	151.00	121.20	132.90	349.23	188.09	118.43
Co$_2$SiO$_4$	40.61	142.6	131.0	133.4			
Ca$_2$SiO$_4$	59.11	120.5	120.5	126.8	308.0	183.2	112.5

† Data from refs 58 and 59.

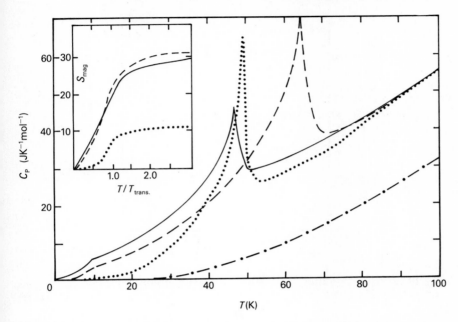

Fig. 10.1. Low-temperature heat capacities of olivines. Solid curve represents Fe_2SiO_4, dotted curve Co_2SiO_4, dashed curve Mn_2SiO_4, dot-dashed curve Mg_2SiO_4. Inset shows total entropy gained in magnetic transition as a function of $T/T_{trans.}$. (Data from refs 58, 59.)

approximately 290 JK^{-1} mol^{-1} at 705 K. Above 750 K the slope is again constant with $C_P = 92 \pm 2$ JK^{-1} mol^{-1}. The evidence is clear and unambiguous for a transition involving a large total enthalpy and entropy increment spread out over a range of more than 100 K. An estimate of the entropy increase associated with the transition can be obtained by plotting C_P against $\ln T$, and measuring the area under the peak associated with the transition. By this method the entropy increase associated with the transition is found to be 16.5 JK^{-1} mol^{-1}.

This entropy change should be compared to the entropy of fusion of PbF_2 which is 16.4 JK^{-1} mol^{-1}, an anomalously low value. Note that the sum of the entropies of transition and fusion is close to the normal entropies of fusion for MX_2 halides (cf. 31.5 JK^{-1} mol^{-1} for $PbCl_2$), suggesting that part of the enthalpy of fusion is absorbed when the anion sub-lattice 'melts' during the transition to ionic conductivity. The ionic conductivity is also plotted in Fig. 10.2. It changes from a magnitude ($\gtrsim 1\Omega^{-1} m^{-1}$) typical of 'normal' salts (e.g. alkali halides) at the melting temperature to a magnitude (several hundred $\Omega^{-1} m^{-1}$) typical of ionic melts, including PbF_2, in just the temperature range of the thermal anomaly. From the magnitude of the

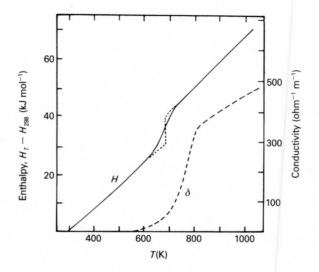

Fig. 10.2. High-temperature heat content (enthalpy) of PbF_2 (solid curve). The ionic conductivity δ is shown on the same temperature scale as a dashed curve. (From ref. 10.)

entropy change, and of the ionic conductivity, the most plausible interpretation is that the solid-state transition represents a gradual disordering, akin to fusion, of the anion sub-lattice of the fluorite structure.

(c) Zero-point entropy

The two examples above share the common feature that the disorder (electronic or ionic) occurs rapidly enough to be detected as an anomalously large absorption of heat in calorimetric measurements of heat capacity or enthalpy. For many materials synthesized in a disordered state at high temperature, the ordering process, though thermodynamically favoured at low temperature, is hindered from proceeding by a high activation energy for bond breaking. The material thus remains in a (generally metastable) disordered state, and retains the configurational entropy arising from this randomness even as $T \rightarrow 0$ K. Because no rearrangement occurs on heating or cooling, no excess energy is absorbed or released, and heat capacity measurements record only the vibrational excitation of the solid. Such a material is an exception to the third law of thermodynamics: it has a non-zero entropy at $T=0$, $S_0^0 > 0$. This can be detected by comparing the entropies of reaction calculated from heat capacity measurements alone (so-called third-law entropies) with those calculated from equilibria at high temperatures, where the disorder represents an equilibrium state (so-called second-law entropies). Table 10.3 lists a number of solids having zero point entropies, the magnitude of S_0^0, and its structural source.

Table 10.3 *Zero-point entropies of some solids*

Material	S_0^0 (JK^{-1}mol^{-1})	Reference	Source of disorder
CO	4.2	(11)	Orientaton of CO molecules
Na$_2$SO$_4\cdot$10H$_2$O	5.8	(11)	Proton disorder involving two positions
Zn$_2$TiO$_4$ (spinel)	4.0	(12)	Randomness of Zn and Ti on octahedral sites
NaAlSi$_3$O$_8$ (high albite)	~18.7	(13)	Al, Si disorder on tetrahedral sites

10.2.4 The interpretation of lattice heat capacities

A crystal containing n atoms per formula unit has $3n$ vibrational degrees of freedom. For a discussion of basic principles, see Kieffer.[14] If one assumes that all these vibrations are independent and occur at the same frequency, the *Einstein model* of lattice vibrations results. Although it correctly predicts the low- and high-temperature limits of C_V, this model does not properly describe the behaviour at low T, predicting C_V to be proportional to T^2 rather than to T^3; and at intermediate temperatures it does not allow a good fit to experimental data. The next level of complexity in approximating vibrational properties is the *Debye model*. Here one assumes that the density of vibrational states varies with the square of the vibrational frequency, with a maximum frequency or cutoff occurring when all degrees of freedom are accounted for. This maximum frequency is the Debye frequency v_D and can be related to a Debye temperature θ_D with h Planck's constant, and k Boltzmann's constant. The heat capacity at constant volume is given by

$$\theta_D = hv_D/k \qquad (10.6)$$

$$C_V = 9nN_A kT^3/\theta_D^3 \int_0^{\theta_D} T e^x x^4 (e^x - 1)^2 dx = 3nN_A \, kD(\theta/T) \qquad (10.7)$$

where $x = hv_D/kT = \theta_D/T$. Thus the thermodynamic properties are a function of the ratio of the Debye temperature to the actual temperature, and the integral $D(\theta/T)$, known as the Debye function, is available in tabulations. The proper high-temperature and low-temperature limits obtain, and at low T

$$C_V = \frac{12}{5} nR\pi^4 \left(\frac{T}{\theta}\right)^3. \qquad (10.8)$$

Values of θ_D, obtained from a best fit to calorimetric measurements, are listed in Table 10.4. It is seen that θ_D increases with the strength of the

Table 10.4 *Calorimetric Debye temperatures for some solids*

Solid	$\theta_D(K)$	Reference
Pb	90	(14)
C (diamond)	1910	(14)
KCl	220	(14)
NaCl	280	(14)
MgO	~773	(15)
MnO	~600	(15)
Mg$_2$SiO$_4$ (olivine)	~900	(15)
Fe$_2$SiO$_4$ (olivine)	~700	(15)

bonding in the crystal, and decreases with increasing molar volume for a given structure.

A test of how well the Debye model describes complex solids can be made by calculating the apparent Debye temperature from the heat capacity at various temperatures using eqn. (10.7). If the Debye approximation were completely adequate, θ_D would be constant. One concludes,[14] that the Debye approximation works fairly well for solids of high symmetry containing only one kind of atom (Pb, C), or atoms of similar masses and bonding characteristics (NaF, KCl), but becomes increasingly inadequate as the bond type, coordination, and mass of different atoms becomes quite different (rutile, silicates). This reflects the fact that vibrational frequencies are not uniformly distributed, and that specific covalently bonded groups (e.g. SiO$_4$) retain their vibrational characteristics in crystals. For a few materials, the vibrational density of states and its dispersion across the Brillouin zone are well characterized by infrared and Raman spectroscopy, and by inelastic neutron scattering; and a rigorous calculation of heat capacity can be made (at least in principle). For most materials, this is not the case. Nevertheless, there are important features, not predictable from a Debye model (especially in complex solids), which need explanation. For example, heat capacities (and entropies) may 'cross over'. Thus pyrope garnet (Mg$_3$Al$_2$Si$_3$O$_{12}$) has a higher heat capacity than grossular (Ca$_3$Al$_2$Si$_3$O$_{12}$) at low temperature, but the reverse is true above 400 K. This leads to an anomalously high entropy for pyrope, which has been linked to the low-frequency vibrations of Mg in the 8-coordinated site.[16] Similarly, a number of high-pressure phases (e.g. the perovskite form of CdTiO$_3$) have higher entropies than their low-pressure, less dense forms (e.g. ilmenite CdTiO$_3$).[16] Layer compounds, strongly bonded in only one or two dimensions, also often have anomalous heat capacities and entropies.

An approach to modelling heat capacities of complex materials has recently been developed by Kieffer.[14,17,18] A Debye-like approximation is retained for the low-frequency acoustic modes, but the optic modes are

Thermodynamic aspects of inorganic solid-state chemistry 371

described by two features: an optic continuum with a low- and high-frequency cutoff, and specific Einstein oscillators corresponding to well-defined vibrations, e.g. Si—O stretching. The approach is quite successful in modelling heat capacities and entropies of complex phases, although a single unique model for a given silicate cannot generally be distinguished; rather, several acceptable models give reasonable results.

From an empirical point of view, the high-temperature heat capacity C_P is often expressed by polynomials. To give the proper 'knee' in C_P near the Debye temperature, various equations have been proposed. A form commonly used is

$$C_P = a + bT + cT^{-2} + dT^{-\frac{1}{2}}. \tag{10.9}$$

As with all empirical polynomials, one must be careful when extrapolating outside the range of the original data.

10.3 Thermodynamics of solid-state reactions
10.3.1 General principles

Consider the general reaction

$$a\text{A} + b\text{B} \rightarrow c\text{C} + d\text{D} \tag{10.10}$$

where at least one reactant or product is a solid. For the reaction at atmospheric pressure and a given temperature

$$\Delta G^0 = \Delta H^0 - T\Delta S^0 = -RT\ln K \tag{10.11}$$

with

$$K = a_C^c a_D^d / a_A^a a_B^b. \tag{10.12}$$

The terms within parentheses are activities, related to chemical potentials μ by

$$RT\ln a_i = \mu_i(\text{given state}) - \mu_i(\text{reference state}). \tag{10.13}$$

At high pressure

$$\Delta G_{P,T} = \Delta G_T^0 + \int_{1\,\text{atm.}}^{P} \Delta V \mathrm{d}P \tag{10.14}$$

where ΔV is the volume change for reaction (10.10). If two solid phase assemblages are at equilibrium at a given P, T, then $\Delta G(P,T) = 0$, which leads to the Clausius–Clapeyron equation for the slope of a univariant phase boundary in P–T space

$$\left(\frac{\mathrm{d}P}{\mathrm{d}T}\right)_{\text{equil.}} = \frac{\Delta S(P,T)}{\Delta V(P,T)}. \tag{10.15}$$

If ΔS and ΔV can be approximated as constants independent of P and T, then dP/dT is constant, and the phase boundary is a straight line is P,T space.

Thus to characterize the thermodynamics of solid-state reactions, one needs to determine the following: (a) the enthalpy, entropy, and volume changes for reactions involving pure stoichiometric compounds, and (b) the thermodynamic activities in phases of variable composition, generally substitutional solid solutions (e.g. $(Mg_xFe_{1-x})_2SiO_4$) or compounds with a homogeneity range (variable cation to anion ratio, e.g. $Fe_{1-x}O$). In general these parameters will all be functions of temperature and pressure, though useful simplifying assumptions can often be made.

A further complication inherent to solids is the difficulty in attaining (and proving one has attained) equilibrium. Consequently, reactions between phases ideally should be 'reversed', phase boundaries approached from high and low T, high and low P, and reactant and product phase assemblages. One must also pay attention to possible equilibria within a single phase, e.g. cation distributions, non-stoichiometry, or defects—all of which may affect thermodynamic properties. In general such problems of equilibrium and metastability are more serious for the solid state than for liquids and gases, because of both lower diffusion rates and problems of nucleation. This is why solid-state thermodynamics often involves measurements at high temperature, especially when dealing with oxides.

10.3.2 Experimental approaches and examples

(a) Oxidation–reduction equilibria

For systems containing an easily reducible metal cation, equilibrium with the metal and an appropriate gas phase provides a useful means of measuring activities and free energies in a binary or ternary oxide (or chalcogenide) phase. Consider the equilibrium

$$NiO \rightleftharpoons Ni + \tfrac{1}{2}O_2 \qquad (10.16)$$

for which

$$\Delta G^0_{16} = -RT\ln K = -RT\ln(pO_2)^{\frac{1}{2}} a_{Ni}/a_{NiO}. \qquad (10.17)$$

For pure NiO in equilibrium with pure Ni, $a_{Ni} = a_{NiO} = 1$, and a knowledge of the equilibrium value of pO_2 at a given T gives ΔG^0 directly, while the temperature dependence of pO_2 can be used to derive values of ΔH^0 and ΔS^0. The value of pO_2 in equilibrium with both metal and metal oxide can be measured in two ways: by gas mixing experiments, or by a solid-state electrochemical cell. In the former, a value of pO_2 is set by mixing oxygen with an inert gas for $pO_2 = 1-10^{-4}$ atm., or, to attain lower values of pO_2, by using a subsidiary gas-phase equilibrium such as

$$H_2O \rightleftharpoons H_2 + \tfrac{1}{2}O_2, \qquad pO_2 = \{K_{18}(pH_2O)/(pH_2)^{-1/2}\} \qquad (10.18)$$

or

$$CO_2 \rightleftharpoons CO + \tfrac{1}{2}O_2, \quad pO_2 = \{K_{19}(pCO_2)/(pCO)^{-1/2}\}. \quad (10.19)$$

The ratio in which the major gases (CO and CO_2, or H_2O and H_2) are mixed then determines the partial pressure (fugacity) of oxygen. This ratio is varied until a value is found for which the solid sample equilibrated with the gas mixture contains both metal and oxide.[19]

The electrochemical method depends on the use of a cell using a solid electrolyte (typically a CaO-stabilized zirconia) which conducts by the movement of oxide ions.[20] For the Ni/NiO equilibrium, such a cell may be written as

$$Pt|Ni,NiO||ZrO_2(CaO)||air|Pt \quad (10.20)$$

The left-hand side of the cell has its pO_2 fixed by the Ni/NiO equilibrium, while the right-hand side has $pO_2 = 0.21$ atm. The half-cell reactions may be written as

$$\text{Left:} \quad O^= + Ni \rightarrow NiO + 2e^- \quad (10.21)$$

$$\text{Right:} \quad \tfrac{1}{2}O_2 + 2e^- \rightarrow O^= \quad (10.22)$$

and the cell reaction

$$Ni + \tfrac{1}{2}O_2 \rightarrow NiO \quad (10.23)$$

with

$$\Delta G^0 = -nFE^0 = -2FE^0 \quad (10.24)$$

where n is the number of electrons transferred, F is the Faraday constant, E^0 the measured cell voltage.

Both the gas equilibration method and the solid-cell method can be used to find the free energies of ternary compounds and activities in solid solutions.[19] Consider, as an example of the former, the reaction

$$2NiO \text{ (rocksalt)} + SiO_2 \text{ (quartz)} = Ni_2SiO_4 \text{ (olivine)}. \quad (10.25)$$

Its free energy can be measured by equilibration with a CO/CO_2 mixture, such that Ni metal, SiO_2 (quartz), and Ni_2SiO_4 (olivine) are present. The reaction is then

$$Ni_2SiO_4 + 2CO \rightleftharpoons 2Ni + SiO_2 + 2CO_2, \quad K_{26} = \left(\frac{pCO_2}{pCO}\right)^2 \quad (10.26)$$

while, for the reaction

$$NiO + 2CO \rightleftharpoons 2NI + 2CO_2, \quad K_{27} = (p^*CO_2/p^*CO)^2. \quad (10.27)$$

Therefore for reactions (10.25) above,

$$K_{25} = \left\{\frac{(p^*CO_2)(pCO)}{(p^*CO)(pCO_2)}\right\}^2 \quad (10.28)$$

and $\Delta G^0 = -RT\ln K_{25}$.

Electrochemically, the same reaction could be measured by the cell

$$\text{Pt}|\text{Ni}_2\text{SiO}_4,\text{Ni},\text{SiO}_2||\text{ZrO}_2(\text{CaO})||\text{Ni},\text{NiO}|\text{Pt} \qquad (10.29)$$

with half-cell reactions

$$\text{Left:} \quad 2\text{Ni} + 2\text{O}^= \rightleftarrows 2\text{NiO} + 2e^- \qquad (10.30)$$

$$\text{Right:} \quad \text{Ni}_2\text{SiO}_4 + 2e^- \rightleftarrows 2\text{Ni} + \text{SiO}_2 + 2\text{O}^= \qquad (10.31)$$

The cell reaction is therefore the reverse of reaction (10.25). The activity of NiO in an NiO–MgO solid solutions (MgO is not easily reduced) can be measured similarly. By gas equilibration, the appropriate reaction, forming just a trace of metal so the solid-solution composition does not change, would be

$$\text{NiO(in MgO)} + \text{CO} \rightarrow \text{Ni} + \text{CO}_2 \qquad (10.32)$$

$$K_{32} = (p\text{CO}_2)(p\text{CO})^{-1}(a_{\text{NiO}})^{-1} \qquad (10.33)$$

where the activity of nickel oxide in the solid solution is now less than unity. Combining this with reaction (10.27), one gets

$$a_{\text{NiO}} = \frac{(p^*\text{CO}_2)(p\text{CO})}{(p^*\text{CO})(p\text{CO}_2)}. \qquad (10.34)$$

Electrochemically, one can use the cell

$$\text{Pt}|\text{Ni},\text{NiO(in MgO)}||\text{ZrO}_2(\text{CaO})||\text{Ni},\text{NiO}|\text{Pt} \qquad (10.35)$$

with overall cell reaction

$$\text{NiO (pure)} \rightleftarrows \text{NiO (in solid solution in MgO)} \qquad (10.36)$$

with

$$\Delta G^0_{37} = -nFE^0 = RT\ln a_{\text{NiO}}. \qquad (10.37)$$

Values of free energies and chemical trends are discussed in Section 10.3.3.

(b) Vapour pressure measurements

If a constituent is relatively volatile, its vapour pressure can be used to probe its thermodynamic activity in the solid state. Measurement of vapour pressure fall into two classes: dynamic measurements in which one monitors weight loss or the amount of volatile species in a carrier gas stream as a function of time,[21] and effusion measurements in which equilibrium in an almost closed system is attained and the vapour is sampled through a pinhole.[22] The detection and measurement of vapour species by mass spectrometry, spectroscopic techniques, or chemical analysis determines the sensitivity and accuracy of vapour pressure measurements.

Such measurements have proved especially useful in semiconductors, where one or both species in a binary system are often volatile, while the solid phases can show considerable homogeneity ranges.

(c) High-pressure phase equilibria

Table 10.5 lists the characteristics of several types of high-pressure apparatus. In general, as P increases, the available sample volume decreases, and at pressures greater than about 100 kbar one can seldom produce more than a few milligrams of sample, especially at high temperature. The appropriate calibration of T and P when both are high is a subject of ongoing development and controversy. The ability to perform measurements (X-ray, conductivity, spectroscopy) at high P and T is improving and is crucial to understanding the actual state of materials under these extreme conditions.

Table 10.5 *High-pressure apparatus*

Apparatus	Ref.	Sample volume (approximate)	Pressure range (kbar)	Temperature range (approximate) (K)
Autoclaves	(23)	0.1 to 10 l	0–1	200–600
Externally heated cold-seal pressure vessels	(23)	10–500 cm^3	0–5	300–1200
Internally heated gas vessels	(23,24)	5–100 cm^3	0–15	300–1800
Piston cylinder solid-media apparatus	(23,25)	0.01–0.1 cm^3	0–40	300–2000
Multiple anvil solid-media apparatus	(23,26)	10^{-3}–10^{-2} cm^3	0–250	300–2000
Diamond anvil cell (laser-heated)	(27)	10^{-6}–10^{-3} cm^3	0–3000	200–2500
Shock apparatus (dynamic methods, pressure for very short time only)	(28)	variable	0–2000	300–3000

To properly constrain a phase boundary, it must be reversed, i.e. the low-pressure phase transformed into the high-pressure phase with increasing pressure, and the high-pressure phase transformed into the low-pressure one with decreasing pressure. Many if not most of the studies reported in the literature fail to meet these criteria, and represent synthesis diagrams rather than equilibrium phase diagrams. Such synthesis diagrams may reflect very considerable metastability, hysteresis, and slow reaction rates; and thermochemical data derived from them may be seriously in error.

As an example of the calculation of thermodynamic parameters from a phase boundary, consider the reactions

$$Ni_2SiO_4(\text{olivine}) \rightarrow Ni_2SiO_4 \text{ (spinel)}. \quad (10.38)$$

Ma[29] using carefully reversed experiments determined that the phase boundary fits the equation

$$P = 21.1 + 0.0118T \quad (10.39)$$

in the range 923–1473 K (P measured in kbar, T in K). From this, one can calculate thermodynamic parameters as follows. First, if one ignores the difference in heat capacity, thermal expansion, and compressibility between olivine and spinel, one can assume that the volume change ΔV for reaction (10.38) is a constant, with a value, obtained from lattice parameters under ambient conditions, of -3.417 ml mol^{-1}. At a point on the phase boundary, 973 K, 31.6 kbar, the free energy $\Delta G_{973\,K,\ 31.6\,kbar}$ is zero. The free-energy change at the temperature and atmospheric pressure, $\Delta G^0_{973\,K,\ 1\,bar}$, is given by

$$\Delta G^0 - \int_{1\,bar}^{31.6\,kbar} \Delta V dP \approx -\Delta V \times P_{\text{trans}}. \quad (10.40)$$

$$\Delta G^0 = -(-3.417) \times 31.6 = 107.75 \text{ ml kbar mol}^{-1} = 10\,920 \text{ J mol}^{-1}.$$

From the Clausius-Clapeyron equation (eqn 10.15), ΔS at the phase boundary is given by

$$\Delta S = \left(\frac{dP}{dT}\right) \Delta V = 0.118 \text{ K}^{-1} \text{kbar} \times -3.417 \text{ ml mol}^{-1}$$
$$= -0.0403 \text{ ml kbar K}^{-1} \text{mol}^{-1} = -4.08 \text{ JK}^{-1} \text{mol}^{-1}. \quad (10.41)$$

Neglecting $(\partial \Delta V / \partial P)_T$ and $(\partial \Delta V / \partial T)_p$ means that $\Delta S = \Delta S^0$ and

$$\Delta H^0 = \Delta G^0 + T\Delta S^0 = 10\,920 - 4.08 \times 973 = 6950 \text{ J mol}^{-1} \text{ (at 973 K)}. \quad (10.42)$$

However, the olivine and spinel phases do not have identical compressibilities and thermal expansions; the denser phase generally has lower values of these parameters (see Table 10.6). In general one must know the simultaneous effect of pressure and temperature on volume, but a simplification to the equation of state can be made by assuming that the effects of P and T can be separated. An approximate equation of state[30] for the volume of a phase, is

$$V = V^0_{298}\{1 + a(T - 298)\}\{KK^{-1}P + 1\}^{-1/K}. \quad (10.43)$$

The term in the first brackets arises from assuming a volume expansion. The term in the second brackets describes the compression according to the Murnaghan equation of state, which is generally adequate to pressures of 150–200 kbar. The integral $\int_{1\,atm.}^{P} V dP$ at constant temperature is then given by, for a given phase

Thermodynamic aspects of inorganic solid-state chemistry 377

Table 10.6 *Equation of state parameters used in calculating thermodynamic values for the olivine–spinel transition. (From ref. 30.)*

Phase	Molar volume V^0_{298} ml mol^{-1}	Thermal expansion coefficient a (K^{-1})	Bulk modulus K (kbar)	$\partial K/\partial P = K'$
Mg$_2$SiO$_4$ (olivine)	43.79	3.8×10^{-5}	1280	4 (or 5.2)
Mf$_2$SiO$_4$ (spinel)	39.65	2.6×10^{-5}	2130	4
Fe$_2$SiO$_4$ (olivine)	46.39	3.2×10^{-5}	1240	4
Fe$_2$SiO$_4$ (spinel)	42.04	2.3×10^{-5}	1970	4
Co$_2$SiO$_4$ (olivine)	44.53	3.7×10^{-5}	1660	4
Co$_2$SiO$_4$ (spinel)	40.62	2.6×10^{-5}	2060	4
Ni$_2$SiO$_4$ (olivine)	42.61	3.4×10^{-5}	1490	4
Ni$_2$SiO$_4$ (spinel)	39.19	2.6×10^{-5}	2230	4

$$\int_{1\,\text{atm.}}^{P} V\,dP = V^0_{298}[1 - a(T-298)][1 - K/K']^{-1}[K/K'][\{PK'/(K+1)\}^{1-(1/K')} - 1]. \tag{10.44}$$

Using eqn (10.43), for both olivine and spinel, with the data in Table 10.6 one gets $\Delta V = -3.461$ ml mol^{-1} at 973 K and 31.6 kbar. From the Clausius–Clapeyron equation, then, $\Delta S_{973\,\text{K},\,31.6\,\text{kbar}} = -4.14$ JK^{-1} mol^{-1}. The entropy change of the reaction at atmospheric pressure, since $(\partial S/\partial P)_T = (\partial V/\partial T)_P$ is then given by

$$\Delta S^0_{973} = \Delta S^P_{973} + \int_P^{1\,\text{atm.}} (\partial \Delta V/\partial T)_P dP \tag{10.45}$$

and

$$(\partial \Delta V/\partial T)_P = V^0_{298}(\text{sp})a(\text{sp}) - V^0_{298}(\text{ol})a(\text{ol}) \tag{10.46}$$

$$\int_{31.6}^{0.001} (\partial \Delta V/\partial T)_P dP = 31.6\,\text{kbar} \times \{(39.193 \times 2.6 \times 10^{-5}) - (43.610 \times 3.4 \times 10^{-5})\}$$
$$= 0.00322\,\text{ml kbar K}^{-1}\,\text{mol}^{-1}$$
$$= 0.326\,\text{JK}^{-1}\,\text{mol}^{-1}. \tag{10.47}$$

Then $\Delta S^0_{973} = -4.14 + 0.326 = -3.81$ JK^{-1} mol^{-1}. From eqn (10.44), ΔG^0_{973} is calculated to be 8517 J mol^{-1}, so ΔH^0 is $8517 \pm 973 \times -3.81) = 4810$ J mol^{-1}.

A calorimetric determination[30] of ΔH^0_{973} gives 5983 ± 3000 J mol^{-1}, which is in the general range of both calculations. Figure 10.3 shows the phase boundary calculated using the calorimetric data for the Ni$_2$SiO$_4$ olivine–spinel transition compared with the phase equilibria. The two agree within the experimental error of both sets of measurements. The values of ΔH and ΔS are discussed further in Section 10.3.3.

(d) Reaction calorimetry

Heat capacity measurements, discussed in Section 10.2 above, permit the characterization of the thermal properties of a single substance. Reaction

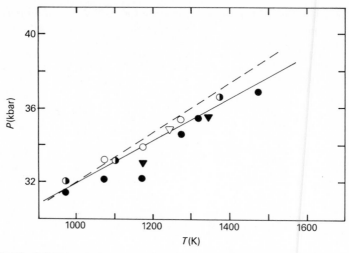

Fig. 10.3. The Ni_2SiO_4 olivine-spinel transition. Points are phase equilibrium data of Ma.[29] Solid symbols represent olivine present after run at P, T; open symbols represent spinel present. Circles represent olivine as starting material, triangles represent spinel (proof of reversed equilibrium). Solid line is best fit to experimental points, dashed line is calculated from thermochemical data in Tables 10.6 and 10.9.

calorimetry, which measures the heat of a chemical reaction, can be used to relate the thermodynamic properties of a given substance to those of other substances, and thus obtain heats of formation, of mixing, and of solid-state reactions. These methods can be classified into two groups: direct methods in which the reaction studied by calorimetry is the one of interest, and indirect methods in which a thermochemical cycle is used to obtain the enthalpy of interest. In the former category is combustion calorimetry, in which the enthalpy of formation of oxides is obtained by burning the metals in oxygen,[31] or the heat of formation of fluorides is measured by burning the metals in a fluorine atmosphere.[32] The absorption or release of oxygen by non-stoichiometric oxides can also be studied by this method,[33] as can the reaction of metals with hydrogen.[34] By combining such calorimetric measurements with free-energy measurements and a knowledge of the composition of the phase as a function of the partial pressure of oxygen or hydrogen, the partial molar entropy of solution of the gas in the non-stoichiometric oxide or hydride can be obtained with fairly high accuracy. This lends insight into solubility mechanisms and defect chemistry,[34] including the role of hysteresis and complex intergrowths of phases on a submicroscopic scale.[35]

The heats of mixing in molten salts and metal alloys can often be measured directly by calorimetry at high temperature,[36,37] as can the heats of formation of some solid alloys.[38] Viscous melts, glasses, and many crystalline solid solutions or compounds cannot be studied by direct reaction calorimetry because they react far too slowly. Instead, indirect methods, such as solution calorimetry, are used. The solvent in such measurements can be an aqueous solution (hydrofluoric acid for oxides and silicates),[4] a molten metal (liquid tin for alloys,[39] liquid copper or a nickel–manganese alloy for carbides and borides[40]), or a molten salt or oxide melt (lead borate, alkali borate, or alkali molybdate or tungstate for oxides and silicates).[41]

As an example, the heats of solution in molten $2PbO \cdot B_2O_3$ near 700°C of a number of oxides are given in Table 10.7. Using these values we can construct the following thermodynamic cycle:

$$SiO_2 \text{ (quartz)} \rightarrow \text{dilute solutin,} \quad \Delta H = -3.3 \text{ kJ} \quad (10.48)$$

$$Mg_2SiO_4 \text{ (olivine)} \rightarrow \text{dilute solution,} \quad \Delta H = 67.3 \text{ kJ} \quad (10.49)$$

$$\text{dilute solution} \rightarrow 2MgSiO_3 \text{ (ortho-enstatite),} \quad \Delta H = -70.0 \text{ kJ.} \quad (10.50)$$

Therefore for

$$Mg_2SiO_4 \text{ (olivine)} + SiO_2 \text{ (quartz)} \rightarrow 2MgSiO_3 \text{ (ortho-enstatite),}$$

$$\Delta H = -3.3 + 67.3 - 70.0 = -6.0 \text{ kJ} \quad (10.51)$$

with an error of $\{(0.2)^2 + (2 \times 0.8)^2 + (2.3)^2\}^{\frac{1}{2}}$ or ± 2.8 kJ. Similarly, for the phase transformation

$$Mg_2SiO_4 \text{ (olivine)} \rightarrow Mg_2SiO_4 \text{ (modified spinel),} \quad (10.52)$$
$$\Delta H = 67.3 - 37.2 = 30.1 \pm 2.9 \text{ kJ.}$$

The first calculation confirms the observation, known from the phase diagram, that $MgSiO_3$ (pyroxene) is stable relative to a mixture of Mg_2SiO_4 (olivine) and SiO_2 (quartz) at atmospheric pressure (since the ΔS of reaction (10.51) is expected to be small). The second calculation shows that the modified spinel structure is indeed metastable at atmospheric pressure, and supports the high-pressure data which give this phase a stability field above about 130 kbar.[42]

Table 10.7 *Enthalpies of solution in molten $2PbO \cdot B_2O_3$ near 973 K*

Oxide	$\Delta H_{\text{sol.}}$ (kJ mol^{-1})†
SiO_2 (quartz)	-3.3 ± 0.2
$MgSiO_3$ (ortho-enstatite)	35.0 ± 0.8
Mg_2SiO_4 (olivine)	67.3 ± 2.3
Mg_2SiO_4 (modified spinel)	37.2 ± 1.7

†Values taken from recent work in Navrotsky's laboratory and previous data in literature.

10.3.3 Systematics and trends

(a) Formation of binary compounds from elements

Table 10.8 shows data for the enthalpies and entropies of formation, of some oxides and chalcogenides from the elements. These data illustrate several general points.

(1) When compound formation involves the consumption of a gas (oxygen, halogens), a large negative entropy change (in the range -90 ± 20 JK^{-1} gram-atom^{-1}) is seen. When the reaction involves all condensed phases, ΔS^0 is small and can be either positive or negative.

(2) ΔH^0 in general parallels the lattice energy of the crystal, being most negative for the most ionic compound having the smallest lattice parameter. The ease of reduction of the compound to metal is generally determined by this value of ΔH^0 since, for a given anion, ΔS^0 of formation does not vary too much.

(3) Plots of ΔG^0 of formation vs. T are generally straight lines, whose slopes are determined by ΔS^0, and which show changes in slope when either reactants or products undergo a phase change. This is shown in Fig. 10.4, which can also be used to find the equilibrium partial pressure for reduction, in analogy to eqn (10.16).

In Table 10.8, the compound 'FeO' is listed as Fe$_{0.95}$O; in fact it has a homogeneity range which varies with temperature, total pressure, and oxygen partial pressure. The phenomenon of non-stoichiometry is a general

Table 10.8 *Thermodynamics of formation of binary compounds from the elements*

	ΔH^0_{298} (kJ mol^{-1})	ΔS^0_{298} (JK^{-1} mol^{-1})	Structure
MgO	-601.7†	-104.6	rocksalt
CaO	-635.1	-103.3	rocksalt
BaO	-558.6	-97.5	rocksalt
MnO	-385.3	-71.1	rocksalt
CaO	-238.9	-78.2	rocksalt
Fe$_{0.95}$O	-266.1	-75.3	rocksalt
NiO	-239.7	-93.3	rocksalt
ZnO	-348.5	-100.0	wurtzite
MgS	-347.2		rocksalt
MgSe	-272.3		rocksalt
MgTe	-209.2		wurtzite
CaS	-460.2	-17.1	rocksalt
BaS	-443.5	-7.5	rocksalt
MnS	-213.8	$+12.6$	rocksalt
FeS	-100.8	-1.3	nickel arsenide
ZnS	-205.9	$+13.8$	zincblende
NaF	-576.5	-101.4	rocksalt
NaCl	-411.3	-90.9	rocksalt

†From standard tabulations; see Section 10.5.

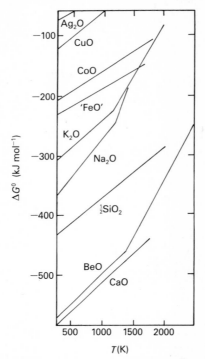

Fig. 10.4. Standard free-energy of formation (kJ mol^{-1}) of binary oxides as a function of temperature.

one; the free-energy relations which determine the homogeneity range are shown in Fig. 10.5. For a very stable binary compound (such as NaCl) the free-energy curve shows a deep and sharp minimum, and the resulting homogeneity range—defined by the common tangents to the free-energy curves for phases A, B, and $A_{1-x}B_{1+x}$—is small. For a less stable compound (e.g. NiS), the free-energy curve shows a broader and shallower minimum, resulting in wider homogeneity ranges. This argument rationalizes the observation that stable ionic compounds (e.g. Mg_2SiO_4) generally show small-to-negligible deviations from stoichiometry, while compounds of relatively low stability (e.g. sulfides, intermetallic alloys) show extensive solid-solution ranges. The exact forms of the free-energy curves are governed by the detailed energetics of substitution and/or defect formation.

(b) Formation of ternary compounds from binary compounds

Consider the enthalpy of the reaction:

$$AO + B_aO_b = AB_aO_{1+b} \qquad (10.53)$$

with A = Ca, Mg, Mn, Fe, Co, Ni, and B = Si, W, C, S (see Fig. 10.6). The following patterns are evident. All enthalpies of formation are in the range 0

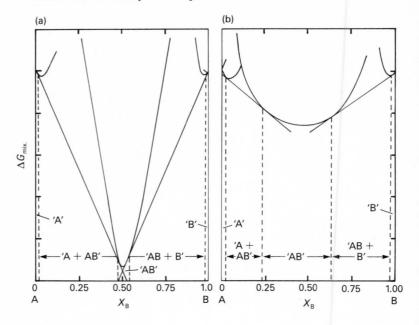

Fig. 10.5. Schematic isothermal free energy relations for a system A–B which has a binary compound for (a) very stable compound with small homogeneity range, and (b) less stable compound with wide homogeneity range. In both cases, the tangents to the free energy curves determine the extent of solubility.

to -300 kJ, with the aluminates, silicates, and germanates in the less negative portion of that range. Thus the stabilization of the ternary compounds is generally considerably less in magnitude than the formation enthalpies of the binary compounds. The entropies of formation, referred to solid binary oxides, are also quite small (from -15 to $+15$ JK^{-1} mol^{-1}). For a given B, the enthalpies of formation become more exothermic in the series Ni, Co, Fe, Mn, Mg, Ca, that is, with increasing basicity of the oxide AO. For a given A, the enthalpies of formation become more exothermic in the series Al, Si, Ge, W, C, S, that is, with increasing acidity of the oxide Al_2O_3, SiO_2, GeO_2, WO_3, CO_2, SO_3. Thus, the most stable compounds form when the most complete transfer of oxide ion from base to acid occurs, and the ternary structures then contain well-defined covalently bonded anions (SiO_4, GeO_4, WO_4, CO_3, SO_4). For compounds between more similar binary oxides (e.g. $Al_2O_3 + SiO_2$, $CuO + Fe_2O_3$, $Fe_2O_3 + TiO_2$) the enthalpies of formation are much smaller in magnitude and in the above three cases are actually endothermic. Thus the stabilities of $3Al_2O_3 \cdot 2SiO_2$ (mullite), $CuFe_2O_4$ (spinel), and Fe_2TiO_5 (pseudobrookite) derive from their entropies (configurational

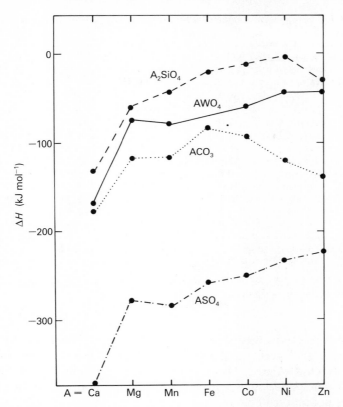

Fig. 10.6. Enthalpies of formation of silicates (A_2SiO_4), tungstates (AWO_4), carbonates (ACO_3), and sulfates (ASO_4) from the binary oxides (AO) plus $2SiO_2$, WO_3, CO_2, or SO_3.

and/or vibrational) and at low temperature these compounds generally are metastable or decompose.

(c) Phase transitions

Table 10.9 shows values of ΔH^0_{1000}, ΔS^0_{1000}, and the transition pressure at 1273 K for some high-pressure transitions. Several general points can be made.

(1) Even for a transition involving the same two structures (e.g. olivine→spinel) ΔS^0 and ΔV^0 vary by more than a factor of three for different compounds (e.g. Mg_2SiO_4, Fe_2SiO_4, Co_2SiO_4, Ni_2SiO_4). The ΔH, $T\Delta S$, and $P\Delta V$ terms may assume equal importance in determining the transition pressure; and the last two terms, although they can be estimated from known transitions, cannot be assumed to have constant values of ΔS^0 and ΔV^0.

Table 10.9 *Thermodynamic parameters for some high-pressure phase transformations*

Reaction	Phases (see key below)	ΔH^0 (kJ mol^{-1})	ΔS^0 (JK^{-1} mol^{-1})	ΔV^0 (ml mol^{-1})	$P_{trans.}$ (kbar) at 1273 K
Mg$_2$SiO$_4$	($\alpha \to \beta$)	30.0	-9.8	-3.24	137
Fe$_2$SiO$_4$	($\alpha \to \gamma$)	2.9	-20.0	-4.35	60
Co$_2$SiO$_4$	($\alpha \to \beta$)	9.0	-9.1	-2.90	69
Co$_2$SiO$_4$	($\beta \to \gamma$)	2.3	-4.1	-1.02	71
Ni$_2$SiO$_4$	($\alpha \to \gamma$)	6.0	-5.8	-3.52	32
CdTiO$_3$	(il $\to$ per)	15.0	$+14.2$	-2.35	~ 0.001
CaGeO$_3$	(wo $\to$ ga)	-0.8	-7.9	-5.97	20
CaGeO$_3$	(ga $\to$ per)	51.0	$+15.1$	-5.35	62
ZnO	(z $\to$ rs)	24.5	$+0.5$	-2.70	90
SiO$_2$	(q $\to$ co)	1.3	-5.0	-2.05	28
SiO$_2$	(co $\to$ st)	49.0	-4.8	-6.63	85

Key: α = olivine, β = modified spinel, γ = spinel, il = ilmenite, per = perovskite, wo = wollastonite, ga = garnet, z = zincite (distorted wurtzite), rs = rocksalt, q = quartz, co = coesite, st = stishovite (rutile).

Values of ΔH^0, ΔS^0 are near 1000 K, values of ΔV^0 are at 298 K. Data from ref. 30 and more recent work of Navrotsky's group.

(2) For transitions involving coordination number-increases, ΔV can be large (about ten per cent), and ΔS^0 is relatively small and can even be positive (ZnO, CdTiO$_3$, CaGeO$_3$). This is related to lattice vibrations and the longer metal–oxygen bonds in the phase with higher coordination number (as discussed in Section 10.2.4).

(3) ΔH^0 for such high-pressure phase transitions is in the range of 0–50 kJ. An approximate upper limit on ΔH^0 can be placed as follows. If the accessible pressure range is 0–1500 kbar, and the maximum volume contraction in a solid–solid phase transition is on the order of -10 ml mol^{-1}, the maximum value of $\Delta G^0 \approx P\Delta V^0$ is 1.5×10^4 ml kbar, or 152 kJ. Since, for such a transition, ΔS^0 is likely to be small because of the vibrational effects discussed above (see also eqn (10.16)), this also represents the maximum value of ΔH^0. Similarly, for a phase transition with increasing temperature at constant P, ΔS^0 must be positive and is unlikely to be greater than 20 JK^{-1} mol^{-1} for a solid–solid transition. If the highest accessible temperature is of the order of 3000 K, then the maximum value of ΔH^0 would be 60 kJ. Thus, phases energetically unstable by greater than about 150 kJ are unlikely to be stabilized by either pressure or temperature, and, in most cases (see Table 10.8), ΔH^0 is in fact much smaller than these limits.

(4) The variation of ΔH^0 for a given transition for various compounds can generally be correlated with bonding factors. Thus, for example, variations in the energetics of the olivine–spinel transition has been discussed in terms of bond lengths and ligand field effects.[30]

10.4 Solid solutions and order–disorder phenomena

10.4.1 'Simple' substitutional solid solutions

A solid solution of given composition will only be stable if its free energy is less than that of an equivalent mechanical mixture of its components, or of any possible exsolution products. This is shown schematically in Fig. 10.7. The free energy of mixing is composed of an enthalpy term (generally positive and destabilizing if no ordering phenomena occur), and an entropy term (generally positive and often approximated by the configurational entropy of random mixing). At high temperature (dot-dashed curve in Fig. 10.7(a)), the single phase is stable over the whole composition range, but at lower temperature (solid curve in Fig. 10.7(a)), a region develops where the system can lower its free energy by separating into two phases, whose compositions are given by the points of common tangency (see the dashed line in Fig. 10.7(a)). The result is that complete miscibility occurs at high temperature where the $T\Delta S$ term dominates, but immiscibility develops at lower temperatures. The exsolved region, defining a solvus or miscibility gap, is shown as the solid curve in Fig. 10.7(b).

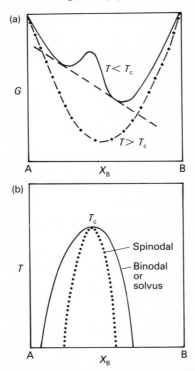

Fig. 10.7. Schematic representation of (a) free energy of mixing, and (b) solvus and spinodal in a binary system where the end-members are isostructural.

In the region between the two minima in the solid curve in Fig. 10.7(a), not only is the single phase metastable with respect to exsolution to two phases of different composition, but it can also lower its face energy by any minute compositional fluctuation, since $(\partial^2 G/\partial X^2) > 0$. This region defines the *spinodal*, which has great significance to phase equilibria and decomposition mechanisms in solid state reactions (see dotted curve in Fig. 10.7(b)).

One can represent the heat of mixing of a binary solid solution by, in order of increasing complexity, a simple regular solution (eqn (10.54)), a subregular or two-parameter Margules fit (eqn (10.55)) which leads to an asymmetric solvus, or a generalized polynomial in the mol fraction X (eqn (10.56)),

$$\Delta H = WX(1-X) \tag{10.54}$$

$$\Delta H = X)1-X)(W_1 X + W_2(1-X)) \tag{10.55}$$

$$\Delta H = X(1-X)(\sum_0^n a_n X^n). \tag{10.56}$$

The entropy of mixing, assuming random substitution, is given (per mol of species being mixed) by

$$\Delta S = -R\{X \ln X + (1-X) \cdot \ln(1-X)\}. \tag{10.57}$$

If ΔH is zero, and ΔS is given by eqn. (10.57), the solution is ideal; and one obtains thermodynamic activities in accord with Raoult's Law

$$a_i = X_i. \tag{10.58}$$

One can extend this definition of ideality carefully for statistically random mixing over more than one set of sites. Thus, for the case of random mixing of Fe and Mg in olivine, ideality would imply

$$a_{Mg_2SiO_4} = X^2_{Mg_2SiO_4} \tag{10.59}$$

but

$$a_{MgSi_{0.5}O_2} = X_{MgSi_{0.5}O_2} \tag{10.60}$$

Other, more complex cases, where species are distributed on several nonequivalent sites, can also be treated.[44]

For the regular solution case, one can derive, if eqns. (10.54) and (10.57) hold

$$RT \ln a_i = W/(1-X_i)^2. \tag{10.61}$$

Equations for activities can be derived for the more complex cases as well. The activities are constrained by the Gibbs–Duhen equation:

$$\sum_i X_i \mathrm{d} \ln a_i = 0. \tag{10.62}$$

Thus, for a binary system, once the activity of one component is known, the activity of the other can be calculated.

If one is interested in the solid solubility among end-members of different structures, the above treatment must be modified. The two end-members no longer fall on the same free energy surface in P–T–composition space; rather each phase defines its own surface. Complete solid solubility in general cannot occur, but the solubility at a given T and P is governed by free-energy relations shown in Fig. 10.8(a) which generally results in a phase diagram, shown schematically in Fig. 10.8(b). Since complete solid solubility is not possible, the solid regions terminate in a modified eutectic at high temperature as shown. The thermodynamic properties of such a system can be described by sets of mixing parameters (eqns (10.54)–(10.56)) for each phase, plus free-energy terms describing the transformation of each end-member to the opposite structure, or cryptomodification.[45]

In isostructural systems four main factors on an atomic scale affect the range and stability of solid solution; these are size difference, charge,

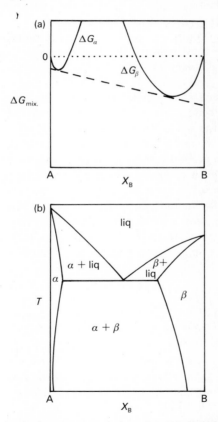

Fig. 10.8. (a) Schematic free-energy relations at a given temperature in the system A–B when A has the structure α, and B the structure β. The two phases then follow different free-energy curves, as shown. (b) Schematic phase relations for this system

covalency, and specific electronic factors. The first is usually the most important factor; it results in a strain energy which is necessarily larger the greater the difference in size. For a given size difference, it is often easier to put a smaller atom into a larger host lattice than vice versa, resulting in asymmetric solubility relations and heats of mixing in binary systems.

Theoretical models of solid solution formation have met with considerable success in alloy and alkali–halide systems. A rigorous calculation of the partition function for ionic crystals—when the nearest neighbours of the cations (namely the anions) remain the same but the next nearest neighbours change—immediately gives rise to difficulties in the definition of a pair-interaction energy. Thus the rigorous approach which was successful for alloys is somewhat difficult for ionic crystals, especially those of complex structure. In such cases, most theoretical models calculate a form of strain energy, and sometimes allow for local relaxation about a given ion. Examples of such models applied to ionic crystals include the work of Fancher and Barsch,[46] Urusov,[47] Driessens,[48] and Catlow et al.[49]

However, for many complex crystals, the ions being mixed occupy only a small fraction of the total volume of the structure. If one compares solid solubility in CaO–MgO (very limited), $CaCO_3$–$MgCO_3$ (quite extensive), and $Ca_3Al_2Si_3O_{12}$–$Mg_3Al_2Si_3O_{12}$ (complete), one is drawn to the conclusion that greater ionic size mismatch can be tolerated by a structure in which the ions being mixed are embedded in a matrix which can itself slightly change geometry to absorb the strain. In a thermodynamic sense, the volume of a phase is a more convenient parameter for correlation than any individual bond length or radius. With these considerations in mind, Davies and Navrotsky[45] developed a correlation between the excess free energy, in a regular or sub-regular solution model, and the volume mismatch for a variety of oxide, chalcogenide, halide, and silicate systems. The use of a volume mismatch term rather than a bond mismatch enables one to simultaneously consider many diverse systems in which divalent ions are being mixed (oxides, chalcogenides, spinels, garnets, olivines, other silicates). The correlation gives

$$W = (100.8\Delta V - 0.4) \text{ kJ mol}^{-1}. \qquad (10.63)$$

Within this correlation are points for cation mixing on 4-, 6-, and 9-coordinated sites, and for anion mixing. Thus, once the effect of coordination number is included in the molar volume of the phases involved, it does not further explicitly affect the thermodynamics of mixing correlations.

The alkali–halide systems as a group show much smaller positive deviations from ideality than the oxide and chalcogenide systems. This confirms the expectation that more highly charged ions mix less easily than ions of lower charge. A few points (e.g. Cr_2O_3–Al_2O_3, TiO_2–SnO_2) for trivalent and tetravalent ions also support this trend by showing more positive heats of mixing than predicted for divalent systems of comparable volume difference.

Thermodynamic aspects of inorganic solid-state chemistry 389

These correlations, though empirical, allow one to predict the magnitudes of deviations from ideality in a wide range of ionic solid solutions.

10.4.2 Order–disorder and complex systems

If the attractions between unlike atoms in a solid solution are stronger than between like atoms, ordering may occur. A classical case is the copper–gold system, whose phase diagram and thermodynamic mixing parameters at high temperature are shown in Fig. 10.9. Similar thermodynamic behaviour, including negative heats of mixing, was recently found in the NiO–MgO solid-solution series.[51] The tendency towards order in this system can be interpreted in terms of the electron configuration of the Ni^{2+} ion. A model was proposed in which Ni^{2+} adopts a compressed octahedral coordination resulting in an ordered alternation of Ni and Mg cations.[51] Even more pronounced ordering (including intermediate compound formation) has been seen in the systems CuO–NiO and CuO–MgO.

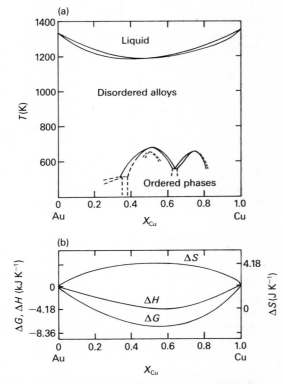

Fig. 10.9. (a) Phase diagram, and (b) thermodynamic mixing properties at 800 K, Cu–Au alloys.[50]

A number of complex solid solutions, including the plagioclase and alkali feldspars, show order–disorder phenomena which may be coupled with changes in symmetry. In the $CaAl_2Si_2O_8$–$NaAlSi_3O_8$ (anorthite–albite) series, peristerite exsolution (at 5–25 mol per cent anorthite) may represent a 'conditional spinodal'[52] while the $I\bar{1}$ to $P\bar{1}$ transformation at compositions near 60 per cent anorthite may be a transition of second- or higher-order. The monoclinic–triclinic inversion in analbite–sanidine ($NaAlSi_3O_8$–$KAlSi_3O_8$) solid solution may also represent a higher-order phase-transition. The formation of dolomite $(Ca,Mg)_2(CO_3)_2$ in the system $CaCO_3$–$MgCO_3$, and its disordering at high temperature, is another example of complex behaviour.[53]

Though the thermodynamics of such solid solutions have often been treated empirically by regular and sub-regular solution models, a real understanding of the driving forces of ordering and exsolution, and of kinetic behaviour, must be based on proper treatments of the coupled symmetry changes and disordering phenomena. Several approaches in the recent literature point in this direction.[52-4] Indeed, the application of structural techniques (notably high-resolution electron microscopy) with resolution in the 3–50 Å range, to solid solutions of complex solids leads to the conclusion that random homogeneous solid solutions may be far less common than previously realized; and complex ordering phenomena leading to submicroscopic ordering and/or exsolved domains occurs quite frequently. Since such complex structures may differ very little in free energy from the metastable homogeneous solid solution, it is not clear to what extent their complexity is reflected in the measured thermodynamic properties. It is clear, however, that the application of simple thermodynamic models (regular and sub-regular formulations, for example) is a matter of empirical convenience rather than a representation of the actual atomic interactions and structural state. The combination of thermodynamic and microstructural characterization of solids of the same thermal history promises to be an expanding and fruitful area for future work.

10.5 Textbooks and tabulations of data

The following textbooks are recommended because they deal in sufficient detail with the thermodynamics of solids:

Bodsworth, C. and Appleton, A. S. *Problems in applied thermodynamics*. Longmans, Green and Co., London (1965).

Darken, L. S. and Gurry, R. W. *Physical chemistry of metals*. McGraw-Hill Book Co. New York (1953).

Denbigh, K. *The principles of chemical equilibrium* (2nd edn). Cambridge University Press, Cambridge, England (1966).

Gaskell, D. R. *Introduction to metallurgical thermodynamics*. McGraw-Hill Book Co., New York (1973).

Kubaschewski, O., Evans, E. L., and Alcock, C. B. *Metallurgical thermochemistry* (4th edn). Pergamon Press, New York (1967).

Lewis, G. N. and Randall, M. (rev. by K. S. Pitzer and L. Brewer) *Thermodynamics* (2nd edn). McGraw-Hill Book Co., New York (1961).
Schmalzried, H. and Navrotsky, A. *Festkörperthermodynamik*. Verlag Chemie, Weinheim, Germany (1975).
Wagner, C. *Thermodynamics of alloys*. Addison-Wesley Pub. Co., Reading, Mass., USA (1952).

The following are tabulations of thermodynamic data.

Coughlin, J. P. *Contributions to the data on theoretical metallurgy* XII. *Heats and free energies of formation of inorganic oxides*. US Bur. Mines Bull. 542 (1954).
Elliott, J. F. and Gleiser, M. *Thermochemistry for steelmaking* Vol. 1. Addison-Wesley Publ. Co., Reading, Mass., USA (1960).
Hultgren, R. et al. *Selected values of the thermodynamic properties of the elements*. Am. Soc. Met. Metals Park, Ohio, USA (1973).
Hultgren, R. et al. *Selected values of the thermodynamic properties of binary alloys*. Am. Soc. Met. Metals Park, Ohio, USA (1973).
Kelley, K. K. *Contributions to the data on theoretical metallurgy* XIII. *High-temperature heat capacity and entropy data for the elements and inorganic compounds*. US Bur. Mines Bull. 584 (1960).
Kelley, K. K. and King, E. G. *Contributions to the data on theoretical metallurgy* XIV. *Entropies of the elements and inorganic compounds*. US Bur. Mines Bull. 592 (1961).
Kubaschewski, O. et al. *Metallurgical thermochemistry*. Pergamon Press, Oxford (1967). Handbook of Chemistry and Physics (52nd edn). Chemical Rubber Publishing Co., Cleveland, Ohio (1972). Landolt-Börnstein, II. Band 4, Teil, Springer-Verlag, Berlin (1961).
Kubaschewski, O. *The thermodynamic properties of double oxides*. Nat. Phys. Lab. DCS Rep. 7 (1970).
Robie, R. A., Heminway, B. S., and Fisher, J. R. *Thermodynamic properties of minerals and related substances at 298, 15 K and 1 atm. and at high temperatures*. US Geol. Survey Bull. 1452 (1978).
Rosenquist, T. *Thermochemical data for metallurgists*. Tapir-Verlag, Trondheim, Norway (1970).
Rossini, F. D. et al. *Selected values of chemical thermodynamic properties*. US Nat. Bur. Stand. Circ. 500 (1952).
Samsonov, G. V. *The oxide handbook* (Englische Übersetzung). IFL–Plenum, New York (1973).
Stiehl, D. R. and Prophet, H. *JANAF thermochemical tables* (2nd edn) NSRDS-NBS 37. US Nat. Bur. Stand., Washington D.C. (1971).
Wagman, D. D. et al. *Selected values of chemical thermodynamic properties—tables for the first thirty-four elements* NBS Tech. Note 270–3. US Natn. Bur. Stand., Washington D.C. (1968).
Wagman, D. D. et al. *Selected values of chemical thermodynamic properties—tables for elements 35 through 53* NBS Tech. Note 270–4. US Nat. Bur. Stand., Washington D.C. (1969).

The following are tabulations of phase-diagram information.

Elliot, J. F. and Gleiser, M. *Thermochemistry for steelmaking*. Vols 1 and 2. Addison-Wesley Publ. Co., Reading, Mass., USA (1960).
Hansen, M. *Constitution of binary alloys*. McGraw-Hill Book Co., Inc., New York (1958). First Supplement (1965). Second Supplement (1969).

Levin, E. M., Robbins, C. R., and McMurdie, H. F. *Phase diagrams for ceramists.* Am. Ceram. Soc., Columbus, Ohio, USA (1964). Supplement (1969). Vol. III (1975). Vol. IV (1980).

Morey, G. W. *Data of geochemistry, phase equilibrium relations of the common rock-forming oxides except water* (6th edn) US Geol. Survey, Paper 440-L (1964).

Toropov, N. A. et al. *Handbook of phase diagrams of silicate system.* Vol.1 Binary systems; Vol.2 Metal oxygen compounds. US Dept. Comm., Nat. Tech. Inf. Serv., Springfield, Va. (1972).

10.6 References

1. Finger, L. W. and Hazen, R. M. *Trans. Am. Cryst. Assoc.* **15**, 93–105 (1979).
2. Westrum, E. F., Jr. *J. chem. Ed.* **39**, 443–55 (1962).
3. Westrum, E. F., Jr. *Colloq. Internationale du C.N.R.S. No. 201—Thermochemic*, 103–18 (1972).
4. Robie, R. A. and Hemingway, B. S. *US Geol. Survey Profess. Paper* **755**, 1–32 (1972).
5. McNaughton, J. L. and Mortimer, C. T. *Int. Rev. Sci. phys. Chem.* (Series 2) **10**, 1–44 (1975).
6. Stout, N. D., Mar, R. W., and Boo, W. O. *High Temp. Sci.* **5**, 241–51 (1973).
7. Bonnell, D. W., Chaudhuri, A. K., Ford, L. A., and Margrave, J. L. *High Temp. Sci.* **3**, 203–10 (1970).
8. Gronvold, F. *Acta Chem. Scand.* **26**, 2216–22 (1972).
9. Holm, J. L., Kleppa, O. J., and Westrum, E. F., Jr. *Geochim. Cosmochim. Acta* **31**, 2289–307 (1967).
10. Derrington, C. R., O'Keefe, M., and Navrotsky, A. *Solid State Comm.* **18**, 47–9 (1976).
11. Lewis, G. N. and Randall, M. *Thermodynamics* (2nd edn) rev. K. S. Pitzer and L. Brewer, pp. 133–4. McGraw-Hill Book Co., New York (1961).
12. Jacob, K. T. and Alcock, C. B. *High Temp. High Press.* **7**, 433–9 (1975).
13. Holm, J. L. and Kleppa, O. J. *Am. Mineral.* **53**, 123–33 (1967).
14. Kieffer, S. W. *Rev. Geophys. Space Phys.* **17**, 1–19 (1979).
15. Watanabe, H. In *High pressure research in geophysics* (eds S. Akimoto and M. H. Munghnani), pp. 441–64. Center for Academic Publications, Tokyo (1981).
16. Navrotsky, A. *Geophys. Res. Lett.* **7**, 709–11 (1980).
17. Kieffer, S. W. *Rev. Geophys. Space Phys.* **17**, 35–59 (1979).
18. Kieffer, S. W. *Rev. Geophys. Space Phys.* **18**, 862–86 (1980).
19. Muan, A. *Proc. Brit. Ceram. Soc.* **8**, 103–12 (1967).
20. Schmalzried, H. and Pelton, A. D. *Ann. Rev. mat. Sci.* **2**, 143–80 (1972).
21. Kitchener, J. A. and Ignatowics, S. *Trans. Far. Soc.* **147**, 1278–86 (1951).
22. Chatillon, C., Pattoret, A., and Drowart, J. *High Temp. High Press.* **7**, 119–48 (1975).
23. Rooymans, C. J. M. In *Preparative methods in solid-state chemistry* (ed. P. Hagenmuller), pp. 72–133. Academic Press, New York (1972).
24. Holloway, J. R. In *Research Techniques for High Pressure and High Temperature* (ed. J. C. Ulmer), pp. 217–58. Springer-Verlag, New York (1971).
25. Boyd, F. R. and England, J. L. *J. Geophys. Res.* **65**, 741–8 (1960).
26. Endo, S. and Ito, S. In *High pressure research in Geophysics* (eds S. Akimoto and M. H. Munghnani), pp. 3–12. Center for Academic Publications, Tokyo (1981).
27. Mao, H. K. and Bell, P. M. *Science* **200**, 1145–8 (1978).

28. Batsanow, S. S. In *Preparative methods in solid-state chemistry* (ed. P. Hagenmuller), pp. 133–46. Academic Press, New York (1972).
29. Ma, C. B. *J. Geophys. Res.* **79,** 3321–4 (1974).
30. Navrotsky, A., Pintchovski, F. S., and Akimoto, S. *Phys. Earth Planet. Interiors* **19,** 275–92 (1979).
31. Boyle, B. J., King, E. G., and Conway, K. C. *J. Am. chem. Soc.* **76,** 3835–7 (1954).
32. Margrave, J. L. and Kybett, B. *Rev. Sci. Instr.* **37,** 673–6 (1966).
33. Boureau, G. and Gerdanian, P. *Canad. metall. Q.* **13,** 339–43 (1974).
34. Kleppa, O. J., Melnichak, M. E., and Charlu, T. V. *J. Chem. Thermom.* **5,** 595–602 (1973).
35. Inaba, H., Navrotsky, A., and Eyring, L. *J. solid-state Chem.* **37,** 67–76 (1981).
36. Castanet, R. *J. Chem. Thermom.* **11,** 787–91 (1979).
37. Kleppa, O. J. In *Thermodynamics of minerals and melts* (eds R. C. Newton, A. Navrotsky, and B. J. Wood), pp. 179–88. Springer-Verlag, New York (1981).
38. Pratt, J. B. and Chua, K. S. *Thermochim. Acta* **8,** 409–21 (1974).
39. Orr, R. I. *Acta Met.* **8,** 489–93 (1960).
40. Kleppa, O. J. and Sato, S. *J. Chem. Thermom.* **14,** 133–43 (1982).
41. Navrotsky, A. *Phys. Chem. Minerals* **2,** 89–104 (1977).
42. Suito, K. In *High-pressure research application to geophysics* (eds M. Manghnani and S. Akimoto), pp. 255–66. Academic Press, New York (1977).
43. Jantzen, C. M. F. and Herman, H. In *Phase diagrams, materials science and technology* Vol. V (ed. A. M. Alper), pp. 127–85. Academic Press, New York (1978).
44. Kerrick, D. M. and Darken, L. S. *Geochim. Cosmochim. Acta* **39,** 1431–42 (1975).
45. Davies, P. K. and Navrotsky, A. *J. solid-state Chem.* In press (1982).
46. Fancher, D. L. and Barsch, G. R. *J. phys. Chem. Solids* **30,** 2503–16 (1969).
47. Urusov, V. S. *Geskhimiya* **4,** 510–24 (1970).
48. Driessens, F. C. M. *Ber. Bunsenges. phys. Chem.* **72,** 764–73 (1968).
49. Catlow, C. R., Fender, B. E., and Hampson, P. J. *J. chem. Soc., Faraday Trans.* (II) **73,** 9110 (1977).
50. Hultgren, R., Desai, P. D., Hawkins, D. T., Gleiser, M., and Kelley, K. K. *Selected values of the thermodynamic properties of binary alloys,* p. 260. Am. Soc. Met., Metals Park, Ohio, USA (1973).
51. Davies, P. K. and Navrotsky, A. *J. solid-state Chem.* **38,** 264–76 (1981).
52. Carpenter, M. A. *Am. Mineral.* **66,** 553–60 (1981).
53. Navrotsky, A. and Loucks, D. *Phys. Chem. Miner.* **1,** 109–27 (1977).
54. McConnell, J. D. C. *Zeit. Kristall.* **147,** 45–62 (1978).
55. Robie, R. A., Hemingway, B. S., and Takei, H. *Am. Miner.* **67,** 470–82 (1982).
56. King, E. G. *J. Am. chem. Soc.* **79,** 5437–48 (1957).

Index

absolute configuration 172
absorption coefficient 181
absorption length 163
acoustic branches 325
acoustic phonons 164, 285
admittance 313
adsorbate 112
aliovalent dopant ions 247, 253
aluminosilicates, MAS NMR 216
amorphous materials 79, 206, 213, 316
 conduction 309
Anderson-localized 318
Anderson transition 309
anhydrous halides 10
anisotropic and antisymmetric exchange 153
anisotropy
 magnetic 154
 single-ion 153
argon ion sputtering 89
Auger transitions 97
Auger electron spectroscopy (AES) 114

bandgap 280
Beevers–Ross (BR) site 270
benitoite, $BaTiSi_3O_9$ 360
binary compounds formation from
 elements 380
Born–Haber cycle 95, 112
Bragg equation 44, 67
Bravais lattices 53
bremsstrahlung isochromat spectroscopy
 (BIS) 115
bridge circuit 129
Brillouin function 134, 137
Brillouin spectroscopy 330
Brillouin zone 324, 330
brucite, $Mg(OH)_2$ 356

cadmium chemical shifts 214
calorimetry 377
calorimetry combustion 378
calorimetry solution 379
chemical
 diffusion 2
 potentials μ 371
 shift, NMR 192

shift anisotropy 195, 210, 214, 225
shifts in core electron spectra 96
 (vapour phase) transport 8, 27
Cheviel phases 21, 24, 34
circular dichroism 171
Clausius–Clapeyron equation 371
cohesive energy 233
computer simulation 231
conduction band (CB) 280
conductivity 279
 A.C. 312
 ionic 365
conductors, organic 292
congruently melting compound 17
contact 288
configuration interaction 100
contact potential 287
contacts, non-ohmic 314
containers 31, 32
co-ordination compounds, MAS, NMR 211
Coulomb integrals 150
counterion choice 16
cross polarization (CP) 197, 198
crystal field, low symmetry 146
 octahedron 144
 splitting 139
 theory 138
crystal structure prediction 272
Curie constant 134
 law 133
Curie–Weiss law 137, 145, 146, 148, 160
cylindrical mirror analyser 89

Davydov splitting 165
d–d transitions, spin forbidden 177
Debye model 369
Debye–Scherrer method 5, 45
Debye temperature 290, 370
defect clustering 250, 254
 elimination 260
defects, point 232, 247
 extended 232
defect simulations 237
deformation density 66
density of electronic states $N(E)$ 310
demagnetizing factor 124
density of states (DOS) 108

diamagnetism 136
diatomic chain 327, 328
dielectric constant dispersion 334
diffraction theory 48
diffractometer-four circle 52
dimensionality, structural 148
dipolar coupling tensor 191, 194
  decoupling 207
  interaction 191
direct methods 58
dispersion relations 324, 325, 326, 333
disproportionation 16
dynamics simulations 240

effective masses 281
Einstein model of lattice variations 369
electrolysis 20
electron density distribution 51
electron energy loss spectroscopy
    (EELS) 77, 117
electron microscopy 71
  analysis 75
  scanning 74
  transmission 6, 72
electron spectrometers 88
electroneutrality principle 283
energy minimization 273
emission spectrum 182
entropy 364
  of fusion 367
  zero-point 368
ESCA 86
Evans method for determination of magnetic
    susceptibilities 131
exchange, anisotropic 159
  Hamiltonian 151, 159
  integrals 150
  interactions 148
exciton 183, 184, 185
extended X-ray absorption fine structure
    (EXAFS) 116

factor group 348, 345
  analysis 330, 350, 354
  splitting 337
Faraday method 127
fast-ion conductors 265, 279
faujasite zeolites 220
Fermi energy 280
  level 92, 104, 106, 107
ferromagnet 185
fluorite structure 266
fluxes, reactive sintering 19
Fourier summation 57
four-probe conductivity measurements 288
free-energy relations 387

Frenkel disorder 247, 249
Frenkel energies 267

Gerade function 167
Gibbs–Duhen equation 386
glide plane 55
Gouy method 123
gradient techniques 236
grain-boundary effects 314

halide exchange 11
Hall coefficients 304
  effect 301
  mobility 303, 307
Hartman–Hahn condition 199
heat capacity 363
Heisenberg 158
  Hamiltonian 155
high-pressure phase equilibria 375
  spectroscopy 345
  synthesis 25
high temperature series-expansion 156
Hooke's law 332
hopping
  of electrons 311
  random range 312
hydrothermal synthesis 26

incongruently melting phase 18
indicatrix 343
inelastic neutron scattering 330
infrared spectroscopy 330
instrumental resolution function 97
intensity of reflection 50
interatomic potentials 241
intercalation 21, 23
interstitial sites
interstitial clusters 256
interstitial-vacancy cluster 268
Ising model 153, 157

Jahn–Teller distortion 176

K_2PtBr_4 single crystal infrared spectrum 355
K_2PtBr_4 single crystal Raman spectrum 355
Koopmans' theorem 93, 94
Kramers–Kronig relationships 313, 342, 356

Lattice energy 233
  modes 337, 353
  parameter determination 47
  symmetry 53

lifetime broadening 107
linear dichroism 179
linear chains, antiferromagnetic 155, 157
longitudinal modes 324, 331
low energy electron diffraction (LEED) 275
low temperature synthesis 22
Lowenstein's Rule 217, 221, 224
luminescence 182

Madelung potential 94, 110, 244
magic-angle spinning (MAS) 190, 196
magic-angle spinning NMR spectra 202
magnetic Bragg peaks 64
　circular dichroism (MCD) 173
　-dipole transition moment 168
　induction 122
　moment 174
　quantity conversion factors 124
magnetization 124
magnon 187
mean-free path 285, 290
metals, as reducing agents 14
　containers 32
　insulator transitions 110
metathetical reaction 3, 8
microprobe 5, 77
mixed-valence system 104
Miller indices 43
mobility of carriers 279, 284
molar extinction coefficient 165
molten salts 378
Mott–Littleton approximation 238, 239
multiplet splitting core electron 101
　structure in photoemission 103
mutual inductances 129

Nelson–Riley function 46
neutron diffraction 61, 257
　scattering lengths 62
Newton–Raphson procedure 236, 239
non-stoichiometric phase, transport 31
normal modes 323
nuclear magnetic resonance 131, 190

ohmic 288
olivine 274, 364, 376
open-shell photoionization 102
optic branches 325
order–disorder 389
optic phonons 286, 348
oscillator strength 165, 336
oxidation–reduction 372

particle size 79

partition function 322
Patterson method 57
Pauli exclusion principle 150
PbO, orthorhombic 357
Peierls distortion 149
phase
　diagrams 48
　relationships 17
　-sensitive detection 128
　transitions 383
phonon 328
　selection rules 352
photoacoustic spectroscopy 165
photoconductivity 306
photoelectric cross section 98
　effect 84
photoemission energy level diagram 85
　time-scales 99
plasmas 9
plasmons 91, 105, 118
point defects 79, 232, 247
polariton 332
point polarizable ion (PPI) model 234
polarization
　circular 170
　experiments 166
　linear 165
polaron 286
polymorphs 209
primitive unit cell 346
pulsed neutron sources 67

quadrupolar interactions 193, 205
　nuclei 204
quadrupole transition moment 168

raising and lowering operations 141, 142
Raman spectroscopy 329, 338
　selection rules 350
Raoult's Law 386
relectance, infrared 341
reflection spectra 339
reflectivity 180, 181
resolving power 72
Rietveld profile analysis 68
rotational strength 171
Russell–Saunders multiplet 170

scattering factor
　electrons 71
　X-rays 42, 49, 57
Schottkey disorder 247, 249, 252
Schrödinger equation 150
screw axis 54
Seebeck coefficients, measurement of 296, 298, 301, 305
Seebeck effect 293

semiconductor 280, 292
  extrinsic 282
  n-type 282, 294
  p-type 282, 295
semimetal 291
shake-up 93, 99, 100, 109
  long range ordering 262
shear planes 254, 261
shear-plane interaction 265
  orientations 264
shell model 235, 242
site group 349
space groups 56
spallation 61
spin 148
spin–orbit coupling 139, 141, 143
spin–spin coupling 192
spinel 376
spinning side bands 197
spinodal 386
SQUID (Superconducting QUantum Interference Device) 130
Stark effect 180
structure factor F_{kkl} 49
superconductor 293
superexchange interaction 149
superionic conductors 265
superstructure 63
surface defects 276
surface layer spectroscopy 110
  simulations 275
susceptibility 123
symmetry elements 55
synchrotron radiation 88, 116
synthesis from the elements 12
systematic absences 53, 54

ternary compounds formation from binary compounds 381
tetragonal distortion 147
thermal decomposition 9
thermal expansion 363
thermopower 299, 300, 302
time-of-flight neutron diffraction 67

transition-metal oxides 258
transverse modes 324, 331
tungsten bronzes 109

unit cell 42

vacancy clusters 259
  concentrations 264
  mechanism 271, 272
valence band (VB) 280
van der Pauw 289, 305
van Vleck equation 132, 144, 148, 152
vapour phase transport 4
vapour presure measurements 374
vector model for exchange 151
vibrating sample magnetometer 130
vibronic wave-function 168

wave vector 164, 179, 183, 323
Weiss molecular field coefficient 159
Wilkinson's catalyst 212

XPS for valence bands 109
X-rays 39, 42, 46
X-ray absorption 41, 46
  emission 40, 41, 77, 94
  fluorescence 5
  photoelectron spectroscopy 84
  powder techniques 45
  scattering 41, 57
  single crystal methods 51
  tube 40
XY model 156

Zeeman effect—second order 145
Zeeman effect 143, 145, 148, 154, 172, 175, 191
zeolites 217, 219, 224, 274
zero-field splittings 158